Cost Reduction in Product Design

Cost Reduction in Product Design

William Wai-Chung Chow, Ph.D.

Assistant Professor
Department of General Engineering
University of Illinois
Urbana, Illinois

VNR
VAN NOSTRAND REINHOLD COMPANY
NEW YORK CINCINNATI ATLANTA DALLAS SAN FRANCISCO
LONDON TORONTO MELBOURNE

Van Nostrand Reinhold Company Regional Offices:
New York Cincinnati Atlanta Dallas San Francisco

Van Nostrand Reinhold Company International Offices:
London Toronto Melbourne

Library of Congress Catalog Card Number: 77-18130
ISBN: 0-442-21540-1

Manufactured in the United States of America

Published by Van Nostrand Reinhold Company
135 West 50th Street, New York. N. Y. 10020

Published simultaneously in Canada by Van Nostrand Reinhold Ltd.

15 14 13 12 11 10 9 8 7 6 5 4 3 2

Library of Congress Cataloging in Publication Data

Chow, William.
Cost reduction in product design.

Includes index.
1. Engineering economy. 2. Engineering design. I. Title.
TA177.4.C45 620'.004'2 77-18130
ISBN 0-442-21540-1

PREFACE

True design work is an art. Scientific principles and technical experience are merely guidelines, although the correct guidelines can save time and effort in creative thinking. The reader will find that the discussions in this book build on, organize, and supplement intuition.

The purpose of this book is to present up-to-date engineering concepts and methods in cost reduction. Many books have been written on such individual approaches as linear programming, stress analysis, value engineering, engineering economics, etc.; however, few engineers are introduced to the basics of all approaches. This book presents an overall view and serves as a basic reference. It also provides a comparison of different engineering analyses, and different manufacturing methods and materials.

Assume a product so designed that it can function well and can be manufactured; this book points the way to perfecting it in terms of economy, efficiency, and engineering integrity. It emphasizes the distinction between feasible design and optimum design, and discusses various opportunities for improvement, citing as specific examples appliances, instruments, machines, toys, vehicles, furniture, and many other consumer products. It covers processes of manufacturing and assembly, including casting, molding, forming, soldering, bracing, welding, and cementing, and materials, such as multifunctional plastics, structural foams, composite materials, and powder metallurgy. Coverage throughout is based on the potential for cost reduction.

While gathering material for this book, I found an emphasis on cost reduction in every field of engineering and business. It seems that if an innovation is superior in engineering performance and lower in cost, it will be widely accepted and become standard practice. After a while, people habitually accept the idea, and the new generation of engineers may even overlook the real advantage of and reason for, a handbook practice. This book can remind engineers at all levels of the principles that form the basis of standard practices. While things run smoothly for ten or twenty years, many parameters change, and efficiency can erode without being noticed. It is helpful to review things now and then to ensure the highest engineering performance and economic competitiveness.

Cost reduction is sometimes regarded as a "dirty" term. Reputable companies prefer not to use it because it suggests reductions in quality and value. In this book, specific discussions are directed to safety, reliability, and value. Often, the reader will find, cost reduction can be accompanied by an increase in quality. Before a new design is introduced into the market, it should be fully tested for quality and acceptability to the consumer. Ideas that yield only a small advantage or a limited cost reduction are often dismissed because of the costs of the engineering development and the marketing effort. However, such ideas are very useful if a product is developed from scratch.

How to get more for less? Getting more production at the same cost is called *increase productivity*. Maintaining production while cutting cost is *conservation.* In this era of limited resources, productivity and conservation are the main concerns of product designers. Some people see these as ways to increase profit. Others think of them as necessary steps to fight inflation and survive competition. Economists agree that these are the proper ways to fight poverty and upgrade living standards. I, myself, am fascinated by the elegance of optimal design and minimum weight design.

I was fortunate to have Professor R. C. Junivall as my advisor during my graduate studies at the University of Michigan. When working with the Kenner Products Co., I found my experience with Mr. Harry March and Mr. John Mayor extremely beneficial. It was the academic and the industrial environments that convinced me this book would be needed. No doubt, my understanding and reasoning in the area of design is influenced by the people I have met and the articles and books I have read. I would like to thank all those who have contributed in this field of knowledge. I hope this book is one further step toward bringing out the results of all the tremendous research done by all of us.

WILLIAM CHOW
Urbana, Illinois

CONTENTS

Cost Reduction in Product Design

1. ENGINEERING CONCEPTS OF EFFICIENT DESIGN

1.1. THE FORCE-FLOW CONCEPT

Machines are made to transmit forces in such a way that useful jobs can be accomplished. The ability of a particular machine member to endure external force is its strength.

The punch in Figure 1.1 is a tool that directs and concentrates forces to a point. Lines of force can be imagined to flow from the hammer to the end of the punch. They pass along the length of the punch and come out at the tip. The concentration of lines of force represents the stress. Thus, the stress is low at the upper end where the area is large. The stress is very high at the tip where the area approaches zero. The stress at any section perpendicular to the lines of force has to be lower than the yield strength of the material to avoid failure.

$$\text{stress } \sigma = \frac{\text{force}}{\text{area}} \tag{1.1}$$

$$\text{stress } \sigma < \text{yield strength } S_y \tag{1.2}$$

In pure compression and pure tension applications, the forces flow smoothly and rather straight through the member. The force experienced by a punch is pure compression, and that existing in a fishing line is pure tension. Such applications are quite efficient in utilizing the load-bearing capacity of the material. If the forces flow in curves,

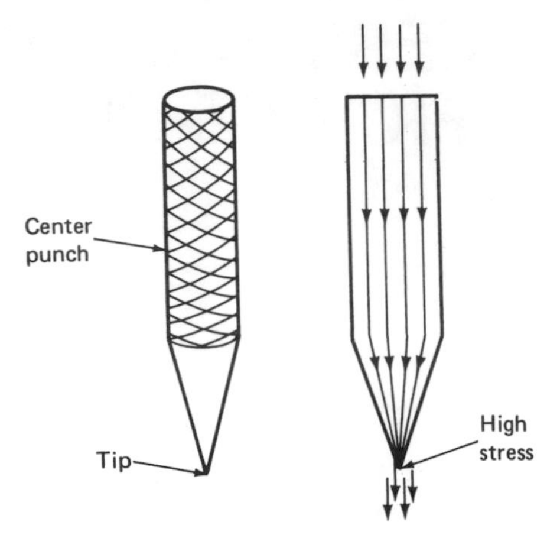

Figure 1.1. Force-flow through a center punch. A center punch is a tool used to produce point indentation on a flat surface. The stress in the punch is proportional to the concentration of force-flow lines and inversely proportional to the sectional area. The stress approaches infinity at the tip where the area approaches zero.

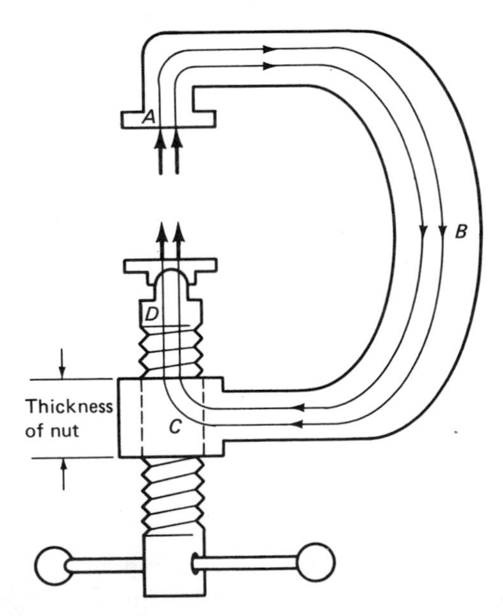

Figure 1.2. Force-flow in a C-clamp. The C-clamp is a commonly used tool that produces clamping force by a screw on one arm. From the force-flow diagram, it can be seen that the regions *A* and *D* are under pure compression, the region *C* is under shear, and the region *B* is under bending and tension.

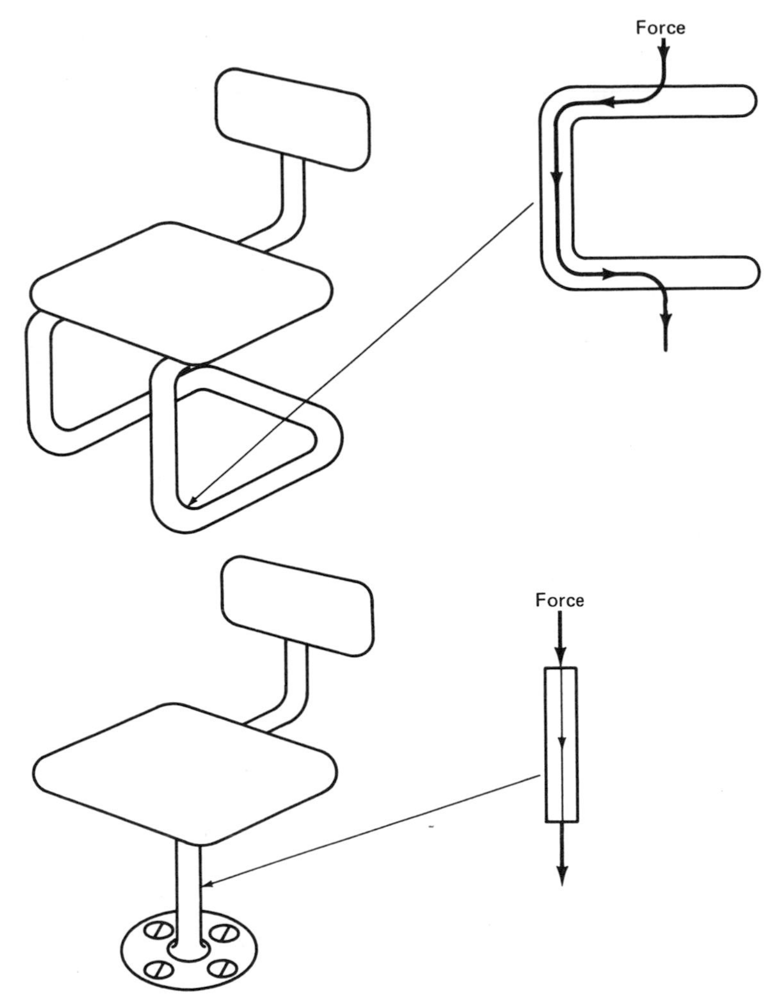

Figure 1.3. Efficiency of load-bearing members related to force-flow patterns.

there is usually bending in the member. Maximum shear stresses exist in planes at 45° to the lines of force.*

Figure 1.2 illustrates the various stresses in different parts of a C-clamp. In region *A* there is pure compression. In region *B* there is bending and axial tension. In region *C* the forces flow from the curved beam into the screw. The forces produce compression in region *D*, similar to that in region *A*.

*The lines of force represent principal stresses. It can be observed from the Mohr Circle that the maximum shear stress is 45° away from the principal stresses.

The C-clamp can fail by compressive yielding in region *A*, bending failure in region *B*, shear failure of screw threads in region *C*, and compressive yielding in region *D*. There should be enough strength to prevent each mode of failure. The force-flow concept is very useful in identifying areas of high stress and potential failure.

To support a load with the least amount of material, lines of force should be allowed to flow directly between points of external forces. Direct lines of force mean no bending and characterize the shortest physical length of a load-carrying member. The two chairs in Figure 1.3 illustrate this point. The chair with direct force-flow is more efficient than the chair with curved force-flow.

Figure 1.4 shows two structures similar in shape but different in ef-

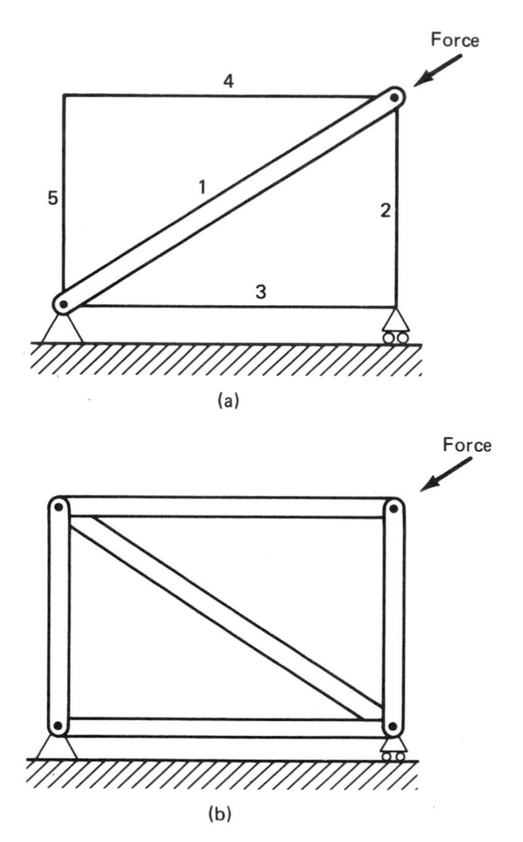

Figure 1.4. Pin-jointed structures with different arrangements of diagonal members. In (a), the diagonal member is placed in line with the external force. The members 2, 3, 4, and 5 are not loaded at all, and can be made very slender. In (b), however, the diagonal member is not in line with the external force, and all members are heavily loaded.

ficiency. In Figure 1.4(a), the diagonal member connects the external force directly to the left ground support. The members 2, 3, 4, and 5 are not loaded, and can be made very slender. In Figure 1.4(b), all five members of the structure are loaded heavily.

1.2. BALANCE OF STRESS AND STRENGTH

In order to prevent failure, the strength of a member has to be greater than the stress in the member. However, a member with excessive strength wastes material. Such a member is "overdesigned." The ratio of strength to stress is called the safety factor. The safety factor should be kept uniform and reasonable throughout the machine.

The component of a machine that has the lowest safety factor is the critical element. For the C-clamp in Figure 1.2, the critical element is B because it has the highest bending moment. Doubling or tripling the strength at sections A, C, or D will not improve the clamp as a whole. A balanced design without excessive overstrength would result if the safety factor of every component in the machine were the same. Balancing strength between sections C and D results in a simplified rule: the thickness of a nut should be between one to two times the diameter of the screw. The ideal situation was embedded in a poem* about a wagon that was so well designed that it ran for 100 years without any trouble. At the end of 100 years, it fell completely to pieces all at once.

Figure 1.5 shows two cantilever beams both of which have the highest bending moments at the fixed ends, which are the critical elements of the beams. If the stresses at the fixed ends are less than the strength of the material, the beams will not fail. Due to their smaller stresses, other elements in the uniform beam have larger safety factors than the critical element. The uniform beam is, therefore, not balanced in stress and strength. The width of the tapered beam decreases correspondingly with the bending moment that decreases linearly from the fixed end to the point of application of the force. The tapered beam, with uniform safety factors, is said to have a balanced design. Both cantilever beams have the same sectional area at the fixed end. They can bear the same maximum external load. Yet, the tapered beam has only half the volume of the uniform beam.

The punch in Figure 1.1 has a uniform load but a decreasing sec-

*"The One-Hoss Shay" ("The Deacon's Masterpiece") by Oliver Wendell Holmes.

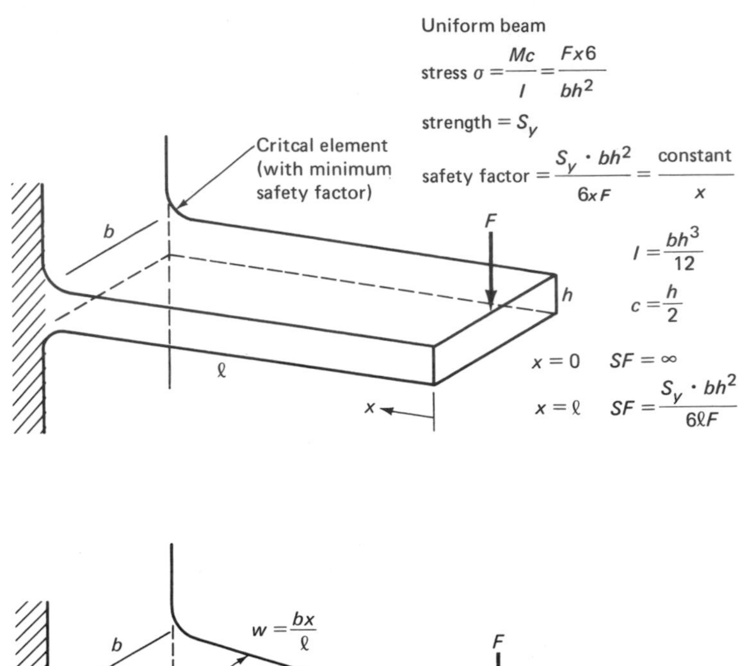

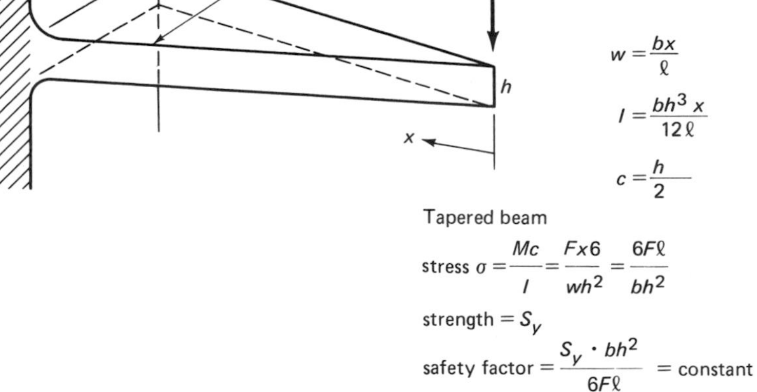

Figure 1.5. Two different cantilever beams with the same load-carrying abilities but significantly different amounts of material. The uniform beam has the critical element at the fixed end. The safety factor is nonuniform throughout the beam and is maximum at the point of application of the external force F. The tapered beam has the same sectional area bh at the fixed end as the uniform beam. The sectional area decreases as the bending moment decreases toward the tip of the beam. This maintains a uniform safety factor and yields a balanced design. The model ignores the stress due to the weight of the beam, the shear stress at the neutral axis, and the difficulty of applying a force to a point because the secondary factors might confuse the main theme.

tional area. A balanced design would call for increasing the yield strength of the material toward the tip. This is usually done by heat-treating the tip of the punch.

$$\text{stress } \sigma = \frac{F \text{ (force)}}{A \text{ (area)}} \tag{1.1}$$

$$\text{safety factor } SF = \frac{S_y \text{ (strength)}}{\sigma} \tag{1.3}$$

$$S_y = SF \cdot \sigma = SF \cdot \frac{F}{A} \tag{1.4}$$

The external force F and the safety factor SF are held constant. For a balanced design, the yield strength S_y should be made inversely proportional to the sectional area A.

The tensile strength of cast iron is about a fourth of its compressive strength. To compensate for this, cast-iron beams should have extra webbing on the tension side. Figure 1.6 shows two cast-iron beams

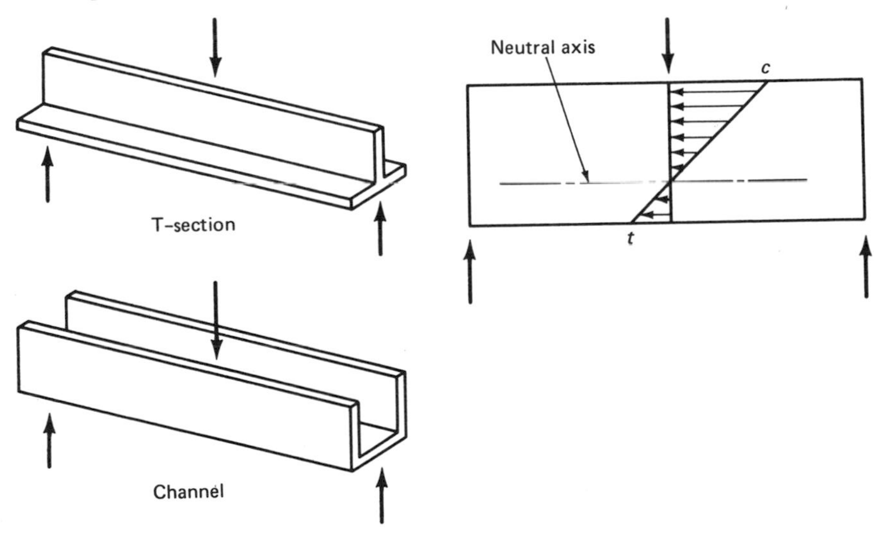

Figure 1.6. Beams under simple bending. If a section is symmetrical about the neutral axis, the tensile and compressive stresses are equal. With an asymmetrical section, the tensile stress can be decreased to match the lower tensile strength of the material. The two cast-iron beams have extra webs on the tension sides. The right-hand drawing shows the distribution of stresses in the beams on the left.

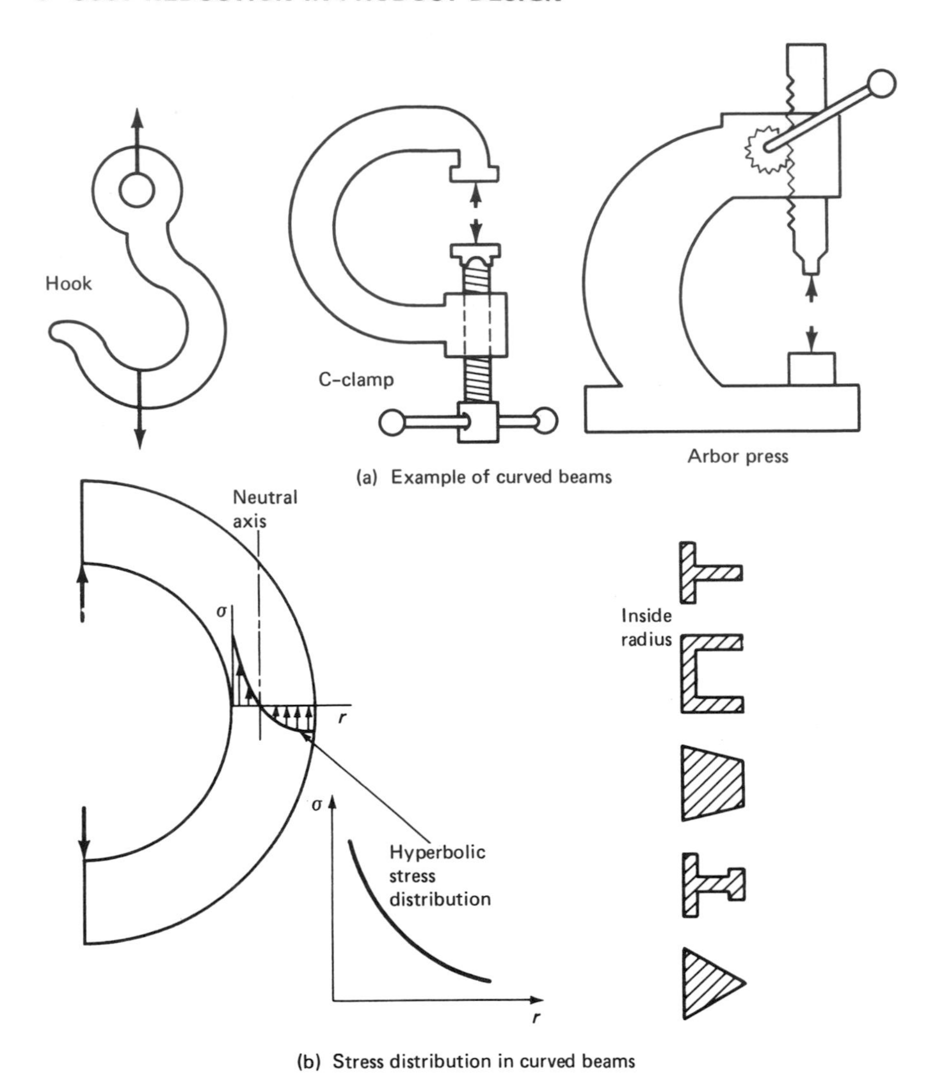

Figure 1.7. Balance of stress and strength for a curved beam. The sections shown in (b), with extra webbing on the inside radius, are desirable. The stress distribution in a curved beam is hyperbolic and the neutral axis does not coincide with the centroidal axis. Since the peak tensile stress at the inside radius is higher, the section should be heavier on the inside radius.

that are balanced in tensile stress and strength at the bottom surface. The geometry of the curved beam (Figure 1.7) dictates unequal tensile and compressive stresses in the beam unless extra webbing is used. The stress distribution in the curved beam is hyperbolic, and the stress is higher at the inside radius. The remedy is to shift the neutral axis toward the inside radius by using a T-section or a channel section with extra webs on the inside radius.

Efficiency is implied in engineering work, and the balance of stress and strength has always offered better products at lower costs. An I-beam is more efficient than a beam with a rectangular cross-section. The material in the I-beam is concentrated at the top and bottom surfaces where stresses are high (Figure 1.8(a)). If the I-beam is used sideways, the material will be concentrated near the neutral axis,

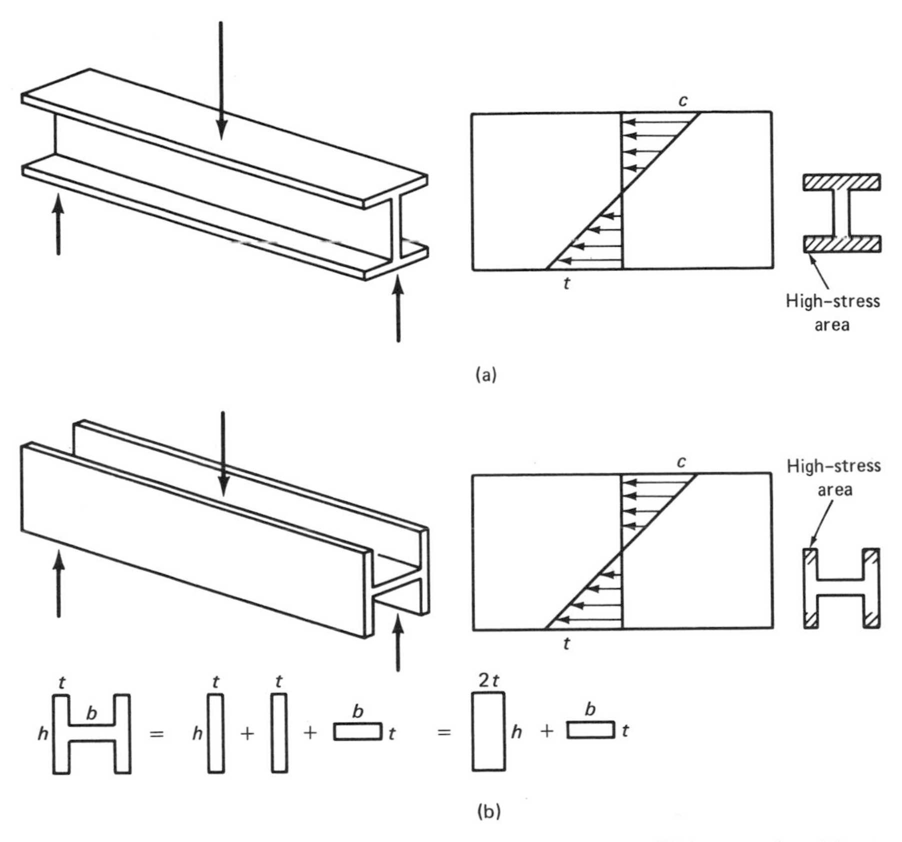

Figure 1.8. The efficient I-beam. In (a), note that the efficiency of an I-beam results from the concentration of material at the areas of high stress. If an I-beam is used sideways, as in (b), it is equivalent to an $h \times 2t$ rectangular section and a $t \times b$ rectangular section. The $t \times b$ section is close to the neutral axis and is under very low stress. When used sideways, therefore, the I-beam is less effective than a rectangular beam of the same depth h.

where the stress is nearly zero (Figure 1.8(b)). When used sideways, the I-beam is less efficient than the rectangular beam.

A quenched glass plate can support four times more bending load than the regular glass plate. Figure 1.9 shows the reason behind this: glass is five times stronger in compression than in tension and practically all failures of glass plates in bending are due to tensile fractures; quenching produces residual compressive stresses on both surfaces of the glass plate. When an external bending load is applied to the glass plate, the residual compressive stress is substracted from the tensile stress caused by the external load. Tensile fracture is thus suppressed. Spherical glass (or acrylic) windows for submarines are stronger if the vessel goes deeper. The material is precompressed by hydrostatic pressure so that tensile stresses are suppressed.

The gear and pinion in Figure 1.10 have equal forces on their teeth

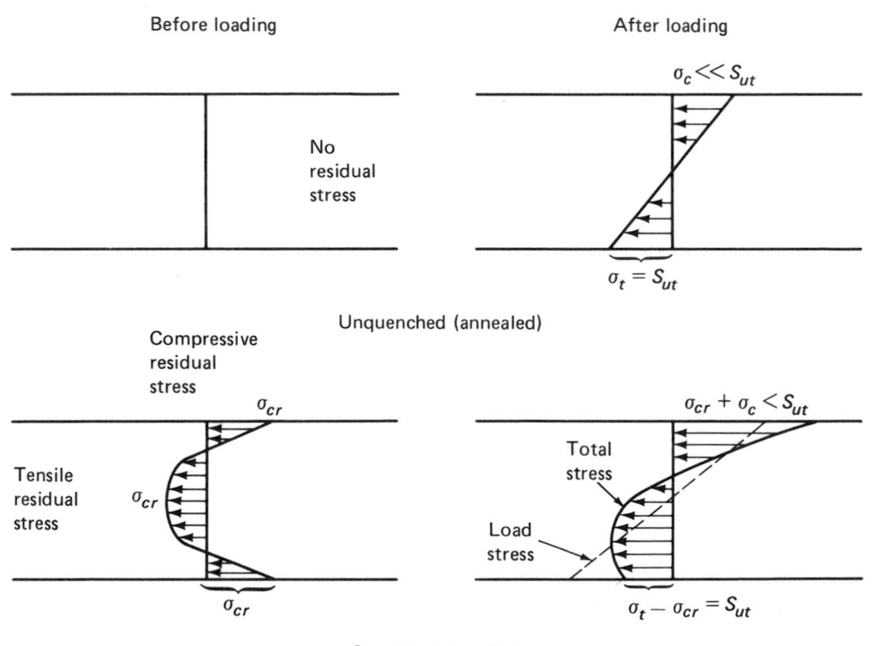

Figure 1.9. Quenched glass is strengthened by residual compressive stress. Since glass is five times stronger in compression than in tension, all failure of a glass plate in bending originated from tensile fracture at the bottom surface. Residual compressive stresses on both surfaces of the glass plate can reduce the tensile stress due to external load. The quenched glass can support four times more bending load than the unquenched glass.

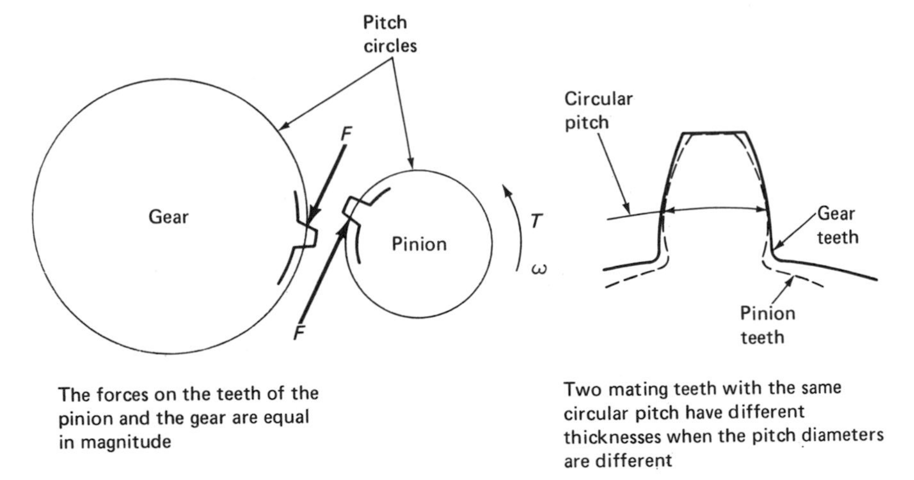

Figure 1.10. Balance of stress and strength in a gear set. The pinion is stressed more frequently because of its higher rotational speed. The pinion teeth are also thinner. In order for the pinion to be as durable as the gear, the pinion has to be made with a stronger material.

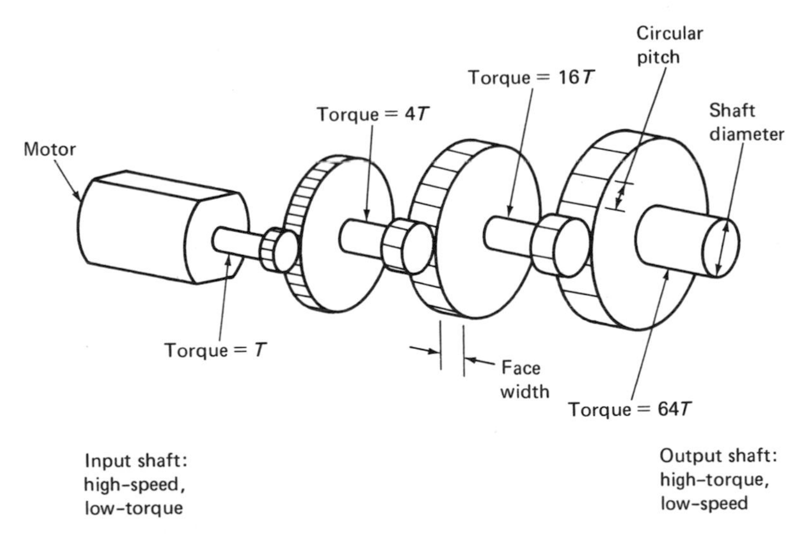

Figure 1.11. Balance of stress and strength in a gear train. Since the torque increases progressively from the motor down the gear train, the face width, the circular pitch, and the shaft diameter have to be increased correspondingly to keep the design balanced.

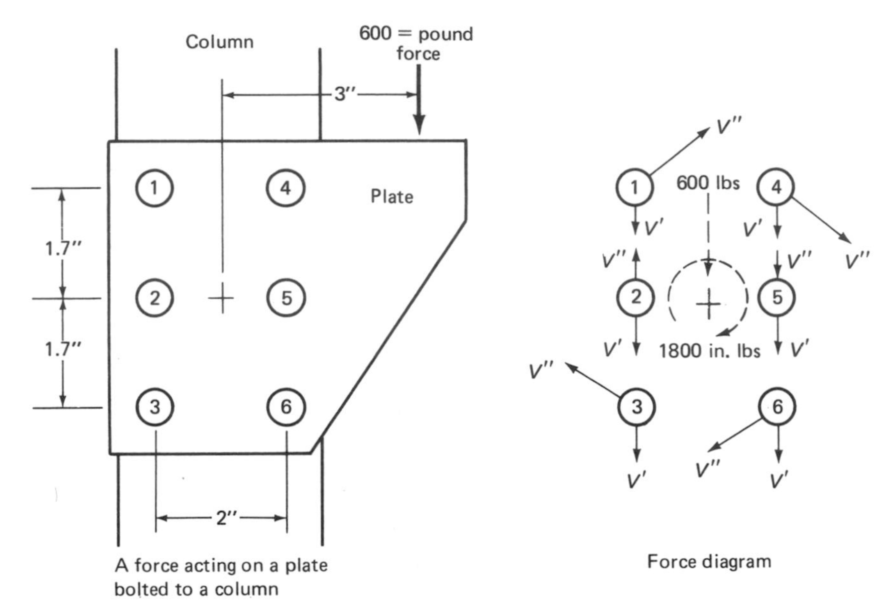

Figure 1.12. Stress distribution in a bolt connection. The eccentric force acting on a system of bolts can be considered as a force and a moment at the center of the system. The 600-lb equivalent force is equally divided among 6 bolts to produce primary shear forces $V' = 100$ lb. The moment (1800 in.-lb) produces secondary shear forces V'' in the bolts proportional to the distance between the bolts and the center of the system. Therefore, the secondary shear force is 200 lbs in bolts 1, 3, 4, and 6, and 100 lbs in bolts 2 and 5. The total load on each bolt is the vector sum of the primary shear force V' and the secondary shear force V''.
For equilibrium of the system,
summation of force: 600 lbs = 100 lbs × 6
summation of moment: 1800 in.-lbs = 200 lbs × 2" × 4 + 100 lbs × 1" × 2

Bolt	1	2	3	4	5	6
V' (pound)	100↓	100↓	100↓	100↓	100↓	100↓
V'' (pound)	200↗	100↑	200↖	200↘	100↓	200↙
V_{total} (pound)	173.2	0	173.2	265	200	265
Area ratio	1.	0.	1.	1.53	1.15	1.53
Diameter ratio	1.	0.	1.	1.24	1.07	1.24

because of their action and reaction relationship. However, the pinion teeth have higher stresses because of their thinner profiles. The pinion teeth are also more vulnerable to fatigue failure because the pinion has the higher rotational speed and the pinion teeth are, therefore, more frequently loaded. If the gear and the pinion are made with the same material and the same accuracy, the pinion will always fail first. Cost reduction can be realized by using slightly less strong material for the gear, such that both the gear and the pinion have the same safety factor. A typical gear set may have a cast-iron gear and a steel pinion.

In a gear train, the motor is usually a high-speed, low-torque driver. The input torque from the motor is progressively increased through the gear train to the output shaft. Figure 1.11 shows an increase in the face width and the circular pitch of the gears. The diameters of the shafts are also increased to keep the balance of strength.

Figure 1.12 shows a bolt connection designed for supporting an eccentric load. As explained in the diagram, the ratio of shear forces on the bolts 1, 2, 3, 4, 5, and 6 is 1:0:1:1.53:1.15:1.53. Bolt 2 can be eliminated because there is no load on it. Bolts 4 and 6 are loaded 53% more than bolts 1 and 3. Bolt 5 is loaded 15% more than bolts 1 and 3. To balance the stress to the strength, the ratios of the diameters of the bolts should be 1:0:1:1.24:1.07:1.24. Theoretically, this balance of stress and strength can be applied to many joining methods. Practically, the extra work involved in installing bolts of different sizes to different holes may offset the savings in the cost of the bolts.

1.3. FAILURE CRITERIA AND DESIGN OBJECTIVES

A system that cannot perform its designed objective is said to fail. Failure does not imply fracture. A fuse is designed to protect electric circuits from being overloaded. If the fuse does not burn in case of an overload, it is a failure. Thus an overdesigned fuse not only wastes material, but also leads to failure.

A micrometer designed to measure accurately would be a failure if the deflection in the measuring arm were excessive. There is no fracture or destruction associated with such failure. Structurally, a C-clamp and a micrometer are similar (Figure 1.13). The former is designed for strength and the latter is designed to measure accurately with no excessive deflection of the curved beam.

The following subsections discuss four common types of load and design strategies.

1.3.1. Design for Static Strength

Design for strength is the same as design against yielding and fracture. The criterion is that the strength of the material must be larger

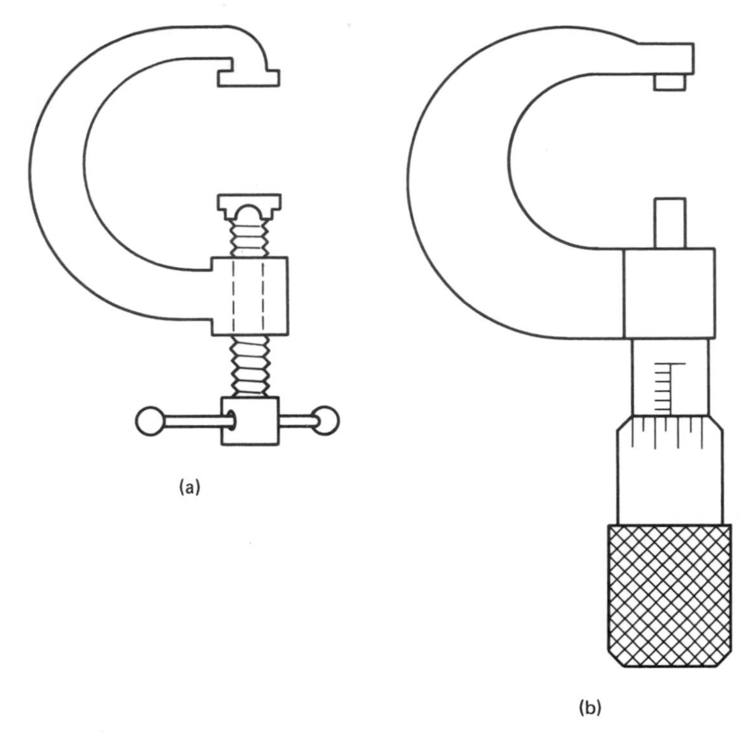

Figure 1.13. Two tools with similar structures but different design objectives. (a) The C-clamp is designed to exert a clamping force, and the micrometer (b) is designed to measure accurately with only minimum deflection in the curved beam.

than the stress σ anywhere in the machine, especially in the critical element.

To prevent yielding,

$$S_y > \sigma \tag{1.2}$$

To prevent fracture,

$$S_{\text{ult}} > \sigma \tag{1.5}$$

where S_y is the yield strength and S_{ult} is the ultimate strength.

Ductile materials usually fail by yielding. Brittle materials usually fail by fracturing.

1.3.2. Design for Deflection

The objective is to design the member to deflect or twist. The load and the stress are both induced by the stiffness of the member.

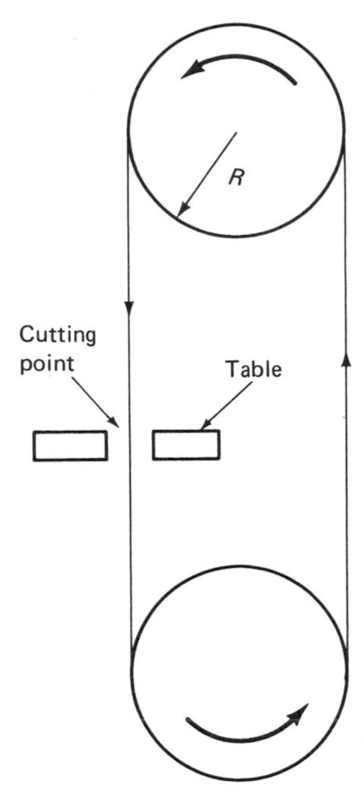

Figure 1.14. A band saw blade. The stress in the band saw blade due to bending is directly proportional to the thickness and inversely proportional to the radius of bending.

$$\text{Strain} = \frac{1}{2}\frac{t}{R}$$

$$\text{Stress} = \frac{E}{2}\frac{t}{R}$$

where t is the blade thickness.

Figure 1.14 shows a band saw blade running between two drums. The minimum radius of the drum is governed by the thickness and the material properties of the blade. If a blade breaks because of bending around the drum, a thicker blade would break even sooner. A thin blade made of a high-strength (large S_y) and low-modulus (small E) material would allow a smaller bent radius. Note that *strong* and *weak* refer to the yield strength or ultimate strength of a material (depending on whether the material is ductile or brittle). *Hard* and *soft* refer to the elastic modulus of the material. Fiber glass, which deflects and bends easily, is very strong in tension.

1.3.3. Design for Rigidity

Design for rigidity (maximum stiffness) is the opposite of design for deflection. This design criterion is usually associated with precise machining and measuring. The stress level involved is very low, and there is no danger of failing by fracture. Cast iron is used a lot in machine tools because of a high modulus-to-strength ratio. Rigidity in structures is important in the prevention of swaying and oscillating. Rigidity and sturdiness are often associated with the feelings of security and strength. When testing a used chair in a garage sale, a buyer will usually perform a rigidity test to judge the condition of the chair. Rigidity is an important factor in the design of consumer products. Note that this factor favors the use of low-strength, high-modulus materials.

1.3.4. Design for Impact

An impact is a load applied suddenly to an object. Impacts can create momentarily peak stresses and subsequent vibrations in the impacted bodies. The design for impact will be treated in detail in the next section.

1.4. IMPACT STRENGTH

It is very unlikely that drinking glasses or china plates will fail under ordinary static loads. However, they will chip or fracture under local or overall impact. Impact is likely to be the most severe load on an object. That is why high-impact styrene* has completely replaced crystal styrene in many applications. Impact design (shockproofing) is important in automobile bumpers, automobile safety devices, power tools, watches, and electric and electronic products. Impact is also the number one factor in designing boxes and containers for packaging.

In an impact situation, the controlling characteristic of a structure is the product of strength and deflection, i.e., its energy capacity. Although some of the kinetic energy is dissipated as heat, sound, and vibration, we can obtain a conservative design by assuming that all the kinetic energy is absorbed by the energy capacities of the im-

*Styrene is the most commonly used thermoplastic for injection molding.

pacted bodies. A specification for impact strength might be: "the object has to survive drops in all directions from a height of 3 feet onto a concrete floor."

If you hit a hammer hard on a steel anvil, the impact force can be as high as 2000 pounds. Applying the same impacting action on a pillow will cause an impact force of about 10 to 20 pounds. The difference in the impact force is due to the fact that a pillow will "give," and thus absorb the impact energy over a longer distance. As the steel anvil will not "give," the impact energy has to be absorbed over a very small distance (such as 0.05 inch), which results in a huge force. Impact energy equals one-half the product of the maximum force and maximum deflection:

$$\text{impact energy} = \tfrac{1}{2} F_{\max} \cdot \delta_{\max} \tag{1.6}$$

For the pillow, 50 in.-lbs = ½ × 20 lbs × 5 in.; for the anvil, 50 in.-lbs = ½ × 2000 lbs × 0.05 in., where the 50 in.-lbs is a reasonable estimate of the kinetic energy in the hammer immediately before impact.

1.4.1. Material Toughness

The energy a unit volume of material absorbs is the area under the stress-strain curve. The elastic part of this energy capacity is called the modulus of resilience R, which is mathematically equal to $\tfrac{1}{2} \cdot (S_y)^2/E$. The energy capacity up to the fracture point of a material is the toughness. Material A in Figure 1.15 has a high yield strength and a high ultimate strength. However, material A is not as tough as material B because of its smaller strain in fracture ϵ_f. Material B has a larger elongation at fracture and, thus, a higher energy-absorption capacity. High-impact styrene is made by adding low-strength ingredients (plasticizers) to crystal styrene. The purpose of this is to increase the elongation at fracture at a small expense of the ultimate strength. The material becomes noticably softer (i.e., with a smaller elastic modulus E). Quenched or cold-worked metallic parts are often annealed to increased ductility and impact strength. Annealing in a controlled elevated temperature allows for increased molecular orientation and increased grain size. Impact strength is an important factor in choosing composite materials. As an example, adding microglass beads to plastics will increase the ultimate strength and decrease the elongation. The toughness of the material

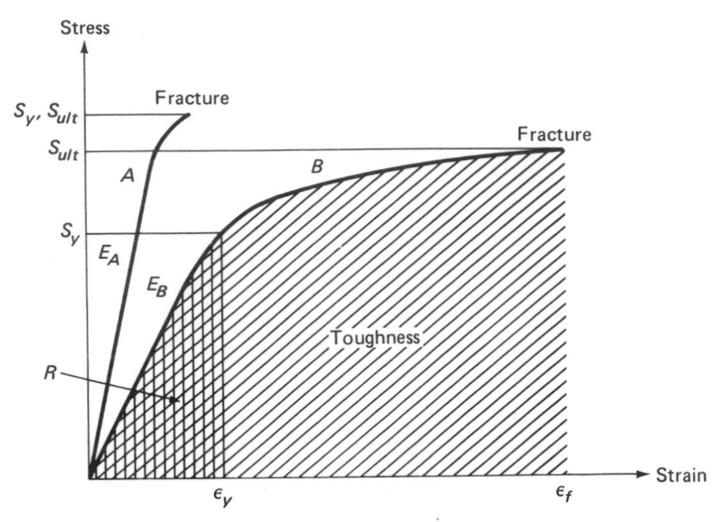

Figure 1.15. Material properties revealed in the stress-strain curve. S_{ult} = ultimate strength; S_y = yield strength; ϵ_f = strain in fracture; ϵ_y = yield strain; E = elastic modulus; and R = modulus of resilience $= S_y^2/2E$.

may be lowered. This composite material would not be suitable for an application where impact strength is needed.

An impact with more energy than the toughness of the material causes fracture. It is found that materials usually are stronger if the duration of the load is short. Hence, the impact toughness might be quite a bit larger than the static toughness obtained from a static stress-strain curve.

1.4.2. Geometric Resilience

A soft object can absorb energy with much lower forces and stresses. Therefore, objects designed to withstand impact should have minimum stiffness. During the early 1970s, many soft drums were placed in front of such highway structures as columns that supported overpasses. If a car crashes into such soft barricades, the soft drums will collapse and absorb the impact energy, leaving the car and the driver relatively unharmed. Another example is the collapsible steering column that is made comparatively soft by means of a net structure (Figure 1.16). Some special floor tiles are made with foam rubber backing. When china plates are dropped on such tiles, they will not shatter.

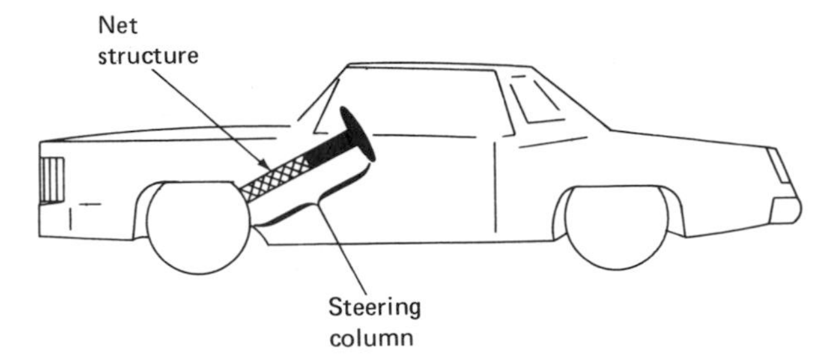

Figure 1.16. The net structure in the steering column of a car reduces impact forces.

The spring rate K can be defined as the ratio of the force F to the deflection δ at any part of an object. The energy of impact will produce a peak force equal to $\sqrt{2KU}$, where U is the impact energy.

$$\text{Impact energy } U = \frac{1}{2} F_{max} \cdot \delta_{max} = \frac{1}{2} \cdot \frac{(F_{max})^2}{K} \tag{1.7}$$

$$\text{Peak force } F_{max} = \sqrt{2KU} \tag{1.8}$$

Therefore, a lower spring rate K will decrease the peak force F_{max}. For the tension element shown in Figure 1.17, the spring rate is $K = AE/l$. To decrease K, a material with a low elastic modulus E can be used, and the bar can be made long and slim. For the cantilever shown in Figure 1.18.

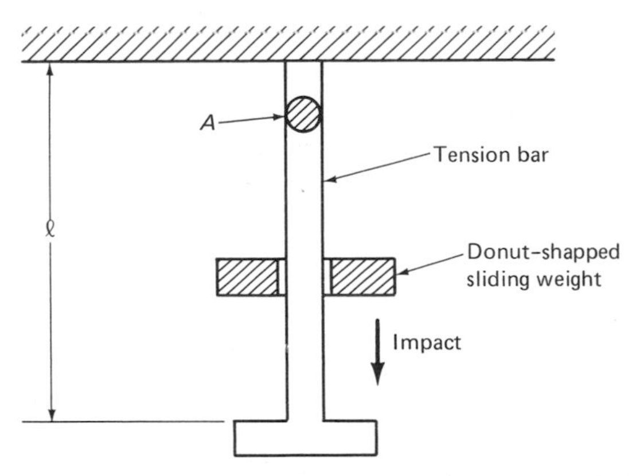

Figure 1.17. A tensile impact. The peak force upon impact is directly proportional to the spring rate K. $K = AE/l$, where A = sectional area, l = length of tension element, and E = elastic modulus of the material.

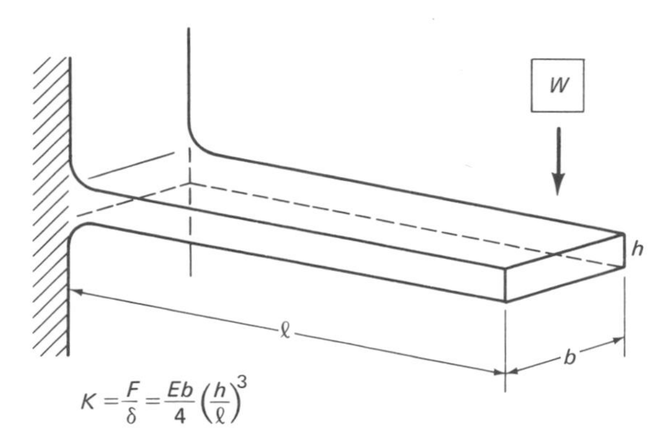

Figure 1.18. Spring rate of a cantilever beam.

$$K = \frac{3EI}{l^3} = \frac{Eb}{4}\left(\frac{h}{l}\right)^3$$

The spring rate K is proportional to the third power of h/l, and is directly proportional to E and b.

1.4.3. Uniform Stress Distribution

Uniform distribution is usually the ideal stress distribution. In such a case, the design is balanced and the material is fully utilized. The eye-bar and threaded rod shown in Figure 1.19 are commonly used to carry tensile loads in the building of bridges, towers, and other structures. The ends of the eye-bar are purposely made stronger than the body because of impact energy considerations. If the body of the eye-bar fails under tension, the energy absorbed would be the toughness times the volume (less the volume of the ends, which is very small compared to the body). If the ends fail first, the body of the eye-bar has no chance to be stressed into the inelastic region. The energy absorbed could be a few percent of the material toughness ($\sigma^2/2E$, where σ is the stress in the tension member when the ends fail). Both the threaded rods in Figure 1.19(b) and (c) are limited by the strength at the threads. The rod in Figure 1.19(c) is stronger in impact and it consumes less material. Since the rod (c) has a slimmer shank, it stretches more upon impact. This elasticity lowers the peak force, in a manner similar to that in the example of the pillow and anvil.

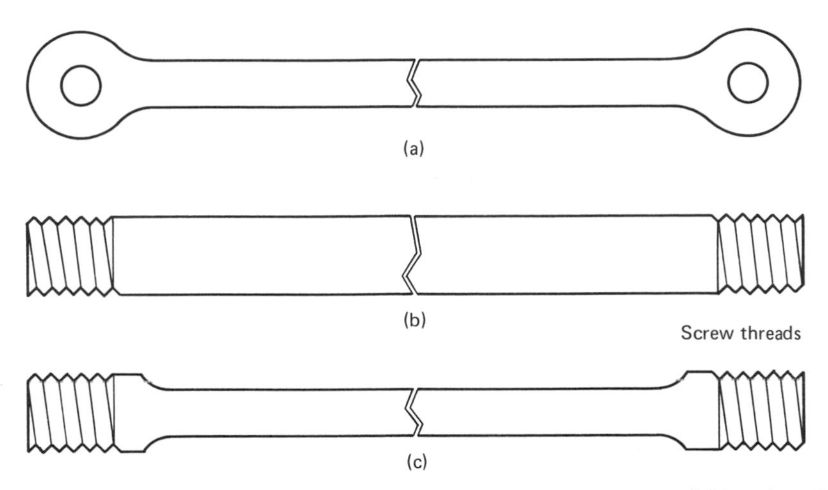

Figure 1.19. Three kinds of tension members. (a) is an eye-bar, (b) is a threaded rod, and (c) is a threaded rod with upset ends. (The actual bars are much longer.)

If the stress is really uniform, the energy capacity of the object is equal to the toughness of the material times the volume. This ideal case is realized in a fishing line. In most practical situations, end connections and geometric arrangements prevent perfectly uniform stress distributions. Both cantilever beams shown in Figure 1.20 have equal static strengths. The uniform cantilever beam can absorb energy which equals 1/9 the modulus of resilience of the material, i.e.,

$$\frac{1}{9}\left(\frac{S_y^2}{2E}\right)$$

The tapered cantilever can absorb 1/3 the modulus of resilience of the material, i.e.,

$$\frac{1}{3}\left(\frac{S_y^2}{2E}\right)$$

The tapered cantilever utilizes only 33% of its material energy capacity because of the nonuniform stress distribution across the section: the stress is zero at the neutral axis of bending. Further improvements can be made by changing the cross-section to an I-section.

To obtain a uniform stress distribution, stress concentration should be avoided, and 100% of the material should be subjected to the

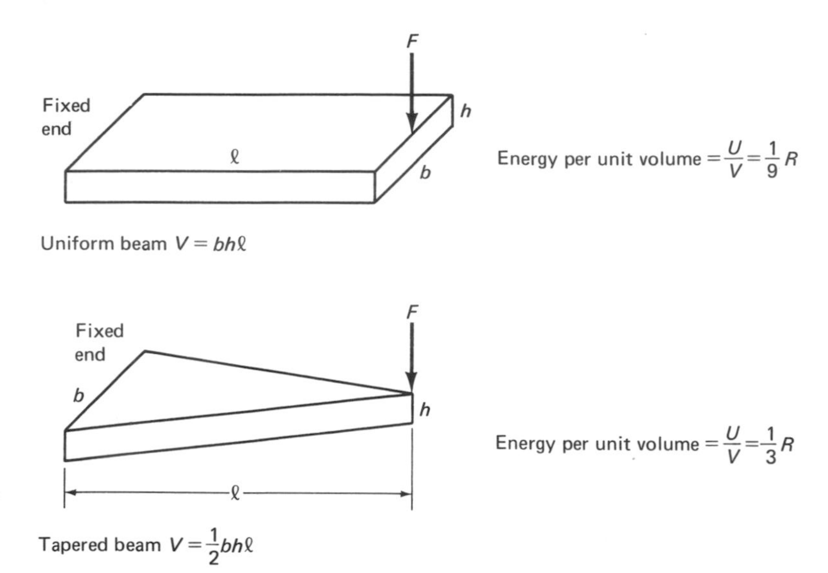

Figure 1.20. Efficiency of material usage in two different cantilever beams. When yielding occurs at the fixed ends of the cantilever beams, the absorbed energy per unit volume of material in the uniform beam is only $1/9$ the modulus of resilience R, but that in the tapered beam is $R/3$.

highest level of stress. Figure 1.21 is a comparison of six different tension members:

(a) a standard specimen with a sectional area A and a length l,
(b) a specimen with nonuniform sectional areas,
(c) a specimen with a stress concentration knotch: the static strength is reduced 50%–30% due to the reduced area and 20% due to the stress concentration factor K_t; the decrease in impact strength is 75%,
(d) a specimen with static strength equal to those of (b) and (c), but with higher impact strength and less material,
(e) an extra resilient specimen, and
(f) an extra stiff specimen.

The static strength of (f) is 200% that of (a). All other specimens have 50% the static strength of (a). The impact-energy capacity per unit volume of the material is the same for (a), (d), (e), and (f). As long as the stress is uniform, all these members are equally efficient in absorbing impact energy. Because there are differences in resilience, (f) has the highest peak force and (e) has the lowest peak force.

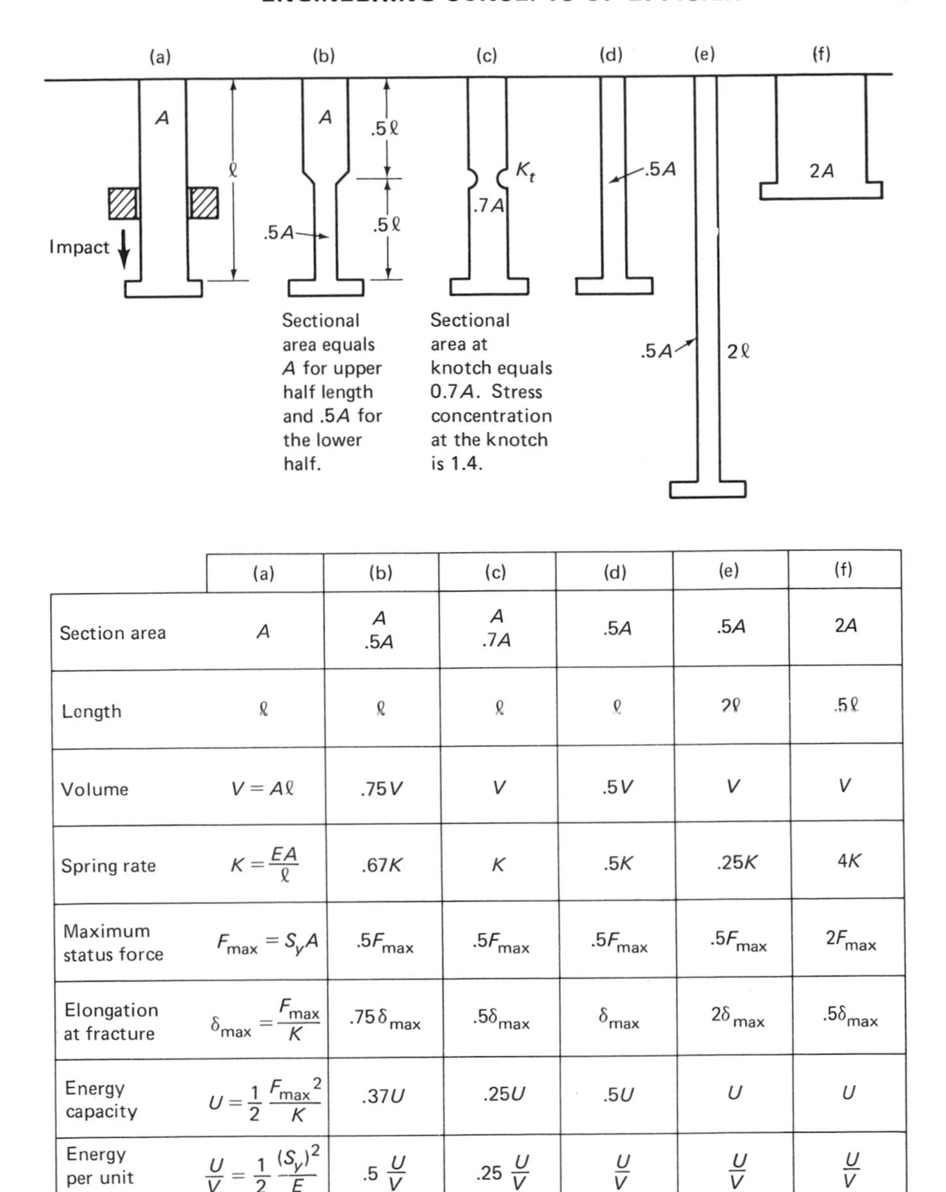

	(a)	(b)	(c)	(d)	(e)	(f)
Section area	A	A $.5A$	A $.7A$	$.5A$	$.5A$	$2A$
Length	ℓ	ℓ	ℓ	ℓ	2ℓ	$.5\ell$
Volume	$V = A\ell$	$.75V$	V	$.5V$	V	V
Spring rate	$K = \frac{EA}{\ell}$	$.67K$	K	$.5K$	$.25K$	$4K$
Maximum status force	$F_{max} = S_y A$	$.5F_{max}$	$.5F_{max}$	$.5F_{max}$	$.5F_{max}$	$2F_{max}$
Elongation at fracture	$\delta_{max} = \frac{F_{max}}{K}$	$.75\delta_{max}$	$.5\delta_{max}$	δ_{max}	$2\delta_{max}$	$.5\delta_{max}$
Energy capacity	$U = \frac{1}{2}\frac{F_{max}^2}{K}$	$.37U$	$.25U$	$.5U$	U	U
Energy per unit volume	$\frac{U}{V} = \frac{1}{2}\frac{(S_y)^2}{E}$	$.5\frac{U}{V}$	$.25\frac{U}{V}$	$\frac{U}{V}$	$\frac{U}{V}$	$\frac{U}{V}$

Figure 1.21. Comparison of energy capacities of six tension bars (for simplicity, the material at the lower end of the bars is ignored).

The largest maximum elongation is in (e) and the smallest maximum elongation is in (c) and (f).

Specimens (c) and (d) illustrate the idea behind high-impact bolt design, which is further described in Figure 1.22. The static tensile strengths of regular and high-impact bolts are the same because the critical area in both is at the root of the screw threads.

$$F = \frac{S_y \cdot A_r}{K_t}$$

where

A_r = root area,
K_t = stress concentration factor, and
S_y = yield strength of material.

With the sectional area of the shank reduced to A_r/K_t, the bolt becomes more resilient as a whole. A slender member, if properly designed, is tougher than a bulky member.

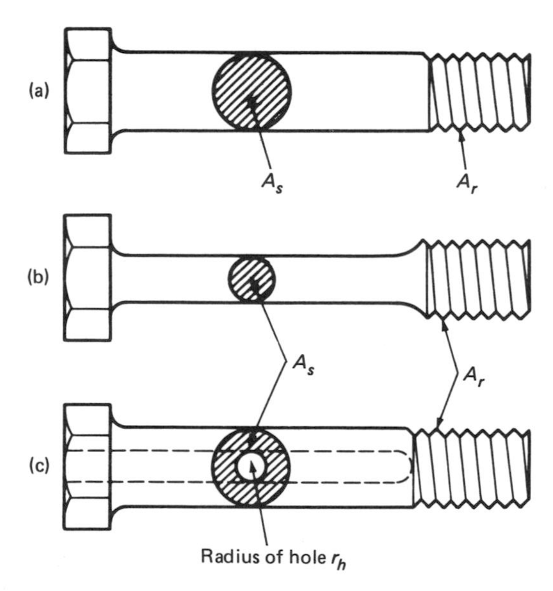

Figure 1.22. Two bolts designed for high-impact strength. (a) is a regular bolt: shank area $A_s >$ root area A_r. (b) is a high-impact bolt with reduced shank: shank area $A_s = A_r/K_t$, where K_t = stress concentration at the thread. (c) is a high-impact bolt with a hole drilled in the shank: shank area $= \pi(r^2 - r_h^2) = A_r/K_t$. The bolts all have the same static strength. Since the shank area is decreased in the bottom two bolts, the material in their shanks is subjected to higher stresses and absorbs more energy. The principle is to subject a large percentage of material to the maximum stress level.

1.4.4. Other Factors in Impact Strength

To reduce impact force, an object should be made as light and soft as possible. Also, the stress should be uniformly distributed, and the material in the object should be tough.

If a 10-ton truck collides with a 2000 pound car, the impact energy will be "unfairly" distributed. The truck contributes ten times as much energy as the car. The total impact energy is then absorbed by the two vehicles in proportion to their softness. Thus, the flimsy car that has contributed little to the impact energy absorbs a large portion of it, and may be badly damaged. The truck, on the other hand, may have its bumper barely scratched by the collision. Many automobile engineers feel that trucks should be made extra soft (by governmental regulations) so that they get a fair share of the impact energy they contribute.

If an impact is elastic, a very small percentage of the impact energy will be dissipated as heat and vibrational energy in the impacting bodies. The rest of the energy will cause the impacted bodies to rebound. The ratio of the retracting velocity to the approaching velocity for the system is called the *coefficient of restitution.* Since most of the energy in an inelastic impact is absorbed, the bodies involved in an inelastic impact have low coefficients of restitution. When a soft bullet is fired into a deer, the impact causes the slug to mushroom. Ammunition experts found that a bullet can be made more devastating if it can be made to mushroom more. The harm is done by the impact shock to the animal and the penetration of a large flattened slug.

Impact can be used to amplify force and power from low-energy and low-force sources. As an example, the hammer amplifies small fist forces to several hundred pounds of driving force. To increase impact, the initial velocity should be high, the mass should be large, the shape should be stiff, and the material should be able to withstand peak stresses produced by the impact.

1.5. TENSION VERSUS COMPRESSION

A suspension bridge supports weight by tension in the steel cables (Figure 1.23). An arch bridge supports weight by compression in the material. Sometimes a designer has to choose between using a tension or a compression member. In this case, the material strength, the geometric stability and the fatigue strength in each member have to be considered.

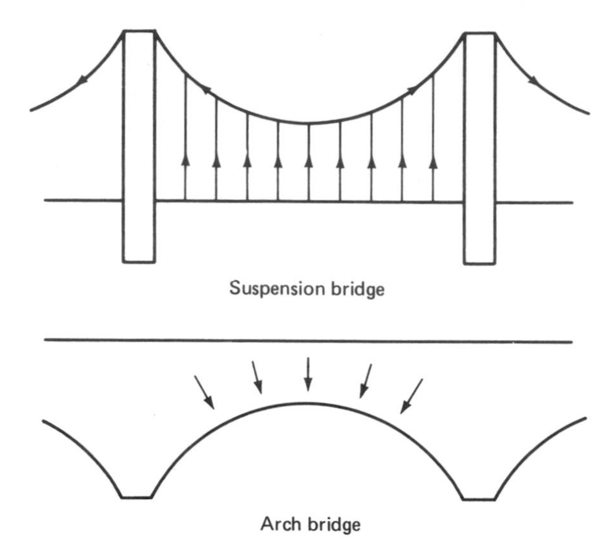

Figure 1.23. A suspension bridge supports weight by tension while an arch bridge supports weight by compression.

Many materials are stronger in compression. Glass rods and concrete bars will fail in tension if bending loads are applied. The ratio of compressive strength to tensile strength is 3.5 for cast iron, 5 for glass, and 10 for concrete. Noticeable exceptions are materials with oriented grain or molecular structures such as steel cables, nylon fibers, natural fibers, and glass fibers. Such materials are stronger in tension than in compression.

Stability is the major advantage of tension members. A member that is slightly curved will straighten out under a tensile load (Figure 1.24(a)). Compression members are unstable and may buckle. The stability of long and slender compression members depends on the elastic modulus E, rather than on the material yield strength S_y. As far as stability goes, high-strength steel members are no better than low-strength steel members.

Though compression members are not as stable as tension members, they will not fail from fatigue or crack propagation as tension members would. If a crack is on the tension side of a beam, it will quickly spread to a gross failure when a bending moment is applied (Figure 1.25(a)). This principle is used in the cutting of glass plates. Impact is applied to the side of the glass plate opposite to the crack in order to break the plate. If the crack is on the compression side of a mem-

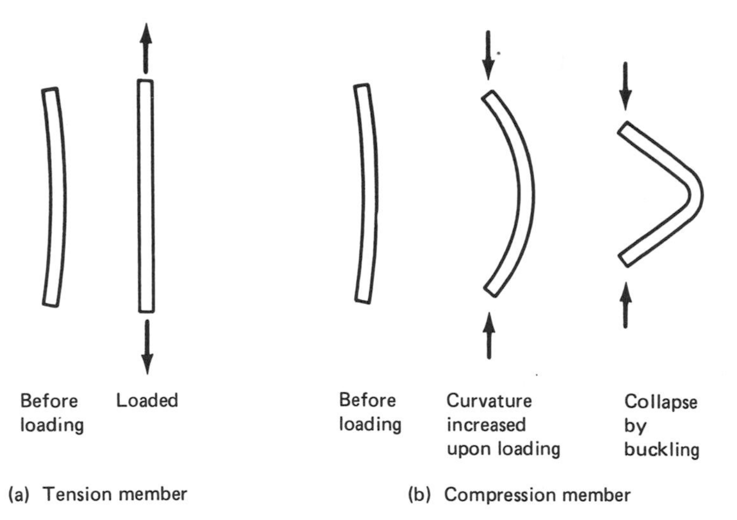

Figure 1.24. Stability in tension member and instability in compression member. A tension member is comparatively stable and is self-straightening. Compression members are usually unstable and may buckle.

ber, a bending moment will close up the crack (Figure 1.25(b)). Figure 1.26 shows that compressive force-flow lines are not interrupted by cracks, but tensile force-flow lines are. The concentration of tensile force-flow lines indicates a severe stress concentration at the tip of the crack.

The above discussion on tension and compression members is summarized in Table 1.1. The following examples will illustrate the design rules.

A concave end-plate for a pressure cylinder experiences compressive stresses and has the danger of buckling (Figure 1.27(a)). A convex

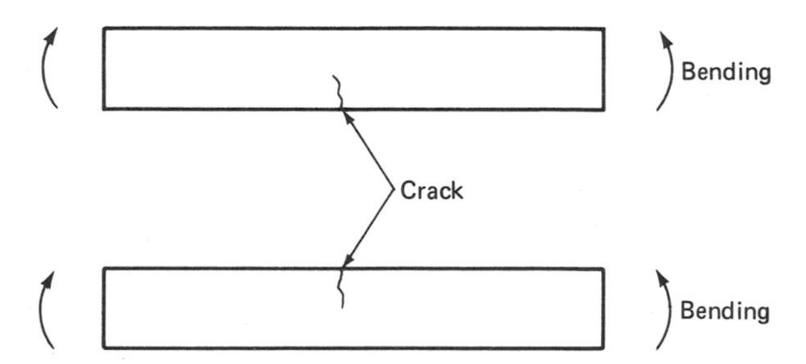

Figure 1.25. Crack propagation and nature of stress. (a) Under bending, a crack on the tension side of a member will propagate until failure. (b) A crack on the compression side will not propagate.

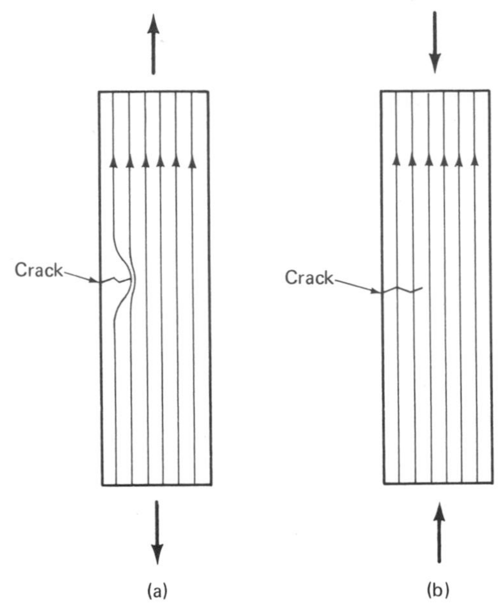

Figure 1.26. Cracks and kinds of load. (a) A crack will interrupt tensile force-flow lines. The concentration of flow lines indicates stress concentration. (b) Compressive force-flow lines are not interrupted by cracks.

TABLE 1.1. COMPARISON OF TENSION AND COMPRESSION MEMBERS.

	Tension member	Compression member
Material strength	Materials usually have lower tensile strength	Materials usually have higher compressive strength
Effect of Poisson's Ratio ν	Sectional area decreases under load by $\Delta A/A = 2\epsilon_l \nu$ ν = Poisson's ratio ϵ_l = longitudinal strain	Sectional area increases under load by $\Delta A/A = 2\epsilon_l \nu$
Crack propagation	Transverse crack easily leads to complete fracture	Cracks will not propagate
Fatigue	Vulnerable to fatigue failure	No danger of fatigue failure
Stability	Stable and self-straightening	May buckle if the member is not rigid enough

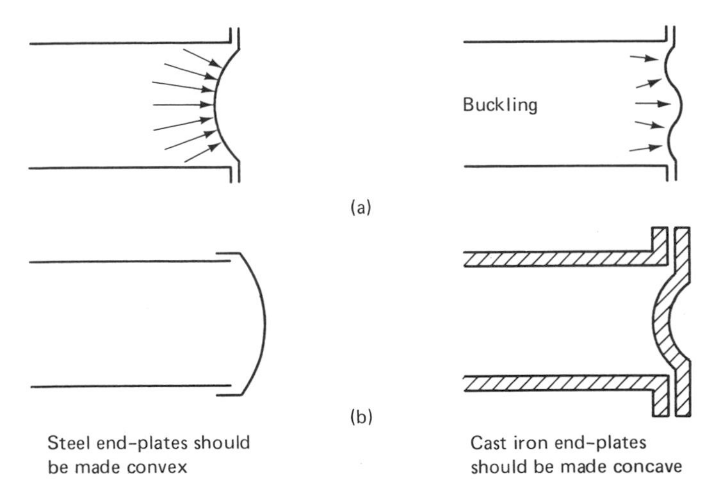

Figure 1.27. Choosing between concave or convex end-plates. (a) The instability of a concave end-plate. (b) The matching of geometries to material properties. A concave end-plate is under compression and may be unstable (may buckle). A convex one is stable, but may fail if the material has low tensile strength.

end-plate is under tensile stress and may fail if the material in the plate has a low tensile strength. Therefore, a steel end plate should be made convex. If cast iron is used, the concave shape is preferable (Figure 1.27(b)). The choice of the shape depends on the strength-to-stiffness (elastic modulus E) ratio of the material. The same argument can be applied to a dam or a bridge.

Figure 1.28 shows a shaft supported by the bearings A and B. A helical gear is located at the point C on the shaft. The force in the gear produces an axial force in the shaft, which tends to move the gear and shaft up toward the bearing A. A thrust bearing at A would produce compression in the section AC in the shaft. A thrust bearing at B would produce tension in the shaft between BC. The location of the thrust bearing depends on the stability of the shaft and the strength of the shaft material. Typically, a thrust bearing at B is better because stability is a great concern in shaft design.

In Figure 1.6, the cast-iron channel is more efficient if the open side is facing up. The same channel, if made of a high-strength steel, should have the open side facing down. The weakness of a cast iron beam is its low tensile strength, whereas the weakness of a high-strength steel beam is its lateral buckling instability.

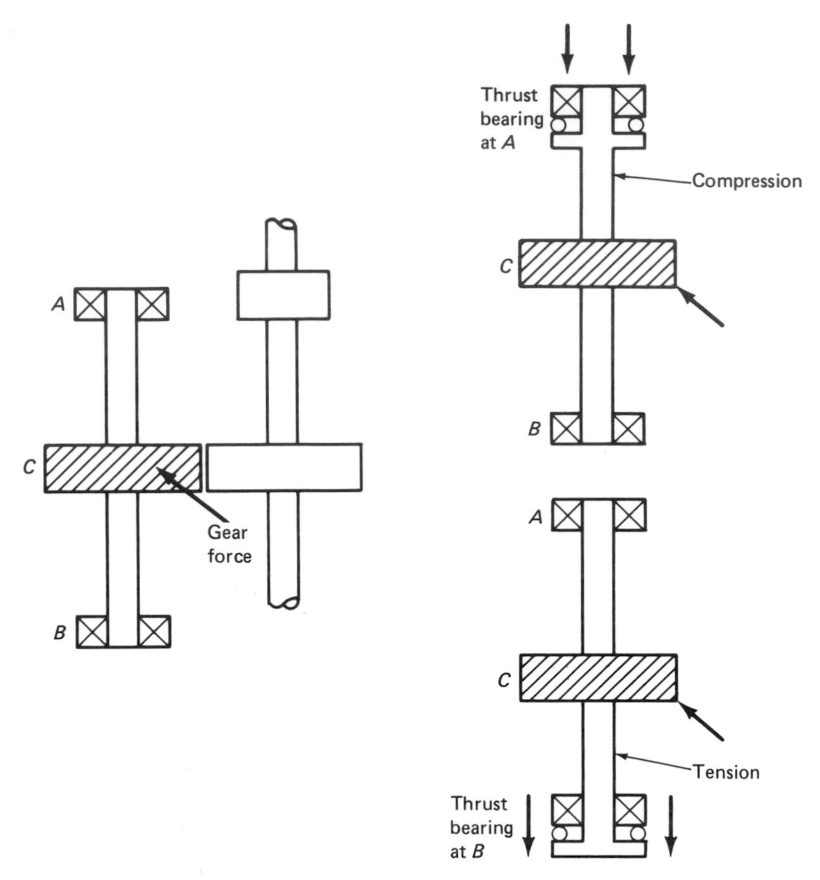

Figure 1.28. Tension and compression in the shaft due to the load at the teeth of a helical gear. A thrust bearing can support axial load in a shaft. Only one thrust bearing should be used because redundant support may produce interference stresses. If the thurst bearing is placed at A, the section AC in the shaft is under compression, section CB is not under axial load. If a thrust bearing is placed at B, the section BC in the shaft is under tension, section AC is not under axial load.

2. MATHEMATICAL METHODS IN COST REDUCTION

If there is more than one possible design for a product, a choice among the feasible designs has to be made. Usually, the one with the minimum cost is chosen. When only a few alternatives exist, the choice is easily made by evaluating and comparing the cost of all alternatives. With a large number of alternatives, such exhaustive evaluation is not practical. A trial-and-error approach not only is time-consuming, but often fails to yield the best design. *Optimization* is a method for determining the best design without guessing.

Optimization has an interesting mathematical implication. For any two points, the number of paths between them is infinite. Only the shortest path is mathematically distinguished from the rest. Optimization in design is also called *design synthesis*, as opposed to the practice of *design analysis*. In a cost-analysis process, the characteristics of the design are specified and the cost is determined from fixed parameters. In synthesis, the design characteristics are variable within some bounds and the cost is minimized. Analysis is simpler than synthesis because it involves direct substitution of known quantities into governing equations. Synthesis deals with such manipulations as mapping constant cost lines or tracing slopes of cost functions.

Many physical laws of nature are in effect representations of ideal solutions. For example, we can make use of the fact that light travels along the shortest path between two points to find an optimum path. In this chapter, we will start with the common, obvious physical models and gradually introduce some techniques of optimization. The aim of this chapter is to let the readers grasp the essentials with the least pain regarding mathematical notations and procedures.

2.1. SHORTEST DISTANCES

The shortest distance is of great interest to a cost-conscious engineer because it represents the minimum travel time, the minimum length of pipes (or wires or roads), and the least energy involved in transportation (gasoline for cars, resistance loss in electricity, viscous loss in fluid). In the simplest form, the shortest distance between two points is a straight line. Real-life problems are more complicated than this. Where there are many nodes (input and output points), the meaning of *distance* is expanded to connote the total length of path multiplied by weighing factors, such as the amount of flow or the traffic. For example, airline companies might like to run the shortest flight routes that can connect several specific cities. This is one of the many problems in *network analysis*, which may appear in a piece of electronic equipment, piping in a hydraulic network, or other applications.

The classical method to determine the shortest distance is the calculus of variation. The paths of shortest distance on a 3-dimensional object are geodesic lines (Figure 2.1). Modern methods in network analysis are found in the field of operations research and electric circuit analysis. Many of the physical laws are helpful in visualizing the

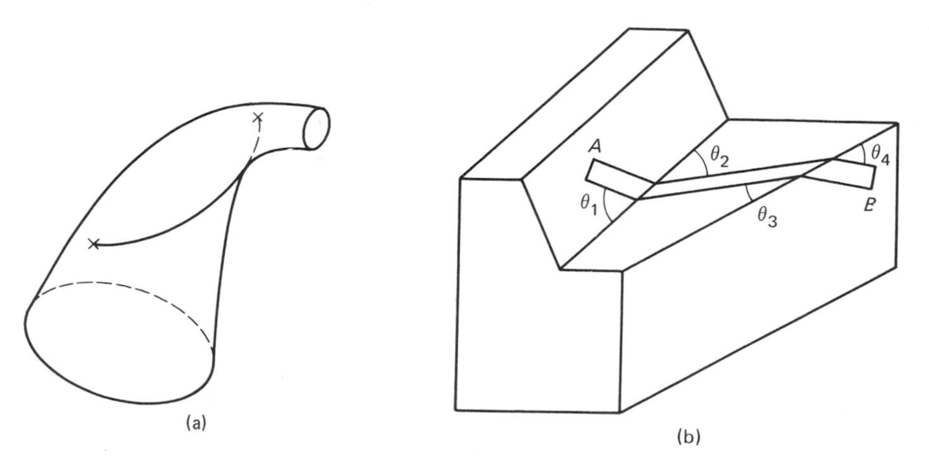

Figure 2.1. Shortest distances on 3-dimensional objects.

mathematics of shortest distances. Examples are: (1) light travels in the shortest path, (2) an equilibrium position is one with the lowest potential energy, and (3) a soap film maintains the smallest surface area because of surface tension.

2.1.1. Light Travels in the Shortest Path

The optical laws of reflection and refraction describe simply how light travels in the shortest path between two points. In the following problem, we are going to find the best location for a hydroelectric power plant to serve two cities A and B. As shown in Figure 2.2, the cities are at distances AR and BQ away from the river. The plant is to be placed so as to minimize the total length of the cable $(AP + PB)$. Assuming the river bank is a mirror and A' is the reflected image of A, the length of the straight line $A'B$ would be the shortest distance between A and B. From the laws of reflection,

$$\theta_A = \theta_B \tag{2.1}$$

$$\frac{AR}{RP} = \frac{BQ}{PQ} = \frac{AR + BQ}{RQ} \quad \text{(similar triangles)} \tag{2.2}$$

$$RP = \frac{AR \cdot RQ}{AR + BQ} \tag{2.3}$$

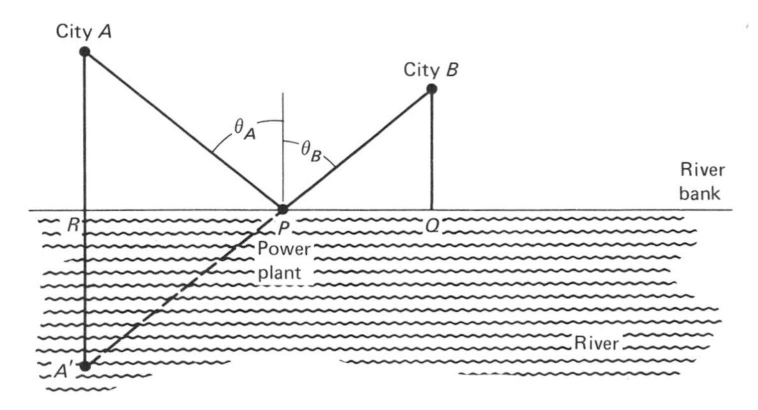

Figure 2.2. Locating the shortest path $(AP + PB)$ by the laws of reflection. A' is the image of A, and θ_A equals θ_B.

There are many other ways to solve the problem—Section 2.2 describes a solution by differential calculus—however, the light analogy is the simplest method in this case.

If the cables in the above problem carry different amounts of electric current, we can apply the laws of refraction to solve it. If we assume city B consumes twice as much power as A, the optimum location of the power plant would shift toward B in order to minimize the transmission power loss. This can be viewed as a refraction problem with the refractive index of 2 for the path PB. The river bank is now the interface between two materials of different refractive indices, as shown in Figure 2.3.

$$\text{the ratio of refractive indices} = 2 = \frac{\sin \theta_A}{\sin \theta_B} \tag{2.4}$$

$$\frac{\sin\left[\tan^{-1}\left(\dfrac{X}{AR}\right)\right]}{\sin\left[\tan^{-1}\left(\dfrac{RQ - X}{BQ}\right)\right]} = 2 \tag{2.5}$$

where X is the design variable, $X = RP$.

The laws of refraction can be used to find the shortest path for an object that travels through different media at different speeds. Figure 2.4(a) shows the shortest path between two points for a hybrid vehicle that travels faster on land than in water. If an aircraft trav-

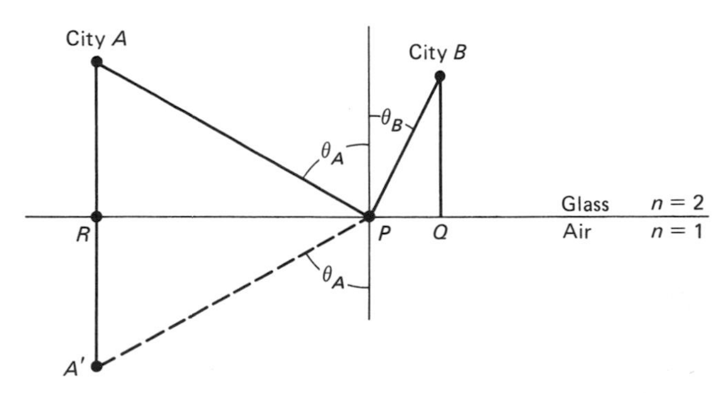

Figure 2.3. The problem in Figure 2.2 modified with a weighing factor of 2 on path PB. The weighing factor is analogous to the refractive index.

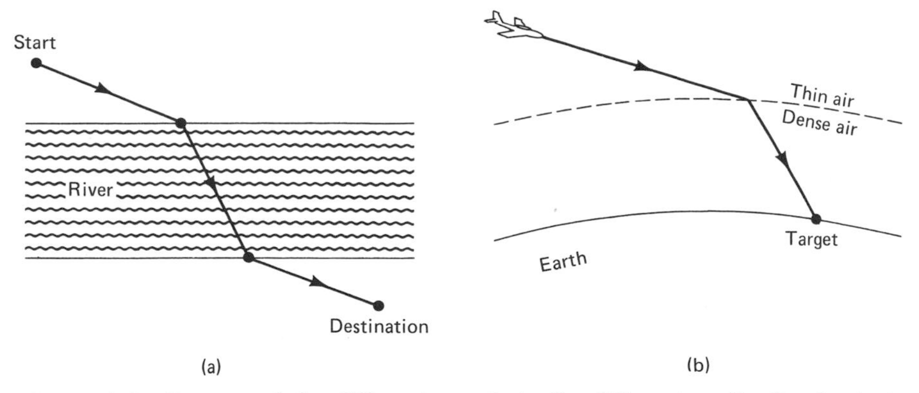

Figure 2.4. Because of the different speeds in the different media, the shortest paths are not straight lines. (a) depicts a hybrid vehicle that can travel faster on land than in water. (b) shows a plane that flies faster in thin air than in dense air.

elling in the upper atmosphere wishes to land on a target some distance away, the shortest path would assume the shape shown in Figure 2.4(b), because the aircraft can travel faster in thin air than in dense air.

2.1.2. An Equilibrium Position is One with the Lowest Potential Energy

The equilibrium position of a set of forces is the position of the system that has the lowest potential energy. A ball moving in a bowl will settle in the bottom of the bowl. Two balls at equilibrium in the bowl will both be off-center, but their overall center of gravity is still at its lowest position. Let w_1 and w_2 be their weights, and h_1 and h_2 be the heights of their centers of gravity from a reference line. When the system is at equilibrium, the potential energy $w_1h_1 + w_2h_2$ is minimum. This idea will be used to select a location for a radio station in the following example. A radio station is to be erected to serve three cities A, B, and C. Because of the differences in population, it is desirable to use the population as a weighing factor for the corresponding distances. The objective is to minimize the sum of products of populations and distances: minimize total distance = minimize (pop. $A \times$ dist. A + pop. $B \times$ dist. B + pop. $C \times$ dist. C). Let us paste a map on a board, drill holes at the locations of the cities, and hang weights proportional to the populations of the cities on three strings,

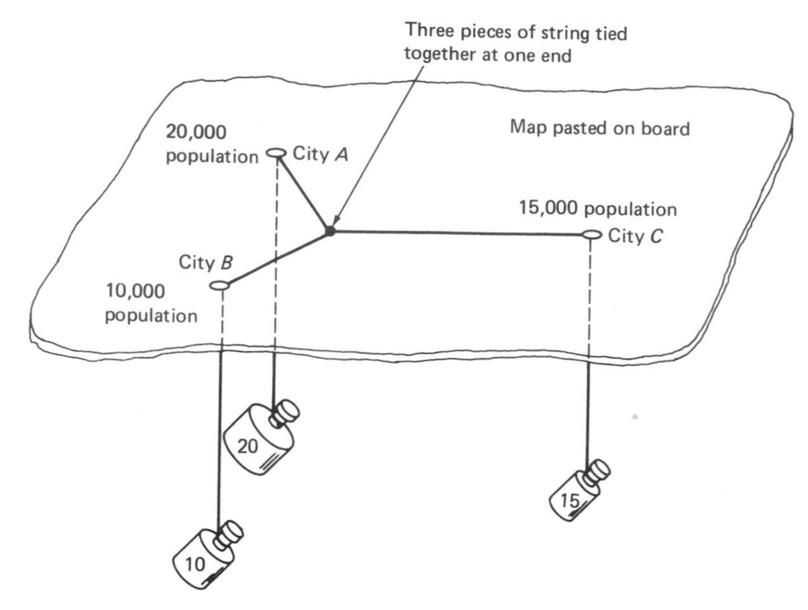

Figure 2.5(a). Strings and weights to determine the location of a radio station. The weights are proportional to the populations of the cities. The knotted end of the strings indicates the best position for the radio station.

as shown in Figure 2.5(a). The weights will settle to an equilibrium position with the knotted end of the strings close to city A. Figure 2.5(b) shows the three forces at equilibrium. The location P_0 indicates the lowest potential energy point. P_1, P_2, and P_3 are contours around P_0 representing positions of progressively higher potential energy. Such contours of potential energy would be actual lines of altitude in the bowl discussed previously. If the knotted end of the three strings is moved to the right of P_0, the potential energy of the 15-pound weight is decreased. Yet this decrease is more than compensated for by the increases in potential energy in the 10- and 20-pound weights. Thus, if external influence is removed, the knotted end will return to P_0. If equal weights are applied to all three strings, the strings would meet at 120° to each other. If the weight on one string were greater than on the other two combined, the intersection point would be at the city with the greatest weight. In order for the forces to be at equilibrium, the sum of the two smaller forces must be equal to, or larger than, the larger force.

The same method can be used to select a site for a home. Consider the following factors: office, five days a week; church, once a week;

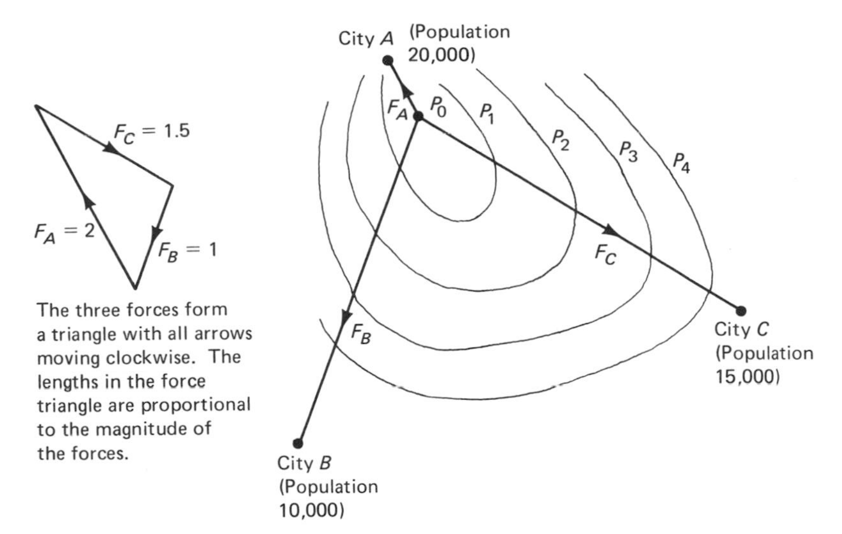

Figure 2.5(b). The equilibrium position of the three forces F_A, F_B, and F_C is the point with the lowest potential energy. Hence, P_0 is the best location for the radio station, where the equal potential energy contours are: $P_0 = 0$, $P_1 = 20$, $P_2 = 60$, $P_3 = 110$, and $P_4 = 160$.

bank, twice a month; supermarket, once a week; school, six days a week. Tying five strings at one end and hanging suitable weights on the other ends of the strings, we can perform the above operation with a suitable map. The knotted end will indicate the position of a house that will minimize travel time for the family.

2.1.3. Shortest Distances on a 3-Dimensional Object (Geodesic Lines)

It is not unusual for an engineer to have to find the shortest distance between two points on a 3-dimensional object (e.g., to design a wiring system inside a car body). In many cases, a rubber band, a piece of string, or a piece of Scotch tape can be used to find the shortest path.

If a rubber band is stretched between two points on a frictionless surface, it will slide over the surface and rest on the shortest path between the points, which will be one of the geodesic lines on the object.

The rubber-band method can be applied to surfaces with two curvatures, such as spheres, ellipsoids, and "animal surfaces." The Scotch-tape method described below is only for surfaces with one curvature,

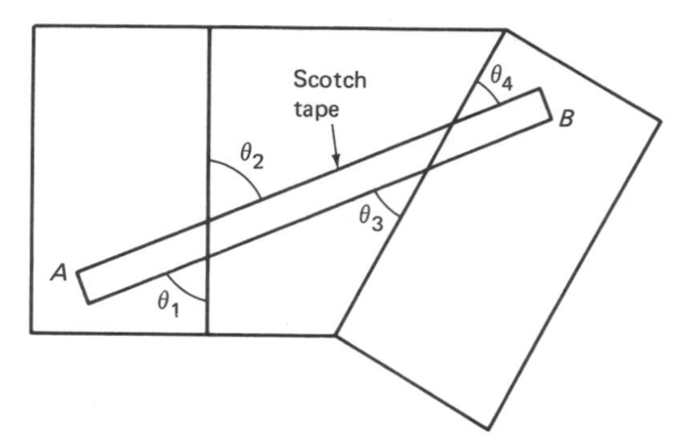

Figure 2.6. The path of the Scotch tape becomes a straight line when the object in Figure 2.1(b) is flattened.

such as cylinders, flat surfaces, straight edges, and cones. If the surface in Figure 2.1(b) is flattened, the Scotch tape connecting points A and B will mark a straight line (Figure 2.6). Since a straight line is the shortest distance, the path of the Scotch tape on the 3-dimensional object has the shortest length. Note that the angle the tape makes with an edge is the same on both sides of the edge ($\theta_1 = \theta_2, \theta_3 = \theta_4$). A piece of Scotch tape would form a helix on a cylinder.

2.1.4. Soap Films Maintain the Smallest Surface Areas

Because of surface tension, a soap film possesses a shape with the minimum surface area. This shape also has the lowest potential energy because the potential energy of surface tension is directly proportional to the area.

Figure 2.7 shows several shapes generated by soap films. These

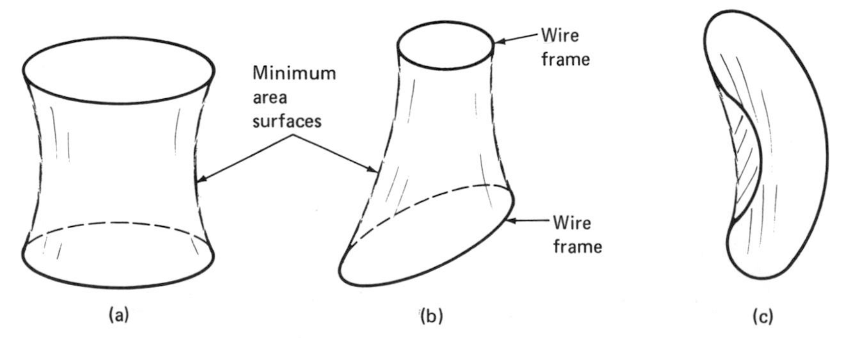

Figure 2.7. Soap films formed on wire frames. (a) and (b) show the minimum surface area of a duct connecting two fixed-end diameters. (c) shows the minimum surface area for a piece of glass to fit a curved window frame.

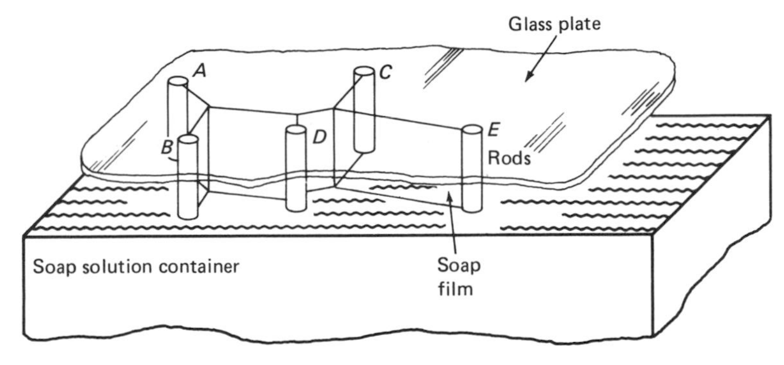

Figure 2.8. Soap-film model for designing a network of pavements to connect five buildings in a factory complex. *A*, *B*, *C*, *D*, and *E* may represent a warehouse, a foundry, a machine shop, an assembly plant, and a business office.

shapes can be determined mathematically by variational calculus. A permanent film surface can be made from plastic dip films, similar to those used for art flowers.

The cylindrical shapes in Figure 2.7 may be used to make connecting ducts with the least material. Such ducts would also result in the least heat loss. The soap-film method can also be used to find the shortest length of pavements. Figure 2.8 shows a way of formulating a pavement network for a factory complex. Note that all the end points are equally weighted, and that the pavement system has several intersection points. It can be proven mathematically that three paths will meet at an intersection at angles of 120° with respect to one another. Figure 2.9 shows that if four or five paths meet at one intersection, the total length of the paths would not be minimum. There are two minimum distance configurations for the system in Figure 2.9(a) and five minimum distance configurations for the system in Figure 2.9(b).

2.1.5. Shortest Total Length of Paths Connecting Several Nodes

Like the pavement problem, we now want to connect several nodes with the shortest total length of paths. Unlike the soap-film method, the paths join only at a node. The smallest number of paths needed to join n nodes together is $(n - 1)$. For the first two nodes, we need one path to connect them. For every additional node in the network, we need one more path. Figure 2.10 shows six nodes to be connected by five paths. The five paths that have the minimum total length are found by the following steps:

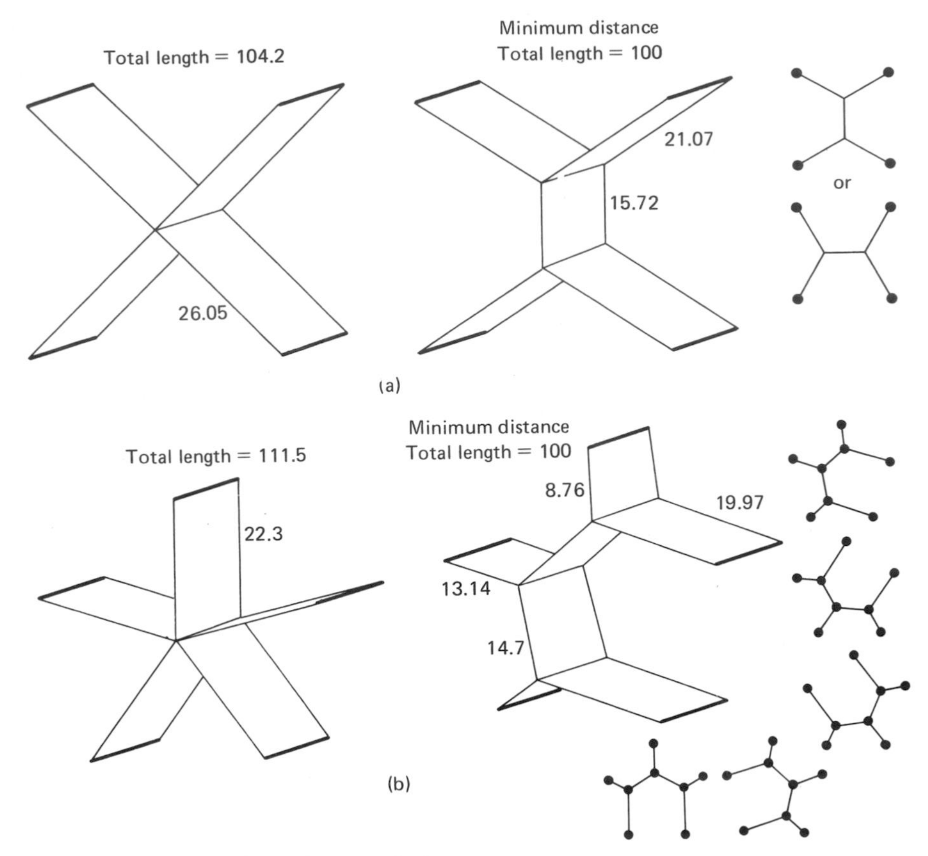

Figure 2.9. Soap-film solution of shortest paths between four and five nodes equally spaced on a circle. Note that when four or five lines meet at one point, the total path lengths are 4.2% and 11.5% longer than the optimum solutions. The optimum solution consists of three lines meeting at 120° to each other. The optimum solution is not unique (Almgren and Taylor [5].)*

*A fee of $130 and a different acknowledgment may be needed.

(1) We arbitrarily pick node *C* as a starting point. Considering all the paths from *C*, we choose the shortest one *CF*.
(2) The two nodes *C* and *F* are now connected. Examining the paths from *C* and *F* to the unconnected nodes, we pick the shortest path *FE*.
(3) The nodes *C*, *E*, and *F* are now connected. We now find *ED* to be the shortest path among all the paths from the connected group to one of the unconnected nodes.

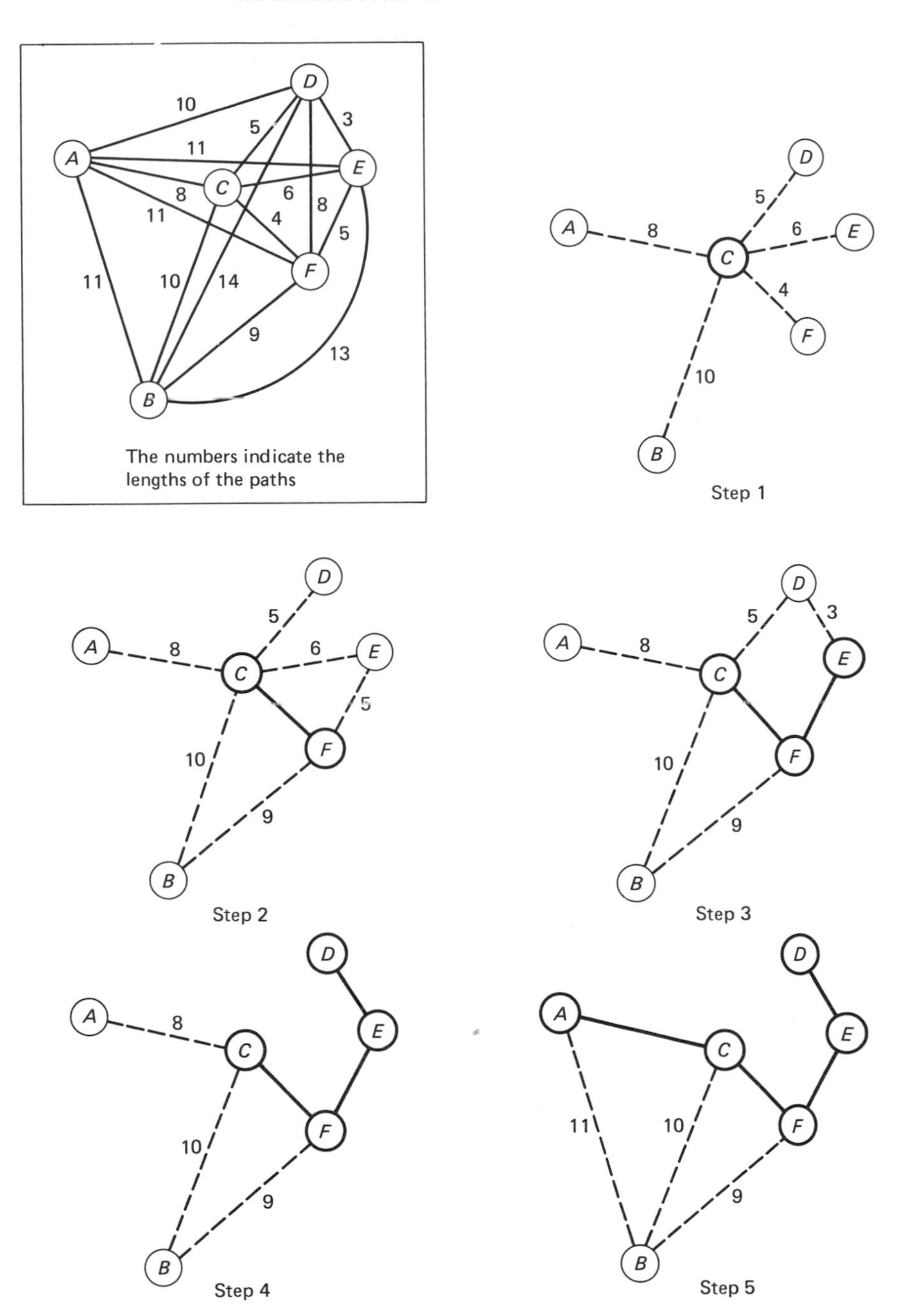

Figure 2.10. A Method to connect six nodes with five paths of minimum total length. There may be more than one pattern that has minimum total length.

(4) The connected nodes are now C, E, F, and D. The shortest path from these four nodes to the remaining nodes A and B is CA.
(5) The only unconnected node is B. The shortest path from the connected group to B is FB.

Now, we have chosen the five paths CF, FE, ED, CA, and FB that connect the given six nodes with the minimum total length. The above algorithm insures that every node is connected to its nearest neighbor, and that the nodes are connected with the fewest lines. The simplicity and power of this method can be appreciated by considering the difficulty in an exhaustive trial method: there are 1206 different ways to connect six nodes with five lines.

2.1.6. The Shortest Route through a Network

Let us try to pass through the maze in Figure 2.11 using the shortest route. The lengths of the path are given. The number marked beside a node is the minimum cumulative distance from A to that node, and is calculated stage by stage as follows:

(1) For node B, the cumulative minimum distance from A is just $AB = 3$. Similarly, the distances from A to nodes C and D are 5 and 6, respectively.
(2) We now want to find the minimum distances from A to nodes E, F, and G. For node E, there are three possibilities:
$$\text{node } E = \text{node } B + BE = 3 + 2 = 5 \checkmark$$
$$\text{node } E = \text{node } C + CE = 5 + 3 = 8$$
$$\text{node } E = \text{node } D + DE = 6 + 12 = 18$$
where the name of a node represents the minimum distance from A to that node. The minimum distance from A to E is, therefore, 5. Doing the same calculations for nodes F and G, we find that the minimum distances for F and G are 8 and 9, respectively.
(3) We repeat the above considerations for nodes H and I. The minimum distance for node H is 14 (through F); that for node I is 10 (through E).
(4) For the end node J, choosing the path IJ will lead to the smallest cumulative distance of 14.

By making progressive judgments from stage to stage, we have narrowed the alternatives down to three simple ones at a time. If the whole system is considered exhaustively, there are eighteen possible

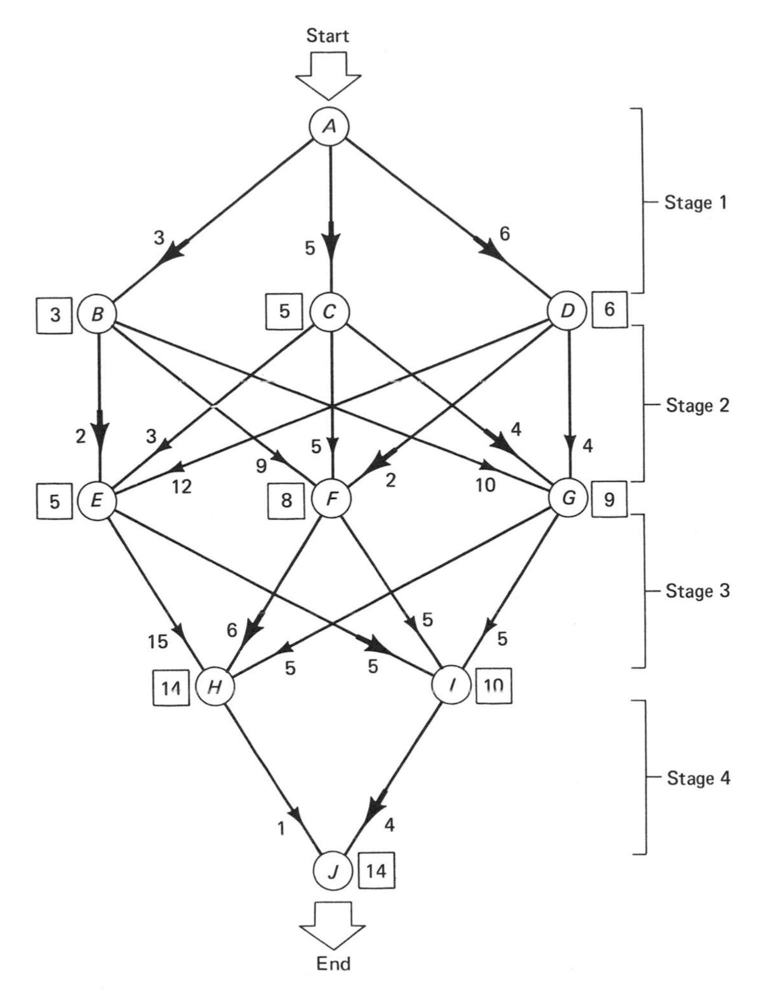

Figure 2.11. Dynamic programming to find the shortest route through a network. From node A to node J, the shortest route is $ABEIJ$.

choices, each involving the sum of four path lengths. The nodes in the above maze are symbolic. They may represent possible strategies for a student to finish a degree in college.

The above method, called dynamic programming, is characterized by sequential solutions in stages. More will be said about dynamic programming later in Section 2.4.2.

Suppose we can travel horizontally through the maze. Let us say we open up paths BC, EF, and HI, and drop out paths BF, BG, DE, EH, FH, and GI, as shown in Figure 2.12. This generalized problem can be solved by a method similar to that in Section 2.1.5. Starting from node A, we can increase the connected nodes one at a time until we arrive at I. Figure 2.12(b) shows several steps in this solution.

(1) Consider all the paths from A. AB is the shortest. Therefore, its length, 3, is marked next to node B.

(2) Consider all the paths from the connected group. The candidates are:

node $D = AD = 6$
node $C =$ node $B + BC = 3 + 1 = 4$ √
node $C = AC = 5$
node $E =$ node $B + BE = 3 + 2 = 5$

where the name of a node denotes its minimum cumulative distance from A. Path BC yields the smallest distance and is chosen. Therefore, 4 is marked beside C.

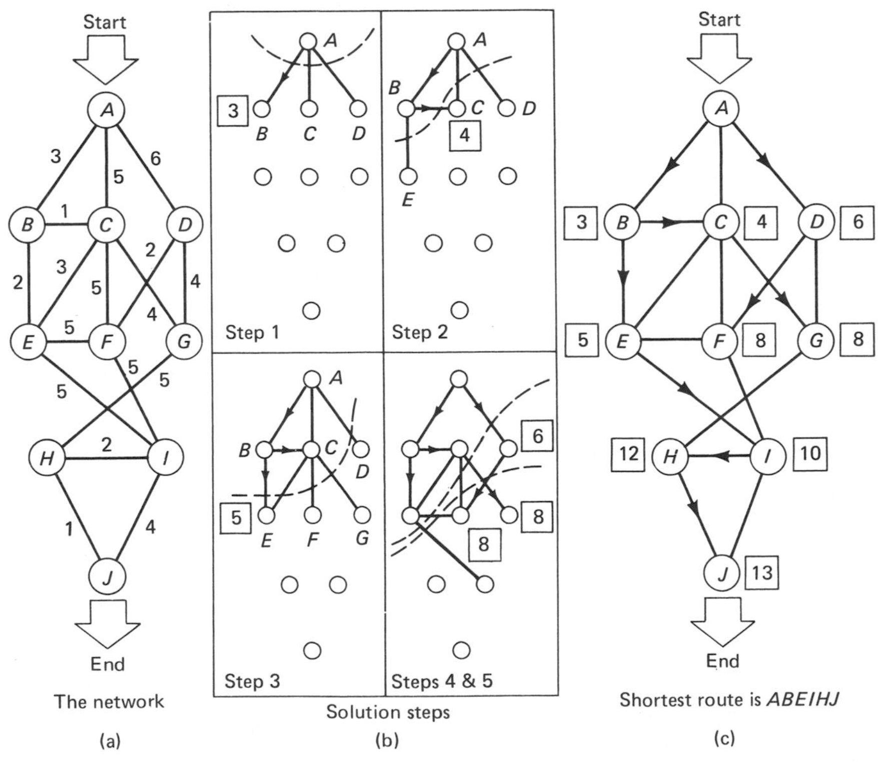

Figure 2.12. Shortest route through a network. The details of the method are explained in the text.

(3) Consider all the paths from the connected group A, B, and C. The alternatives are:

node $D = AD = 6$
node $E =$ node $B + BE = 3 + 2 = 5$ ✓
node $E =$ node $C + CE = 4 + 3 = 7$
node $F =$ node $C + CF = 4 + 5 = 9$
node $G =$ node $C + CG = 4 + 4 = 8$

Path BE is chosen and the number 5 is marked next to node E.

(4) With A, B, C, and E connected, the next successful candidate is node D.

(5) With A, B, C, D, and E connected, there are two nodes equally close to A in their cumulative distances.

node $F =$ node $D + DF = 6 + 2 = 8$ ✓
node $G =$ node $C + CG = 4 + 4 = 8$ ✓

Both are included in the propagating group of connected nodes.

(6) Every additional node connected will be marked with the minimum cumulative length of paths from node A until node I is reached. The shortest route through the network is thus found to be 13 (Figure 2.12(c)).

2.2. Differentiable Cost Functions

If the cost of a product can be expressed as a continuous function of a parameter x, one necessary condition at the point of minimum cost is $(d(\text{cost})/dx) = 0$. On the immediate left of a minimum point, the slope of the cost function is negative. On its immediate right, the slope is positive. Finding the point of minimum cost by differentiating the cost function is simple and useful, although it does require a mathematical relation between the cost and the design variable x that is differentiable. The examples in the next three subsections will illustrate the differentiation method for minimizing costs.

2.2.1. Location of a Hydroelectric Plant

Referring to the hydroelectric plant problem in Section 2.1.1 and Figure 2.2, let $x = RP$

$$AP = \sqrt{x^2 + AR^2}$$

$$PB = \sqrt{(RQ - x)^2 + BQ^2}$$

$$\text{cost} = c(AP + PB) = c\left(\sqrt{x^2 + AR^2} + \sqrt{(RQ - x)^2 + RQ^2}\right) \quad (2.6)$$

where c is the cost per unit length. Equation 2.6 expresses the cost of the electric cables as a function of x. We want to find the points at which $(d(\text{cost})/dx) = 0$ because the minimum cost point $x_{\min}$ will be among them.

$$\frac{d(\text{cost})}{dx} = c\left(\frac{x}{\sqrt{x^2 + AR^2}} + \frac{x - RQ}{\sqrt{(RQ - x)^2 + BQ^2}}\right) = 0$$

$$x^2(BQ^2 - AR^2) + x(2AR^2 \cdot RQ) - AR^2 \cdot RQ^2 = 0$$

$$x = \frac{AR \cdot RQ}{AR + BQ} \quad \text{or} \quad \frac{AR \cdot RQ}{AR - BQ}$$

$$x_{\min} = \frac{AR \cdot RQ}{AR + BQ} \tag{2.7}$$

from direct substitution or double differentiation. The other value of x in Equation 2.7 is an inflection point. $x_{\min}$ and the inflection point are shown in Figure 2.13. The point may be a minimum, inflection or a maximum, depending on whether the second derivative is positive, zero, or negative. In some problems, the differentiation method may produce two local minima. These have to be compared to determine the overall lowest cost.

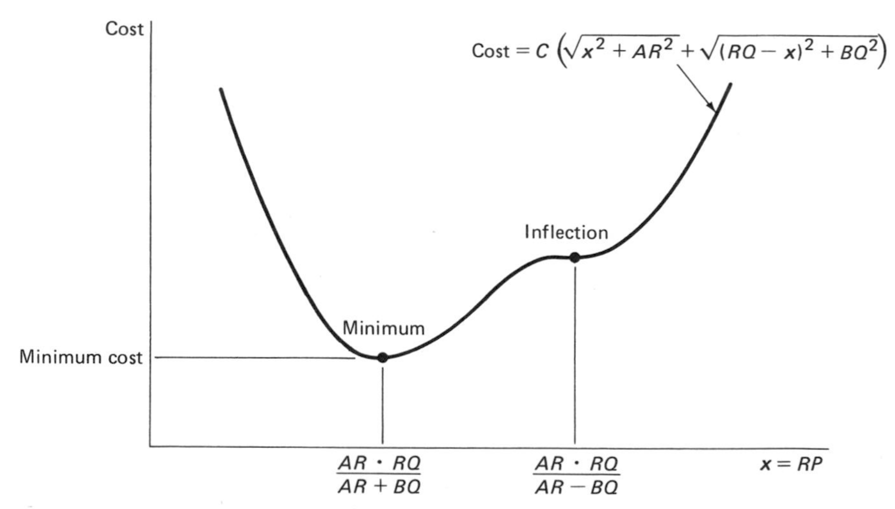

Figure 2.13. Differentiable cost function of one variable. The plot illustrates the cost function for the location of the hydroelectric plant. There are two points with zero slope.

2.2.2. Optimum Shapes of Milk Cartons and Tin Cans

The objective is to contain a certain volume of milk with as little paper as possible. The ratio of the surface areas of a sphere, a cylinder,* and a cube of the same volume is 1 : 1.164 : 1.255. Though this relative ratio is constant, the surface-to-volume ratio for each shape depends on its size. As the size increases, the surface area increases by the square of the length, and the volume increases by the cube of the length. Small packaging is inherently uneconomical.

Though a sphere has a low surface-to-volume ratio, it is not a practical shape for paper containers. The rectangular shape is more realistic for milk cartons. In the following problem, we are going to find the optimum ratio of the height to the width of a paper carton (Figure 2.14).

Let x be the ratio of the height h to the width w of a carton with a square base, and let the volume of the milk to be contained be 28.88 cu in. (1 pint).

$$x = \frac{h}{w} \qquad \text{or} \qquad h = xw \tag{2.8}$$

$$\text{volume} = w^2 \cdot h = w^2(xw) = xw^3 = 28.88$$

$$w = \sqrt[3]{\frac{28.88}{x}} = 3.065x^{-1/3} \tag{2.9}$$

*A cylinder with the minimum surface-to-volume ratio.

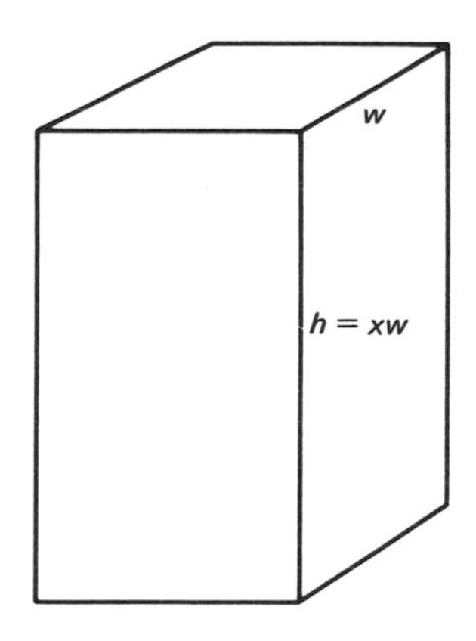

Figure 2.14. Milk carton with a square cross-section.

$$\text{surface area} = 2(w^2) + 4(wh) = 2w^2 + 4xw^2 \tag{2.10}$$

$$\text{(2.9) in (2.10) surface area} = 18.8x^{-2/3} + 37.6x^{1/3} \tag{2.11}$$

At $d(\text{surface area})/dx = 0$, $x = 1$; the surface area = 56.4 sq in.; the surface-to-volume ratio is 1.961. This means that when the height equals the width of the carton, the least amount of paper is used. The cubic carton is, therefore, the most efficient within the set of rectangular boxes. If x is defined as the height-to-diameter ratio for a cylinder, the optimum value for x will also be 1. For a tin can, the minimum surface-to-volume ratio is obtained when its height equals its diameter.

The above analysis assumes no overlapping of material at the seams and rims. To make the problem more practical, we will include lap and edge joints in the milk carton. Suppose the spout, the edge joint, and the lap joint measure 0.7 w, 0.1 w, and 0.55 w, respectively (Figure 2.15). The total height of the paper tube needed will be $(1.35 + x)w$.

$$\text{surface area} = 4(1.35 + x)w^2 \tag{2.12}$$

$$\frac{d}{dx}(\text{surface area}) = 0 \text{ will give } x = 2.7$$

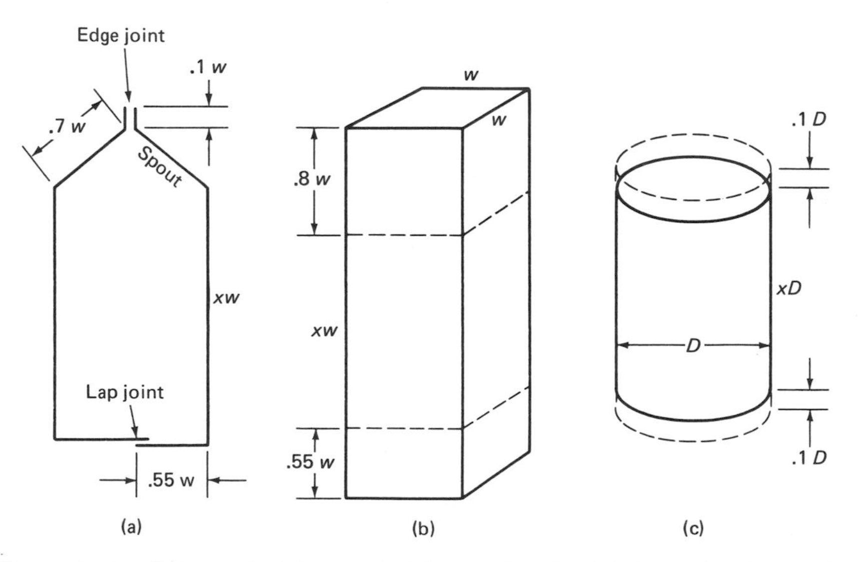

Figure 2.15. Extra material needed at the overlapping joints can be expressed as a percentage of the width w or diameter D. (a) is a milk carton, (b) is the paper tube needed for (a), and (c) is a tin can with rims.

Because of the paper used in the top folds and bottom folds, a taller and narrower shape is more advantageous than a cube. Parallel analysis of a tin can with overlapping rims will yield similar results. If the folding rim is assumed proportional to the diameter, an optimum ratio of height/diameter would be larger than 1 (Figure 2.15(c)).

2.2.3. Optimum Configurations of Structures

The tripod structure in Figure 2.16 supports a pot over a campfire. We are going to optimize the height h of the structure for a given radius R. The objective is to minimize the volume (and consequently, the cost) of material in the three posts. Note that as h is decreased, the axial compression in the posts increases, and the diameter of the posts has to be increased. There is a trade-off between the length and the sectional area of the posts. From the free-body diagram in Figure 2.17

$$\frac{\frac{1}{3}w}{P} = \sin\theta = \frac{h}{l} \tag{2.13}$$

$$A = \frac{P}{S_y}$$

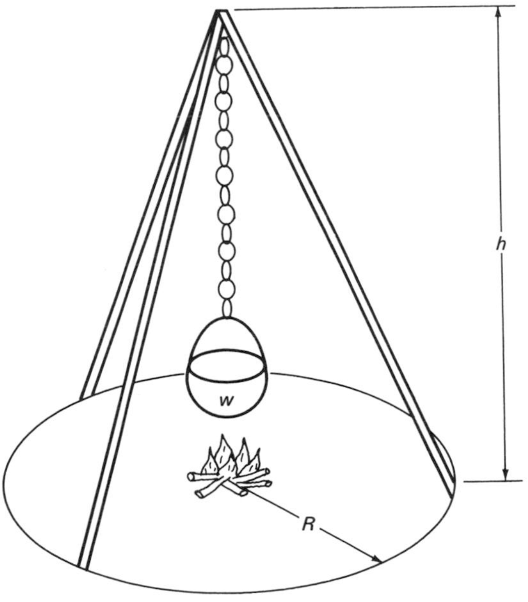

Figure 2.16. A tripod structure: three posts supporting a weight.

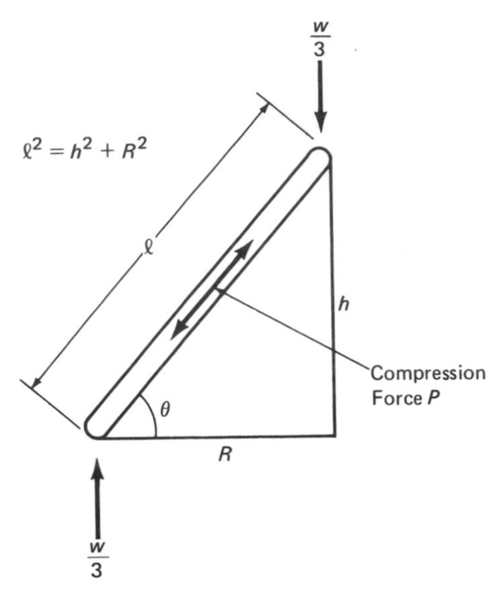

Figure 2.17. Force analysis of post.

$$\text{Volume of material } V = 3 \cdot l \cdot A = 3l\left(\frac{P}{S_y}\right) = \frac{h^2 + R^2}{h} \cdot \frac{w}{S_y} \tag{2.14}$$

where:

h = height of tripod (design variable),
A = sectional area of a post,
P = axial compression in a post,
S_y = yield strength of material,
w = weight of the pot,
l = length of the posts, and
R = distance from the foot of a post to the fire.

Setting $dV/dh = 0$, we get $h = R$. Therefore, to minimize volume,

$$h = R; \quad l = R\sqrt{2}; \quad A = \frac{\sqrt{2}}{3}\frac{w}{S_y}.$$

So far, our differentiation examples have only one parameter. For problems with several parameters, the solution procedures are quite similar. The cost function would be partially differentiated with respect to each of the parameters. With two variables, x and y, the minimum point is among the points that satisfy both of the following equations:

$$\frac{\partial}{\partial x}(\text{cost}) = 0 \quad \text{and} \quad \frac{\partial}{\partial y}(\text{cost}) = 0. \tag{2.15}$$

Let us minimize the volume of material in a hoisting structure with two design variables x and y. The structure (Figure 2.18) is set up to support a pulley at a fixed distance D away from a wall. The tension member is anchored at a point x inches above the center of the pulley. The compression member is pivoted at a point y inches below C. The allowable stress in the tension member equals the yield strength S_y of the material. In the compression member, the allowable stress is 0.5 S_y because of the danger of buckling.
Let

T = the axial tension of the tension member,

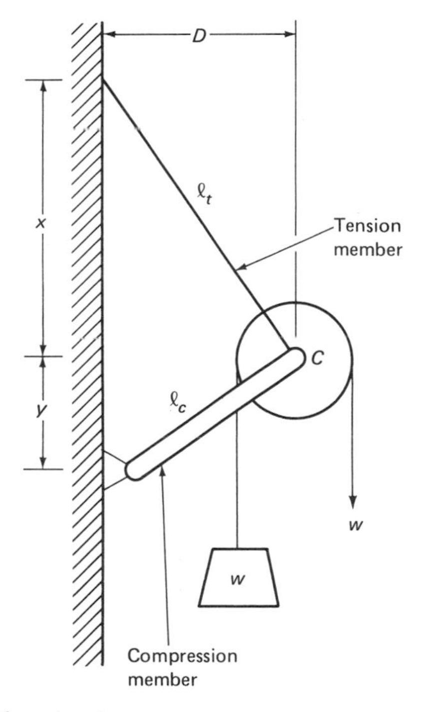

Figure 2.18. A hoisting structure.
The pulley is supported at a distance D away from the wall. The tension member is l_t long, and is anchored at a point x above the center of the pulley. The compression member is l_c long, and is pivoted at a point y below the center of the pulley.

P = compression in the compression member,
A_t = cross-sectional area of the tension member,
A_c = cross-sectional area of the compression member,
l_t = length of the tension member,
l_c = length of the compression member,
w = the weight being hoisted on the structure, and
V = volume of material in the two members.

Then

$$\frac{T}{A_t} = S_y \tag{2.16}$$

$$\frac{P}{A_c} = 0.5\, S_y. \tag{2.17}$$

From the free-body diagram in Figure 2.19,

$$T = \frac{2wl_t}{x + y} \tag{2.18}$$

$$P = \frac{2wl_c}{x + y} \tag{2.19}$$

But

$$V = A_t l_t + A_c l_c$$

$$= \frac{T}{S_y} l_t + \frac{P}{0.5S_y} l_c \qquad \text{(from 2.16 and 2.17)}$$

$$V = \frac{2w}{S_y}\left(\frac{x^2 + 2y^2 + 3D^2}{x + y}\right) \qquad \text{(from 2.18 and 2.19).} \tag{2.20}$$

To find the minimum volume, partial derivatives of V with respect to x and y are set to zero. The resultant equations are then solved. One of the solutions gives the minimum volume.

$$\frac{\partial V}{\partial x} = 0 \text{ gives } x^2 + 2xy - 2y^2 - 3D^2 = 0 \tag{2.21}$$

$$\frac{\partial V}{\partial y} = 0 \text{ gives } -x^2 + 4xy + 2y^2 - 3D^2 = 0 \tag{2.22}$$

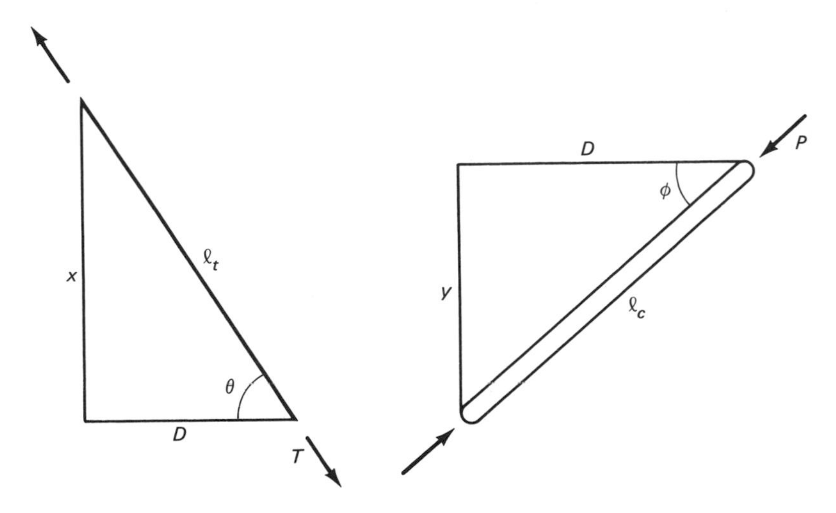

Figure 2.19. Free-body diagrams of the two-member truss.

$$l_c^2 = y^2 + D^2, \; l_t^2 = x^2 + D^2$$

Equilibrium of horizontal and vertical forces:

$$\Sigma F_x = 0 = P \cos \phi - T \cos \theta$$

$$\Sigma F_y = 0 = P \sin \phi + T \sin \theta - 2w$$

$$T = \frac{2w/\cos \theta}{\tan \phi + \tan \theta} = \frac{2wl_t}{x + y}$$

$$P = \frac{2w/\cos \phi}{\tan \phi + \tan \theta} = \frac{2wl_c}{x + y}.$$

Adding 2.21 and 2.22,

$$xy = D^2 \tag{2.23}$$

Let e^2 be the ratio of x to y

$$x = eD$$

$$y = \frac{1}{e} D \tag{2.24}$$

Substitute x and y in 2.21:

$$e^4 - e^2 - 2 = 0. \tag{2.25}$$

$$e = 1.414$$

is the only meaningful solution among the four solutions of 2.25. Substitute e in 2.24

$$x = 1.414D$$

$$y = 0.707D$$

Substitute x and y in 2.20,

$$\text{minimum } V = \frac{5.7wD}{S_y}$$

To prove this is the minimum, Equation 2.20 can be evaluated with

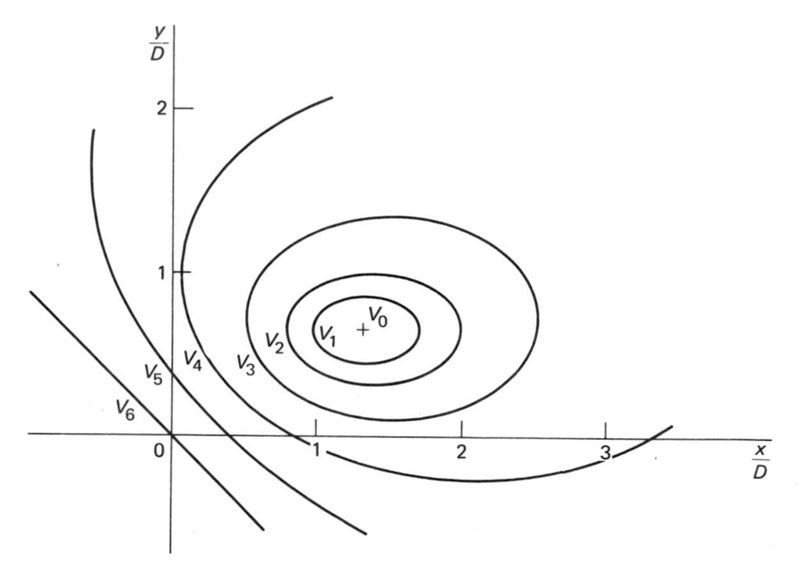

Figure 2.20. Plotting volume as a function of $\frac{x}{D}$ and $\frac{y}{D}$ by Eq. 2.20. The diagram shows lines of equal volume. V_0 is the minimum point. This figure is a two-dimensional analogy of Figure 2.13.

$$\text{Minimum Volume} = V_0 = 5.6577\frac{wD}{S_y}$$

$$V_1 = 5.8\frac{wD}{S_y} \qquad V_4 = 9\frac{wD}{S_y}$$

$$V_2 = 6\frac{wD}{S_y} \qquad V_5 = 16\frac{wD}{S_y}$$

$$V_3 = 6.6\frac{wD}{S_y} \qquad V_6 = \infty$$

any combination of x and y. Regardless of the x and y picked, the volume is always larger than $(5.7wD)/S_y$. Figure 2.20 shows a mapping of x and y into volume. The contour lines shown in the diagram are lines of equal volume. Figure 2.20 differs from Figure 2.13 only in the number of independent variables. For a problem with more than three design variables, visual display is impossible, but the above solution procedures are still valid.

2.3. LINEAR PROGRAMMING

A programming method is usually an algorithm for meeting an objective through an appropriate allocation of limited resources. Linear programming is an allocation technique that uses a linear mathematical model. The objective function and the constraints are formulated as linear expressions.

In Figure 2.13, a minimum exists because the cost function is a curve. In linear systems, the objective function is a "straight line," and an optimum solution would not exist unless there were constraints on the design space. In fact, an optimum solution (if there is any) falls on the boundary between the feasible and infeasible regions.

2.3.1. Mixing of Ingredients

The following is a simple problem to illustrate linear programming. Tin and lead are mixed in different proportions to produce two grades of solder. The maximum quantities available for tin and lead are 240 pounds and 300 pounds, respectively. Each pound of 50/50 solder contains ½ pound of tin and ½ pound of lead, and each pound of 25/75 solder contains ¼ pound of tin and ¾ pound of lead. The profit is 40¢ per pound for 50/50 solder, and 70¢ per pound for 25/75 solder. The sales potential for 25/75 solder is 300 pounds. How much of each grade of solder should be produced so that the total profit is maximized?

Let x and y pounds be the quantities of 50/50 and 25/75 solder produced, respectively. The objective is to maximize the profit P:

$$P = 0.4x + 0.7y \tag{2.26}$$

The design variables x and y are subjected to the following constraints:

$$\tfrac{1}{2}x + \tfrac{1}{4}y \leqslant 240 \qquad \text{(availability of tin)} \tag{2.27}$$

$$\tfrac{1}{2}x + \tfrac{3}{4}y \leqslant 300 \quad \text{(availability of lead)} \tag{2.28}$$

$$y \leqslant 300 \quad \text{(sales potential of 25/75 solder)} \tag{2.29}$$

$$x \geqslant 0 \quad \text{(feasibility)} \tag{2.30}$$

$$y \geqslant 0 \quad \text{(feasibility)} \tag{2.31}$$

In Figure 2.21, the labeled lines represent the equality parts of the first three constraints. Points lying on or to the left of each line would satisfy the inequality constraint represented by that line. The shaded region, *OUVRS* on the graph, lies to the left of all three lines. It is also in the first quadrant of the graph where $x \geqslant 0$ and $y \geqslant 0$. This means that points in this region satisfy all constraints. We will now find out which point in this region of feasible solutions (design space) will give the maximum profit P_{max}.

For clarity, the same feasible region is shown in a separate graph (Figure 2.22), in which the zero profit line $0.4x + 0.7y = 0$ is also drawn. If we move the profit line upward without disturbing its slope, the equation for the new line would be $0.4x + 0.7y = P$. The profit

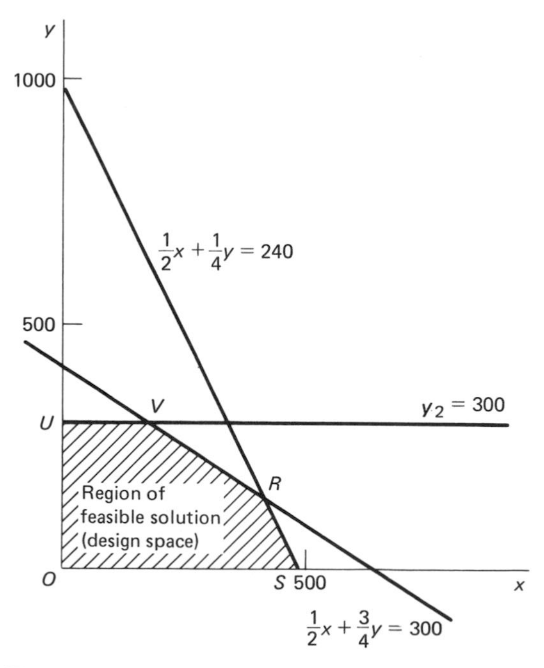

Figure 2.21. The constraints define a feasible region.

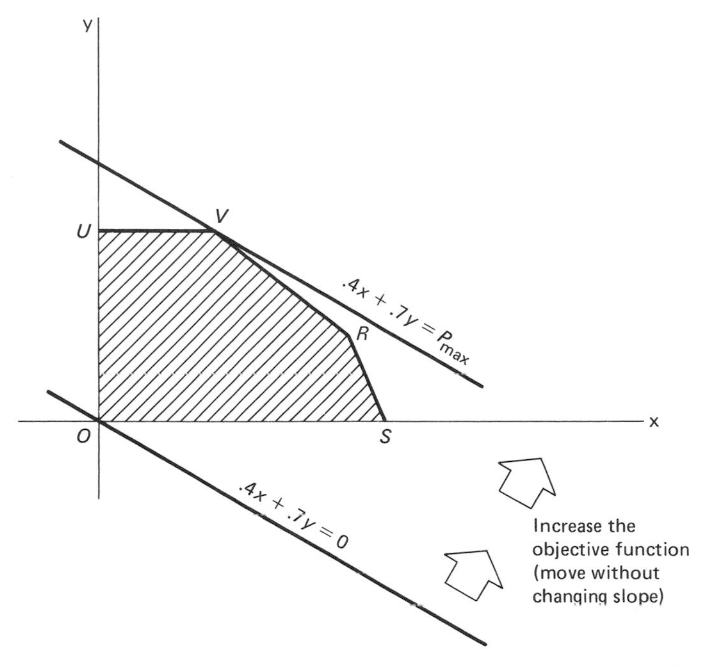

Figure 2.22. The optimum solution falls on an extreme point (corner).

P gets greater and greater as the line is moved farther and farther upward. When the constant profit line touches the point V in the graph, any further movement will push the line off the feasible region. The equation of the constant profit line at V becomes $0.4x + 0.7y = P_{max}$, where P_{max} is the maximum profit. The point V represents the optimum solution that yields the maximum value for the expression $0.4x + 0.7y$ under the given set of constraints.

From Figure 2.22 we can see that the optimum solution V is the intersection point of the constraining lines UV and VR. The equations for the two lines are solved to yield the coordinates of V:

$$x = 150, \quad y = 300 \qquad (2.32)$$

Substituting this pair of values into Equation 2.26:

$$P_{max} = 0.4(150) + 0.7(300) = 270$$

The maximum profit of \$270 would be obtained when 150 pounds of 50/50 solder and 300 pounds of 25/75 solder were produced.

Note that no matter what the objective function happens to be, one of the optimum solutions will fall on an extreme point of the region of feasible solutions. We can understand this by drawing any line $C_1x + C_2y = 0$, and moving the line, without changing its slope, until the right-hand side of the equation attains the maximum value within the region of feasible solutions. The optimum solution could be a line.

If a linear programming problem involves n variables, the set of feasible solutions will be a region in a n-dimensional space. In this region, there are extreme points analogous to the corners (O, U, V, R, S) in Figure 2.21. One of the extreme points is an optimum solution. The Simplex method is an algebraic method of finding the optimum solution by examining the extreme points selectively. Basically, a search-pointer is placed at the origin where the value of the objective function is always 0. It is moved to a neighboring extreme point with a higher value for the objective function. This search-pointer will continue to move along the boundary as long as the objective function is increasing from one extreme point to another. When it arrives at an extreme point whose neighbors all have smaller values for the objective function, an optimum solution is obtained.

2.3.2. The Diet Problem and the Transportation Problem

The diet problem is very similar to the solder problem. Here, the objective is to minimize the cost of obtaining enough nutrients from a variety of foods. Suppose x_1, x_2, x_3, x_4 and c_1, c_2, c_3, c_4 are the quantities and unit costs of four types of foods, the objective would be to minimize the total cost $c_1x_1 + c_2x_2 + c_3x_3 + c_4x_4$. The constraints on the variable x_is are that the total amount of nutrients, vitamins, and minerals contained in them must be greater than the minimum daily requirements of the nutrients, and that the quantities of foods (x_1, x_2, x_3, x_4) are not negative.

Another popular example is the transportation problem, where the cost of the distribution of goods is to be minimized. Say, gasoline is to be moved from three storage tanks to three stations. Let x_{ij} and c_{ij} be the quantity and unit cost of gasoline moved from tank i to station j. The objective would be to find an optimum set of x_{ij}s such that the cost function of $c_{11}x_{11} + c_{12}x_{12} + c_{13}x_{13} + c_{21}x_{21} + c_{22}x_{22} + c_{23}x_{23} + c_{31}x_{31} + c_{32}x_{32} + c_{33}x_{33}$ is minimized. The constraints on x_{ij}s are that the total quantity from a particular tank must be less than the quantity stored in that tank, and that the total quantity supplied to a station must be less than the sales potential of gasoline at that station. Furthermore, for practical reasons, the x_{ij}s must be nonnegative.

2.3.3. The Pin-Jointed Structure

Let us now solve a design problem. Pin-jointed structures are easily adapted to programming solutions because they are statically determinate and their forces, stresses, sectional areas, and volumes of material are governed by linear relationships. We can optimize a structure with redundant members, which would be eliminated by the programming method. The structural layout is initially defined with fixed locations of nodes and external applied forces, but variable sectional areas for the members. By minimizing the volume of material in the structure, the optimum sectional area of each member will also be known. The constraints on the sectional area are the yield strength of the member and the equilibrium of forces at the nodes. A particular member will be eliminated if its optimal area is zero.

Figure 2.23 shows a structure supporting a force F. The locations of the nodes A, B, C, and D and the lengths of the members l_1, l_2, l_3 are fixed. Let A_i be the sectional area, and T_i the force in the ith member. The yield strength of the material is S_y.

Objective function:

$$\text{minimum volume } V_{\min} = 2\sqrt{2}A_1 + \sqrt{5}A_2 + 2\sqrt{2}A_3 \quad (2.33)$$

Constraints: from equilibrium of vertical forces at B

$$\frac{1}{\sqrt{2}}T_1 - \frac{1}{\sqrt{5}}T_2 - \frac{1}{\sqrt{2}}T_3 - F = 0; \quad (2.34)$$

from equilibrium of horizontal forces at B

$$\frac{1}{\sqrt{2}}T_1 + \frac{2}{\sqrt{5}}T_2 + \frac{1}{\sqrt{2}}T_3 = 0; \quad (2.35)$$

for practical reasons

$$A_1 \geqslant 0 \quad (2.36)$$
$$A_2 \geqslant 0 \quad (2.37)$$
$$A_3 \geqslant 0 \quad (2.38)$$

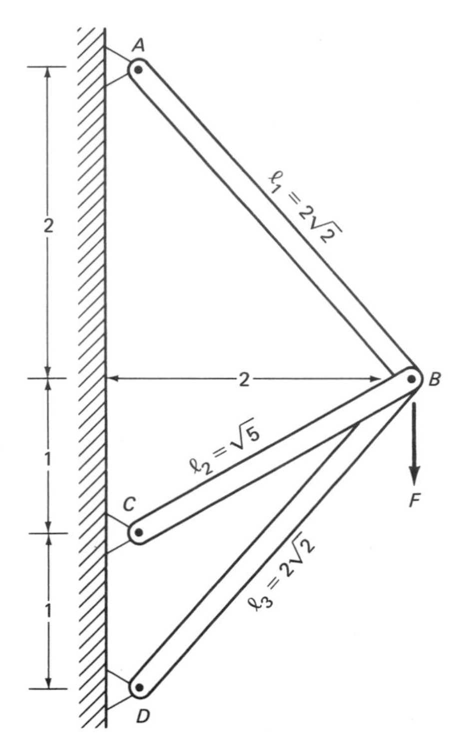

Figure 2.23. The redundant members of this pin-jointed structure will be eliminated by linear programming optimization of material volume.

from the yield strengths of members

$$-A_1 S_y \leqslant T_1 \leqslant A_1 S_y \tag{2.39}$$
$$-A_2 S_y \leqslant T_2 \leqslant A_2 S_y \tag{2.40}$$
$$-A_3 S_y \leqslant T_3 \leqslant A_3 S_y \tag{2.41}$$

Notice that the first two constraints are nodal force equilibrium equations. We have only two independent equations because one of the members is redundant. When there are no redundant members, the number of equations equals the number of members, and the linear system has a unique solution. Figure 2.24 shows a transformation of variables from T_i and A_i into T_i' and T_i''. This transformation converts Equations 2.39, 2.40, and 2.41 into a handier form for linear programming. The new variables T_i' and T_i'' are called slack variables, and they represent the excess strengths of the members. To minimize volume in the original problem means to have the lowest

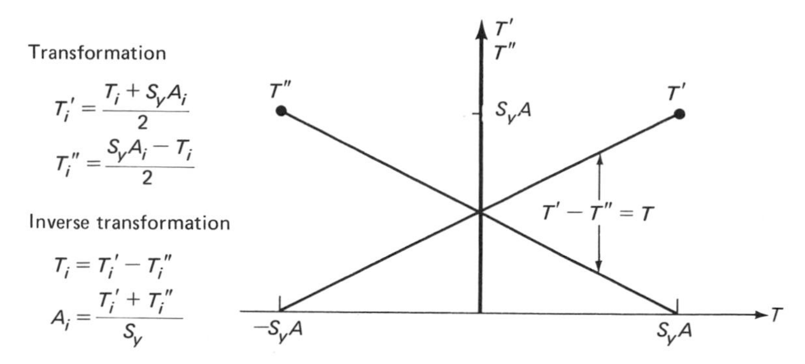

Figure 2.24. Transformation of T and A into T' and T''. T' and T'' are slack variables that represent excess strengths in the members.

Original Equations	Transformed Equations
Objective Function	**Objective Function**
$\text{Vol.} = 2\sqrt{2}A_1 + \sqrt{5}A_2 + 2\sqrt{2}A_3$	$\text{Vol.} = \frac{2\sqrt{2}}{S_y}(T_1' + T_1'') + \frac{\sqrt{5}}{S_y}(T_2' + T_2'') + \frac{2\sqrt{2}}{S_y}(T_3' + T_3'')$
= minimum	= minimum
Constraints	**Constraints**
$\frac{1}{\sqrt{2}}T_1 - \frac{1}{\sqrt{5}}T_2 - \frac{1}{\sqrt{2}}T_3 - F = 0$	$\frac{1}{\sqrt{2}}(T_1' - T_1'') - \frac{1}{\sqrt{5}}(T_2' - T_2'') - \frac{1}{\sqrt{2}}(T_3' - T_3'') - F = 0$
$\frac{1}{\sqrt{2}}T_1 + \frac{2}{\sqrt{5}}T_2 + \frac{1}{\sqrt{2}}T_3 = 0$	$\frac{1}{\sqrt{2}}(T_1' - T_1'') + \frac{2}{\sqrt{5}}(T_2' - T_2'') + \frac{1}{\sqrt{2}}(T_3' - T_3'') = 0$
$A_1 \geqslant 0$	$T_1' \geqslant 0$
$A_2 \geqslant 0$	$T_1'' \geqslant 0$
$A_3 \geqslant 0$	$T_2' \geqslant 0$
$-A_1 S_y \leqslant T_1 \leqslant A_1 S_y$	$T_2'' \geqslant 0$
$-A_2 S_y \leqslant T_2 \leqslant A_2 S_y$	$T_3' \geqslant 0$
$-A_3 S_y \leqslant T_3 \leqslant A_3 S_y$	$T_3'' \geqslant 0$

possible T_is and A_is. In the transformed version, minimizing the volume is the same as finding the smallest set of slack variables.

By solving the transformed problem with the Simplex method, we obtain an optimum set of values:

$$T_1' = \frac{F}{\sqrt{2}} \qquad T_2' = 0 \qquad T_3' = 0$$

$$T_1'' = 0 \qquad T_2'' = 0 \qquad T_3'' = \frac{F}{\sqrt{2}}$$

By the inverse transformation equations in Figure 2.24, the minimum sectional areas and the minimum volume in the original problem are:

$$A_1 = \frac{1}{\sqrt{2}}\frac{F}{S_y} \qquad A_2 = 0 \qquad A_3 = \frac{1}{\sqrt{2}}\frac{F}{S_y} \qquad V_{\min} = \frac{4F}{S_y}$$

The approach described above may not look very simple or elegant. However, with digital computers, this solution method can handle structures with hundreds of members automatically. The minimum weight structures attracted a lot of attention because of their theoretical implications and economic values.

2.4. NONLINEAR PROBLEMS

There are many ways to solve different types of nonlinear problems. Differential calculus, variational calculus, and Langrangian Multipliers are classical ones. Modern methods include linear approximation methods and some thirty or more specific techniques. We will limit our discussion to the search methods, dynamic programming and geometric programming.

2.4.1. Search Methods

Since most nonlinear methods have great limitations, it is not easy to find a suitable method for a nonlinear problem or to formulate a problem into a standard programming format. A search method is an organized way of trial-and-error, which can handle almost any practical problem if enough time is available. A search method is always inferior to an exact method because a problem has to be redone if there is any change in the parameters. It is also difficult to prove that the result is the best possible solution. However, understanding the method and the problem helps to perform an efficient search.

Given a design space, several basic questions can be asked of any search method. Will the search results converge? How fast and in what manner will the results converge? How accurate is the final result? Here we will discuss four important search methods.

1. *Grid Method*. First, a coarse grid sections the design space into squares that are quickly checked. A finer grid is then used to investigate the promising areas. The fineness of the grid is progressively increased until the area to be searched is narrowed down to a pin point (Figure 2.25).

The grid method has the danger of mislocating the target areas for further investigation if the initial grid is too crude. On the other

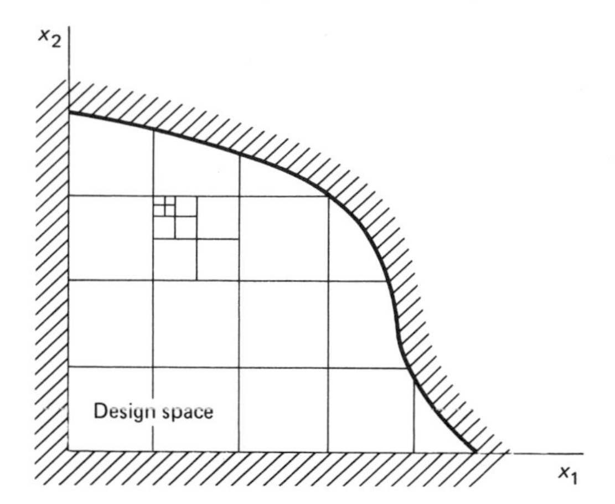

Figure 2.25. The grid method.
First a coarse grid, then a finer grid, and then a still finer grid are applied over the promising area.

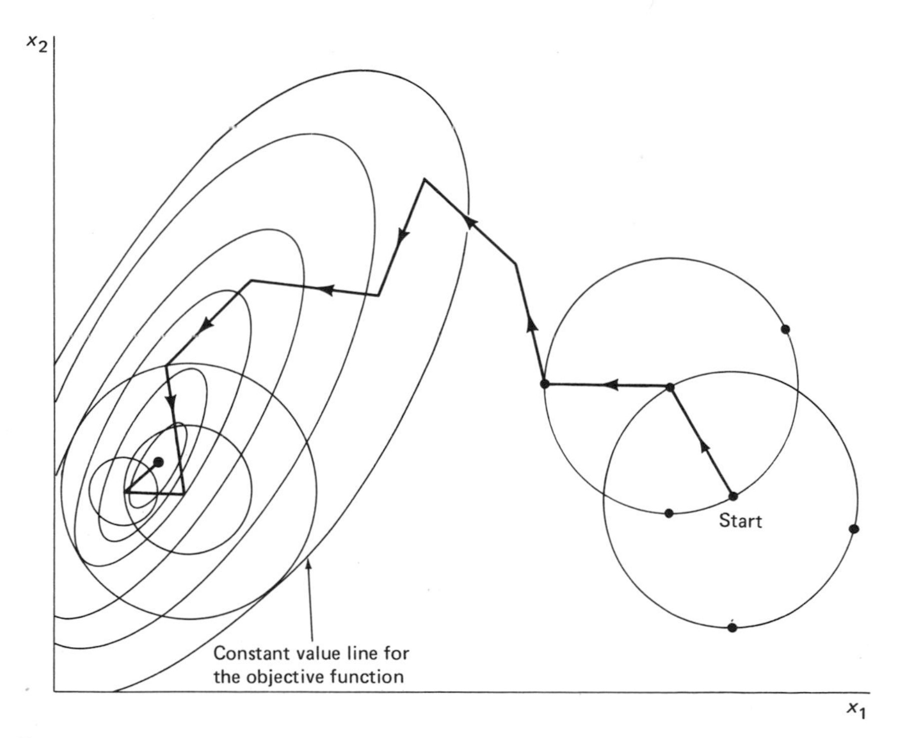

Figure 2.26. Random direction search. The points at a fixed radius are picked randomly and compared to the center. The center is moved toward the optimum value.

hand, a very fine grid, which might be very accurate, is costly in computing time. If the design variables can have only integer values, there will be fewer alternatives, and the grid method would be more efficient.

2. *Random Direction, Progressive Accuracy Search*. The value of the objective function at an arbitrary point is compared with the values of points at a fixed distance away (Figure 2.26). Every time a "better" point is found, the center is moved to the new point. When the points at a fixed radius do not yield a "better" value, the center is roughly at the optimum solution. The radius of search is then decreased, and the whole search process is repeated. The center of search moves closer to the optimum solution as the radius of search is repeatedly decreased to continue the "fine tuning." The search is terminated when the desired accuracy is obtained and no improvement can be gained for a preset number of trials.

3. *Steepest Gradient Search*. The gradient of a point in the design space can be obtained by taking the partial derivative of the objective function with respect to the design variables (Figure 2.27). The

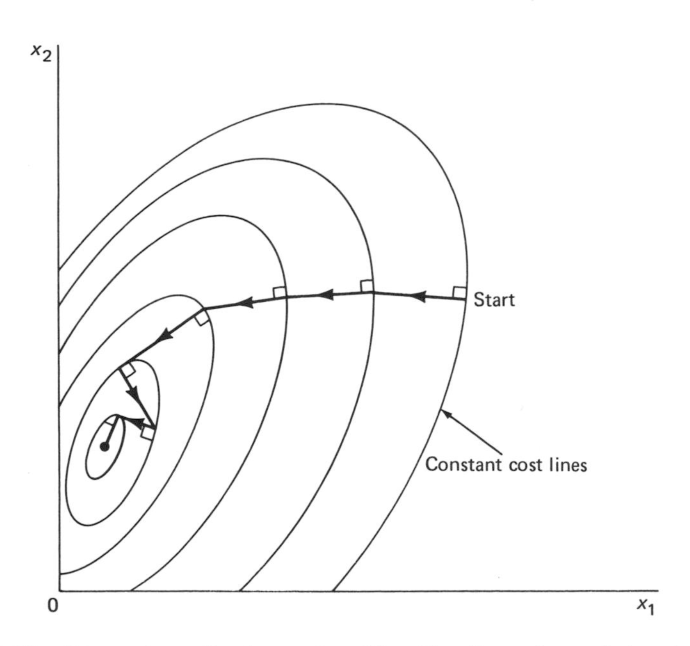

Figure 2.27. Steepest gradient search. The direction of search is set equal to that of the steepest gradient. The steepest gradient is perpendicular to the constant cost lines.

steepest gradient, which is a direction perpendicular to the constant cost lines, replaces the random direction of the previous method. In many respects the steepest gradient search is more efficient than the other two methods discussed above.

4. *The Simplex Search*. The simplex search is like the steepest gradient search and yet it does not require derivatives of the function. The cost (objective function) is evaluated at three points, which are the corners of an equilateral triangle. The point with the highest cost is rejected and a new triangle is formed using the mirror image about the opposite side of the triangle (Figure 2.28(a)). The process is repeated until a mirror image fails to produce an improvement. The triangle is contracted by retaining the corner that has the lowest cost and cutting all sides in half. The search continues until the required accuracy is obtained.

An improvement from the equilateral triangle (by Nelder and Mead [12]) is an irregular triangle that can rotate and expand. The mirror image is replaced by a projection about the midpoint of the opposite side. If an improvement is detected, a projection of twice

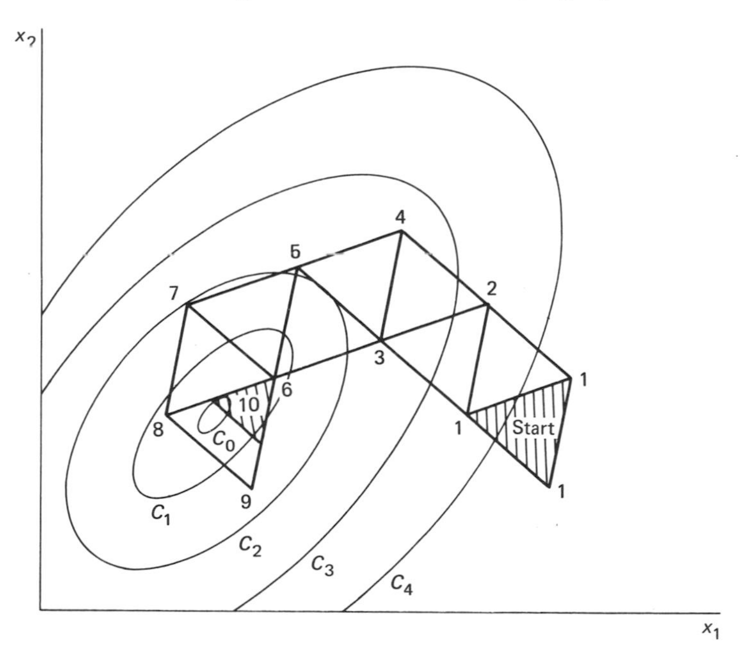

Figure 2.28(a). The Simplex search method with an equilateral triangle. 1, initial triangle; 2 through 9, reflections; 10, contraction. C_0 is the minimum cost, where $C_0 < C_1 < C_2 < C_3 < C_4$.

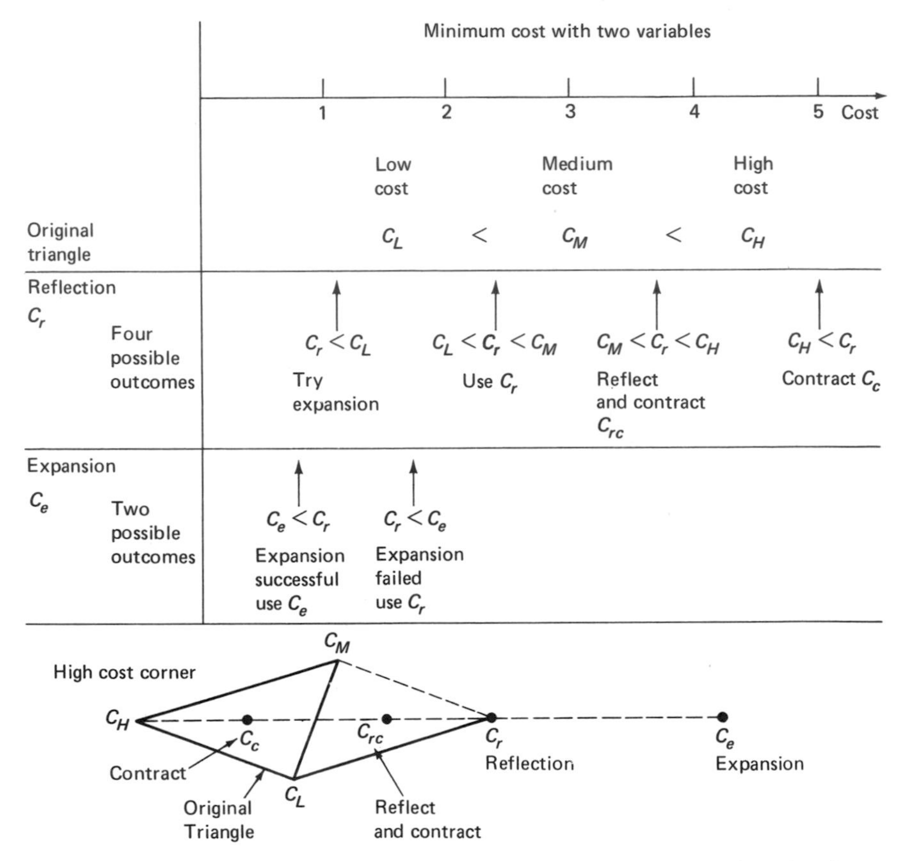

Figure 2.28(b). Improved Simplex search method with irregular triangle.

the length is attempted. If an improvement is not detected, a half length contraction will be taken (Figure 2.28(b)). In this manner, expansion, contraction, and reflection operations are possible at each step. Such a triangle can accelerate the search by taking large steps in the proper direction and avoiding delays at valleys or cycling around the minimum.

We will not get into the details of search methods because they involve computer algorithms and computer languages. The reader is advised to study the suggested references for further information and sample computer programs.

2.4.2. Dynamic Programming

Dynamic programming has been briefly discussed in connection with network analysis in Section 2.1.6. We will now elaborate on this concept of solving multistage problems with sequential decisions. Typical engineering applications of dynamic programming are in: network analysis, allocation of resources to maximize profit, probability and reliability problems, feedback control problems, optimum levels of labor and machinery to meet fluctuations in production levels (soothing operation capacities), structural design, and variational calculus problems.

Strictly speaking, dynamic programming is not a nonlinear programming method because it can solve linear problems as well. Dynamic programming is an optimization concept that may involve different kinds of mathematics. For example, problems traditionally solved by calculus of variation can be replaced by a similar problem with many decision-making stages. So, a continuous problem is approximated by a finite number of steps. Another example is representing interrelated machine components by a network of performance paths. During each stage of the decision process, only the results from the previous stage are considered. Problems with many variables can be simplified by sequential computation. A problem with n variables is segmented into n decision stages, each of which deals with one variable.

Example 1. Two mining machines are to be allocated to four existing coal mines. The mines that receive extra machinery will increase production by the amount shown in Table 2.1.

TABLE 2.1

	Extra production, tons/hr			
No. of extra machines	Mine No. 1	Mine No. 2	Mine No. 3	Mine No. 4
0	0	0	0	0
1	5	4	6	4
2	8	9	7	6

What should be the allocation pattern in order to maximize the extra production?

It takes some imagination to see a sequential connection in the problem. Figure 2.29 shows the symbolic network with four stages, each of which deals with a mine. The decision for each mine is

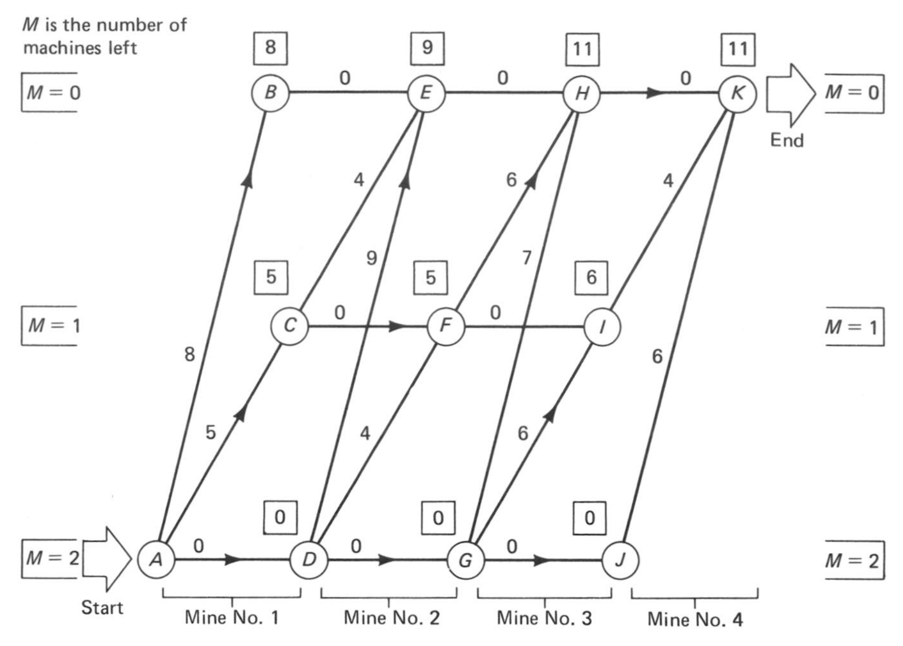

Figure 2.29. Allocating two machines to four mines by dynamic programming. The maximum extra production is 11 tons/hour. The optimum route is ACFHK, representing an allocation of one machine to mine no. 1 and one machine to mine no. 3.

made consecutively. The nodes in each stage represent the number of machines still remaining after the previous stage of allocation. Note that a total of two machines is allocated during the four stages of decision. If two machines are allocated in stage 1, nothing can be allocated in stages 2, 3, and 4. At the starting point, two machines are available. At the end, no machines are left. A path represents the amount of extra production for the corresponding allocation. We want to maximize the total extra production of a route through the network. With the dynamic programming procedure discussed in Section 2.1.6, the maximum extra production is found to be 11 tons/hr with one machine for mine no. 1 and one for mine no. 3. In order to appreciate the solution technique, the reader should check the cumulative extra production for each node. Because this is a small problem, the answer can be found by simple evaluation. If ten machines are to be allocated to four mines according to the data shown in Figure 2.30, the advantage of dynamic programming over other methods becomes obvious. Since real-life problems are often larger than these examples, a graphical representation such as Fig-

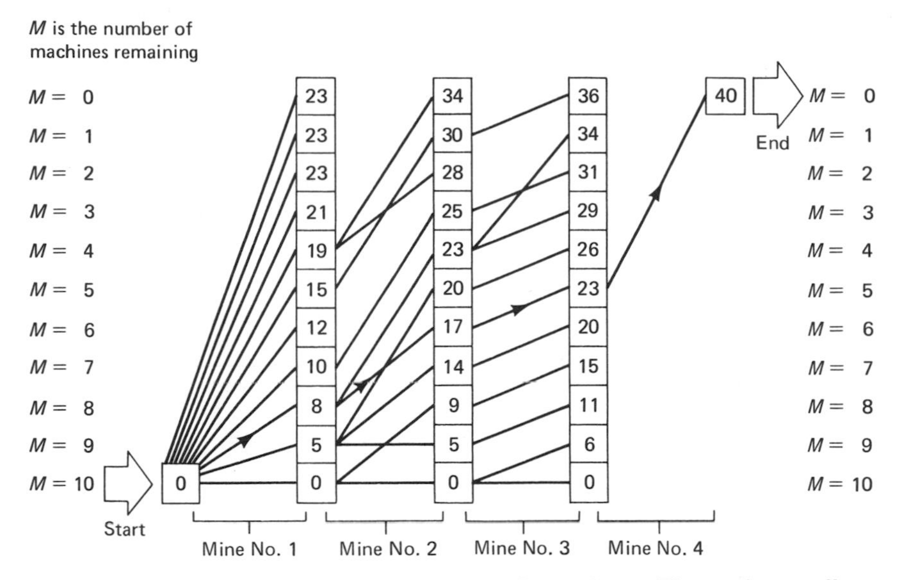

Figure 2.30. Allocation of ten machines to four mines. The optimum allocation pattern is: two machines to mine no. 1, two machines to mine no. 2, one machine to mine no. 3 and five machines to mine no. 4. The increase in production is 40 tons/hour. The numbers in the cells represent the maximum cumulative extra production up to that allocation.

Number of Extra Machines	Extra Production (Ton/Hour)			
	Mine 1	Mine 2	Mine 3	Mine 4
0	0	0	0	0
1	5	4	6	4
2	8	9	7	6
3	10	11	11	10
4	12	15	13	14
5	15	17	16	17
6	19	18	17	20
7	21	19	18	23
8	23	20	22	25
9	23	21	26	27
10	23	22	29	28

ure 2.29 quickly becomes impractical and an iterative computer algorithm will be needed.

Example 2. An engineer wishes to improve the reliability of a machine having four components in series. The overall reliability of this machine is the product of the component reliabilities because the machine can function only if all four components are working. Limited by the budget, the engineer can either select one component for heat-treatment and grinding, or select two components for heat-treatment. The cost of heat-treatment and grinding are assumed to be the same. Table 2.2 shows how much the treatments might improve an individual component.

TABLE 2.2

		Reliability of components			
	Number of treatments	Comp. no. 1	Comp. no 2	Comp. no. 3	Comp. no. 4
No treatment	0	0.8	0.9	0.6	0.7
Heat-treated	1	0.9	0.95	0.7	0.75
Heat-treated and ground	2	0.95	0.99	0.8	0.8

The engineer is in effect trying to allocate two treatments to four components so that the product of component reliabilities is maximized. In Figure 2.31, the stages represent the components; the nodes represent the remaining resources after the previous allocation. A path represents the reliability of a component after a treatment. The number next to each node is the maximum product of reliabilities from node A to that node. We wish to choose the route that maximizes the product of reliabilities at the end point (node K). Referring to Figure 2.31:

(1) Stage 1 (allocations to component no. 1)—the paths to node B, C, and D are just 0.95, 0.9 and 0.8.

(2) Stage 2—For node E, consider three alternatives:

node E = node $B \times BE = 0.95 \times 0.9 = 0.855$ ✓

node E = node $C \times CE = 0.9 \times 0.95 = 0.855$ ✓

node E = node $D \times DE = 0.8 \times 0.99 = 0.792$

Routes ABE and ACE are equally good; the maximum product is 0.855.

For node F, consider two alternatives:

node F = node $C \times CF = 0.9 \times 0.9 = 0.81$ ✓

node F = node $D \times DF = 0.8 \times 0.95 = 0.76$

The maximum product is 0.81.

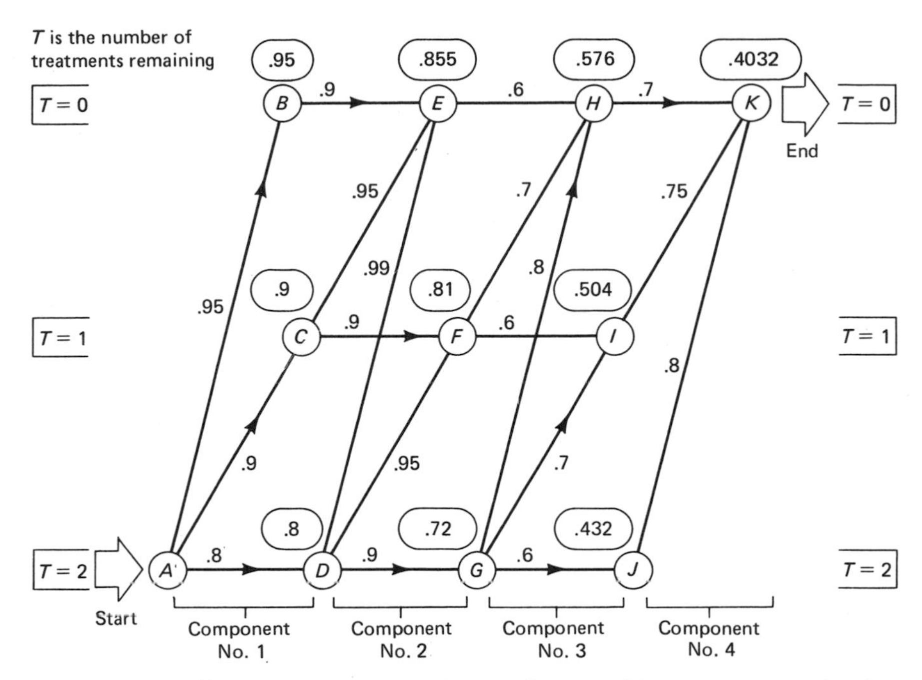

Figure 2.31. Allocating two treatments to four machine components by dynamic programming. Since the components are in series, the values next to the nodes represent products of path contributions. The optimum route is *ADGHK*, which represents heat-treatment and grinding of component no. 3.

For node G, there is only one path DG: node $G = 0.8 \times 0.9 = 0.72$.

(3) The same procedure is repeated for stage 3 (component no. 3). For node H, the best path is GH; the maximum product is 0.576. For node I, the best path is GI with a maximum product of 0.504. For node J, the only path is GI, giving a maximum product of 0.432.

(4) For stage 4

node K = node $H \times HK = 0.4032$ ✓
node K = node $I \times IK = 0.378$
node K = node $J \times JK = 0.3456$

The route with the highest reliability for the network is *ADGHK* (0.4032), which represents heat-treatment and grinding of component no. 3.

The first of the two examples involves the summation of path contributions, while the second involves the multiplication of path contributions. The concept of dynamic programming is independent of the mathematics used. Adding, multiplying, and differentiating are all compatible with this method.

2.4.3. Geometric Programming

Geometric programming is a method designed to minimize cost functions in the form:

$$\text{cost} = \text{constituent } 1 + \text{constituent } 2 + \text{constituent } 3 + \cdots \tag{2.42}$$

where each cost constituent can be expressed as a product of several variables with positive or negative exponents. The cost constituents may be component costs, operation costs, administration costs, etc.

Let us look at a cost function of two variables:

$$\text{cost} = C_1 x_1^a x_2^b + C_2 x_1^c x_2^d + C_3 x_1^e x_2^f \tag{2.43}$$

where $C_1, C_2, C_3, a, b, c, d, e$, and f are all constants.
Based on the theories of geometric programming, the minimum cost is

$$\text{cost}_{\text{min}} = \left(\frac{C_1}{\delta_1}\right)^{\delta_1} \left(\frac{C_2}{\delta_2}\right)^{\delta_2} \left(\frac{C_3}{\delta_3}\right)^{\delta_3} \tag{2.44}$$

where the vector $(\delta_1, \delta_2, \delta_3)$, called the dual vector, has the following properties:

(1) orthogonality—the scalar product of the exponent vector of each variable and the dual vector is zero.

$$a\delta_1 + c\delta_2 + e\delta_3 = 0 \tag{2.45}$$

$$b\delta_1 + d\delta_2 + f\delta_3 = 0 \tag{2.46}$$

(2) normality—

$$\delta_1 + \delta_2 + \delta_3 = 1 \tag{2.47}$$

In this particular case, the dual vector $(\delta_1, \delta_2, \delta_3)$ is unique. The optimum situation also represents the best breakdown of costs of constituents.

Constituent 1 is $\delta_1\%$ of the total minimum cost:

$$\delta_1 (\text{cost}_{\text{min}}) = C_1 x_1^a x_2^b \tag{2.48}$$

Constituent 2 is $\delta_2\%$ of the total minimum cost:

$$\delta_2(\text{cost}_{\min}) = C_2 x_1^c x_2^d \tag{2.49}$$

Constituent 3 is $\delta_3\%$ of the minimum total cost:

$$\delta_3(\text{cost}_{\min}) = C_3 x_1^e x_2^f \tag{2.50}$$

The optimum values of x_1 and x_2 can be solved from Equations 2.48, 2.49, and 2.50, when the dual vector and the minimum cost are known.

Geometric programming originates from the geometric inequality:

$$u_1 + u_2 + u_3 \geqslant \left(\frac{u_1}{\delta_1}\right)^{\delta_1} \left(\frac{u_2}{\delta_2}\right)^{\delta_2} \left(\frac{u_3}{\delta_3}\right)^{\delta_3} \tag{2.51}$$

where $\delta_1 + \delta_2 + \delta_3 = 1$.

The left-hand side of the inequality is called the primal function. It is at least as large as the expression on the right-hand side, its dual function. Applying this inequality to the cost function in Equation 2.43, we have

$$\text{cost} = C_1 x_1^a x_2^b + C_2 x_1^c x_2^d + C_3 x_1^e x_2^f$$

$$\geqslant \left(\frac{C_1 x_1^a x_2^b}{\delta_1}\right)^{\delta_1} \left(\frac{C_2 x_1^c x_2^d}{\delta_2}\right)^{\delta_2} \left(\frac{C_3 x_1^e x_2^f}{\delta_3}\right)^{\delta_3}$$

$$\text{cost} \geqslant \left(\frac{C_1}{\delta_1}\right)^{\delta_1} \left(\frac{C_2}{\delta_2}\right)^{\delta_2} \left(\frac{C_3}{\delta_3}\right)^{\delta_3} x_1^{a\delta_1 + c\delta_2 + e\delta_3} \cdot x_2^{b\delta_1 + d\delta_2 + f\delta_2} \tag{2.52}$$

When the cost is minimum and the dual function is maximum, the two sides are equal. Therefore, if we can find the maximum value of the dual function, the minimum cost is also known. The dual function will reach its maximum value when $(\delta_1, \delta_2, \delta_3)$ satisfies the orthogonality condition in Equations 2.45 and 2.46. The dual function is then independent of x_1 and x_2, and the exponents in the dual function (Equation 2.52) become zero. The inequality Equation 2.52 reduces to Equation 2.44. We now turn to two examples of cost minimization with geometric programming.

Example 1. We will repeat the milk carton problem in Section 2.2.2, but will look at the cost from a different angle. Assuming the cost per unit area of paper is 1, and neglecting the assembly cost:

cost of paper carton = cost of bottom and top + cost of side panels.

From (2.11)

$$\text{cost} = 18.8\,x^{-2/3} + 37.6\,x^{+1/3} \tag{2.11}$$

From (2.45)

$$-\tfrac{2}{3}\delta_1 + \tfrac{1}{3}\delta_2 = 0 \tag{2.53}$$

From (2.47)

$$\delta_1 + \delta_2 = 1 \tag{2.47}$$

Solving the two simultaneous equations 2.53 and 2.47:

$$\delta_1 = \tfrac{1}{3} \qquad \delta_2 = \tfrac{2}{3} \tag{2.54}$$

(2.54) in (2.44)

$$\text{cost}_{min} = (3 \cdot 18.8)^{1/3} \times (\tfrac{3}{2} \cdot 37.6)^{2/3} = 56.4$$

(2.54) in (2.48)

$$\tfrac{1}{3}(56.4) = 18.8\,x^{-2/3}$$

$$x = 1$$

This solution is the same as the solution from differentiation. The optimum relative cost of the top and the bottom to the sides of the milk carton is 1 : 2, as depicted by $\delta_1 : \delta_2$. The ratio of δs always gives the optimum breakdown of the cost constituents.

Example 2. The solution from geometric programming is always a global minimum, as opposed to a point of inflection or a local minimum.

We wish to minimize the cost of spraying herbicide on a farm. The spraying will be done by a spray gun attached to a tractor. The cost of operation consists of the cost of the spray gun, the cost of the tractor, and the cost of labor. Since the cost of herbicide is fixed by the area of the farm, it is not considered a variable.

Let W = width of the spray,
S = speed of the tractor,
the flow rate of the herbicide is proportional to WS,
the cost of the spray gun α (pump capacity)$^{1.5}$ and W

$$\alpha\, W^{2.5} S^{1.5},$$

the cost of the tractor $\alpha\, S$
the cost of labor α time $\alpha\, (WS)^{-1}$.
Combining the three constituents:

$$\text{total cost} = C_1 W^{2 \cdot 5} S^{1 \cdot 5} + C_2 S + C_3 W^{-1} S^{-1} \tag{2.55}$$

where C_1, C_2, and C_3 are cost coefficients.
Using the three simultaneous equations 2.45, 2.46, and 2.47:

$$\delta_1 = .222, \quad \delta_2 = .222, \quad \delta_3 = .555 \tag{2.56}$$

The minimum cost is obtained by substituting 2.56 in 2.44:

$$\text{cost}_{\text{min}} = \left(\frac{C_1}{.222}\right)^{.222} \left(\frac{C_2}{.222}\right)^{.222} \left(\frac{C_3}{.555}\right)^{.555}$$

The ratio of the cost constituents is gun cost : tractor cost : labor cost = $\delta_1 : \delta_2 : \delta_3 = 0.222 : 0.222 : 0.555$. The values of the design variables W and S can be determined by using Equations 2.48 to 2.50.

In the above examples, we note that the size of the dual vector δ is equal to the number of terms in the cost function. In order to get a solution for the dual vector, the number of terms in the cost function must be at least one more than the number of design variables. When

$$\text{no. of terms} - \text{no. of variables} = 1$$

solution to dual function is unique. When

$$\text{no. of terms} - \text{no. of variables} > 1$$

the dual function has more than one solution. The maximum for the dual function is the minimum for the cost function.

In addition to the characteristics discussed above, the geometric programming method can also handle inequality polynominal constraints. The constraints have a form similar to the objective function:

$$K_1 x_1^p x_2^q + K_2 x_1^r x_2^s \leqslant 1 \tag{2.57}$$

2.5. DISCUSSIONS

The following is an optimization approach to a problem:

(1) Define the problem as exactly as possible. Use a suitable mathematical model and express the objective function in terms of the design variables.

(2) Place constraints and limits on the design variables. This is also known as defining the design space. If the design is unique, the design space is a point and no choice is necessary.
(3) Optimize the objective function subjected to all constraints and limits.
(4) Interpret and adapt the solution to the original problem.

There are many optimization methods to handle different kinds of problems. The objective may be to minimize cost, to maximize strength, to maximize service life, to maximize power, etc. A composite objective function can be used when there is more than one goal. If these goals conflict with one another in some way, negative coefficients can be used to make up "penalty functions" as part of the objective function. For a design engineer, the simplest methods are differentiation calculus and linear programming. Other appropriate mathematical tools include: calculus of variation, dynamic programming, network analysis, and geometric programming. For a given problem, there may be several methods that are applicable. Structural synthesis, for example, has been solved with all the methods mentioned above.

A dual problem exists for every optimization primal problem. For example, maximizing profit in a primal problem may become minimizing cost in the dual problem. Constraints define a design space. If an objective function does not have maximum or minimum points, an optimum solution exists only when there are constraints. Sometimes it is advantageous to convert a constrained problem into an unconstrained problem. Equality constraints may be substituted directly into the objective function, while inequality constraints may be converted into penalty functions.

When maximum or minimum points exist in an objective function, the points have a zero rate of change of the objective with respect to all variables. Sometimes several local maxima or minima exist, one of which is the true global solution. With constrained problems, if the maximum or minimum falls within the design space, the constraints could be neglected. If the maximum or minimum is outside the design space, or if there are no extreme values in the objective function, the solution would fall on the boundary of the design space. If the constraining equation at the maximum value of the objective function is parallel to the objective function, there would be a series of equally good optimum solutions.

Modern optimization methods originated in the field of operations research. Their adoption for machine design and structural design started in the late 1950s. To many practical designers, the mathematics tends to become overwhelming. In this chapter, we have concentrated on principles and applications. For formal definitions and mathematical derivations, the readers are advised to read specific books in the list of references.

Finally, we should always use good judgment with optimization problems. The validity of the results depends on how well the model represents the real situation. In a way, the results are implied by the assumptions and the definition of the problem. In optimizing the milk carton, as defined in Section 2.2.2, we can get only an optimum ratio of its height to its width. The mathematics in the problem would not even hint that a cylindrical or spherical container is better. Linear programming structural analysis is seriously restricted by having to fix the nodes. If the nodes are not placed in the best possible locations, the solution will never be really good. Furthermore, some design features, such as color and style, cannot be fully represented in the objective function, and such significant constraints as taste, buying trends, and social and political influences cannot be expressed effectively by mathematics. An optimum solution is optimum only for a specific set of constraints. If the load on a structure is changed, the original solution may be short of the optimum. The study of the range of loads within which the design remains optimum is called sensitivity analysis. An optimum solution usually needs careful adaptation to real-life conditions.

3. ECONOMIC FACTORS INFLUENCING DESIGN

In this chapter, we will discuss a group of topics not directly related to the function of a product, but nevertheless important to the success of the product.

First, we will examine some basic tools of value-engineering that deal with the relationship between performance and cost. A product might have undesirable effects such as accidents, failures, and environmental damage, the absence of which actually increases the product's value indirectly. We will study the safety and reliability requirements of a product, and will discuss briefly the factors affecting recycling.

Then, we will discuss the production operations that contribute significantly to the cost of a product, and our last subject of study will be the evaluation of investment alternatives and the control of production systems.

The common objective of these discussions is to show how product design is influenced by various economic factors. Technology and engineering represent only what can be done. Very often what is actually done is governed by economic factors.

3.1. VALUE IN PRODUCTS

If small eggs sell for 70¢ a dozen and large eggs sell for 85¢ a dozen, which kind of eggs has a better value? Assuming that large eggs and small eggs are equally convenient for cooking, we must reduce the eggs to a common base in order to compare them rationally. Suppose that each large egg weighs 2 oz without the shell and each small egg weighs 1.7 oz without the shell. The "value" of the large

TABLE 3.1. ANALYZING VALUES OF TYPEWRITER OPTIONS

	Speed, word/min	cost, $	value, word/min/$
Manual typewriter	50	50	1
Electric type	additional 30	20	1.5
Electric return	" 10	12	0.83
Electric index	" 2	5	0.4
Electric backspace	" 5	5	1
Tabulator	" 5	2	2.5
Repeated type	" 5	3	1.6
Repeated spacing	" 5	2	2.5
Snap-in ribbon	" 1	2	0.5

egg is 0.282 oz/cent and that of the small egg is 0.291 oz/cent. The small egg has a better value!

Obtaining the maximum performance per unit cost is the basic objective of value-engineering. The value of a product is the ratio of its performance to its cost. Since cost is a measure of effort, then the value of a product is just the ratio of output to input commonly used in engineering studies. In a complicated machine system, every component contributes to the cost and the performance of the entire system. The ratio of performance to cost of each component indicates the relative value of individual components.

For example, in the design of a typewriter, the following options are available: electric type, electric return, electric index (vertical advance without return), electric backspace, tabulator (skip spacing), repeated type, repeated spacing, and snap-in ribbon. Let the speed be chosen to quantify performance. Hypothetical costs and values are shown in Table 3.1. The basic, manual typewriter has a value of 1 word/min/dollar. Those options with a value larger than 1 are more desirable than the basic version. Those options with a value less than 1 cannot be justified on the basis of typing speed only. As the table shows, the tabulator and repeated spacing are the best values. Of course, a broader view of performance includes low noise, tidiness of type, appearance and convenience of the typewriter, and other values more difficult to assess than the typing speed.

3.1.1. Function

It is universally true that in order to do exactly what is needed without wasting effort, one must be able to state one's objective clearly and precisely. The definition of a function in value-engineering is rather unusual. Value and performance are sought from an abstract and conceptual approach without implying any particular mathematical model or physical quantity. As an example, let us examine the bicycle taillight. Its function may be described as: improve safety, draw attention, or illuminate bicycle. The three descriptions all lean toward increasing the safety of a bicycle, but they involve different degrees of abstraction. *Illuminate bicycle* is the most exact description but it implies the use of light bulbs and limits the scope of the solution. *Draw attention* may allow the use of sound, motion, flare, flags, reflectors, bright paint, etc. *Improve safety* may suggest the construction of special bicycle paths, or the use of helmets or bumpers. If an engineer is blocked into the design of a taillight, he might reduce cost by using cheaper materials and easier manufacturing methods. Optimization may lead to a better lens angle and, hence, visibility. Time-and-motion studies may reduce the labor cost by 10% to 20%, but only the abstract analysis of a function can bring out totally different ideas. For this reason, function definition in value-engineering can be considered as a technique to identify the problem. Most other engineering methods are problem-solving methods in which the problems are intuitively assumed. In value-engineering, the emphasis is placed on analyzing the function and not on the hardware that performs the function.

Many products have a variety of functions. There is usually a principal function and a host of secondary functions. To define a function means to find out what the consumer wants the product to do. In this sense, value-analysis is related to marketing research. As an example, to measure the play value of a toy, a test may be conducted for the following aspects: (1) the average length of time the toy can sustain interest in a child, (2) the frequency with which a child returns to the toy after playing with something else, (3) the percentage of children that pick the toy among several competitive alternatives, (4) the percentage of children of a particular age that fail to master the skill needed to play with the toy. Similar consumer testings and ratings have been used regarding the taste of different beers and the comfort of rides in automobiles.

To better understand the term *value,* let us take a closer look at the different kinds of values in a toy. (1) Play value: the play value is the entertainment the toy provides. It may be the excitement of competition, the challenge of doing something, the stimulation of creative instincts, or the satisfaction of imitative yearnings. Toy dolls, comic-book heroes, and soldiers may satisfy and stimulate childish fantasies. Sound effects and visual effects usually increase the play value of toys. (2) Educational value: for a child, learning is often stimulating. The educational value of a toy is also a comfort to the parents. (3) Esteem value: radios shaped like Snoopy or Mickey Mouse may appeal to children more than "real" radios. In everyday life, esteem value is usually associated with brand names and symbolic status. For example, a Cadillac is regarded as more prestigious than a Chevrolet. (4) Reliability value: a reliable toy lasts longer and, therefore, effects more entertainment. (5) Safety value: for the sake of safety, children are not allowed to play with real guns, chain saws, lawn mowers, ovens, and sewing machines. (6) Social value: a child usually wants to possess what his friends have. Toys may shape the behavior of children or dramatize real life. There are toys made to promote heroism, racial harmony, and sexual equality.

After a function is defined or measured, the value is obtained by dividing the function by the cost. At this time, the engineer should consider seriously all the alternatives that could perform the function. There are several questions that may help: (1) How does the product meet the function? (2) Which features of the product do not contribute to the function? (3) What else can do the job? (4) Can an additional component or a change in material improve the performance?

One of the value-engineering approaches consists of three steps: (1) determine the function and state it in the simplest verb-noun form; (2) find the cheapest system to satisfy the function—this system usually represents fairly decent value because of the least cost; (3) compare more advanced systems to the cheapest system by the ratio of performance to cost. This approach is named "Blast, Create, and Refine". The verb-noun definition of function eliminates things that would confuse the real issue. It is amazing that we are sold so many things we think we need that we really don't need. To pose the question "what is the cheapest system?" would be a way to start from the basic and add just enough fancy stuff to make it good. In the pursuit of "the cheapest system," we often discover the real objective of the product and the distinction between necessity and luxury. Thus, a lighter is used to "start fire." The cheapest way is a match—

30 lights/cent. Each alternative will be tabulated with its function and cost. The value of every alternative is compared to that of the cheapest system. In order for a $1 lighter to be as valuable as a match, it has to light 3000 times. A checklist of ways to generate alternative systems is included in Appendix 1.

3.1.2. Lower Cost and Higher Value

Quantitative value is displayed by a performance-versus-cost plot. Figure 3.1 shows such a plot for the conductive value of metals. Silver and copper are recognized as good conductors. They have the highest conductivity, but their costs render their conductive values lower than those of aluminum, iron, and steel. Aluminum becomes the best conductor when the cost of material is taken into consid-

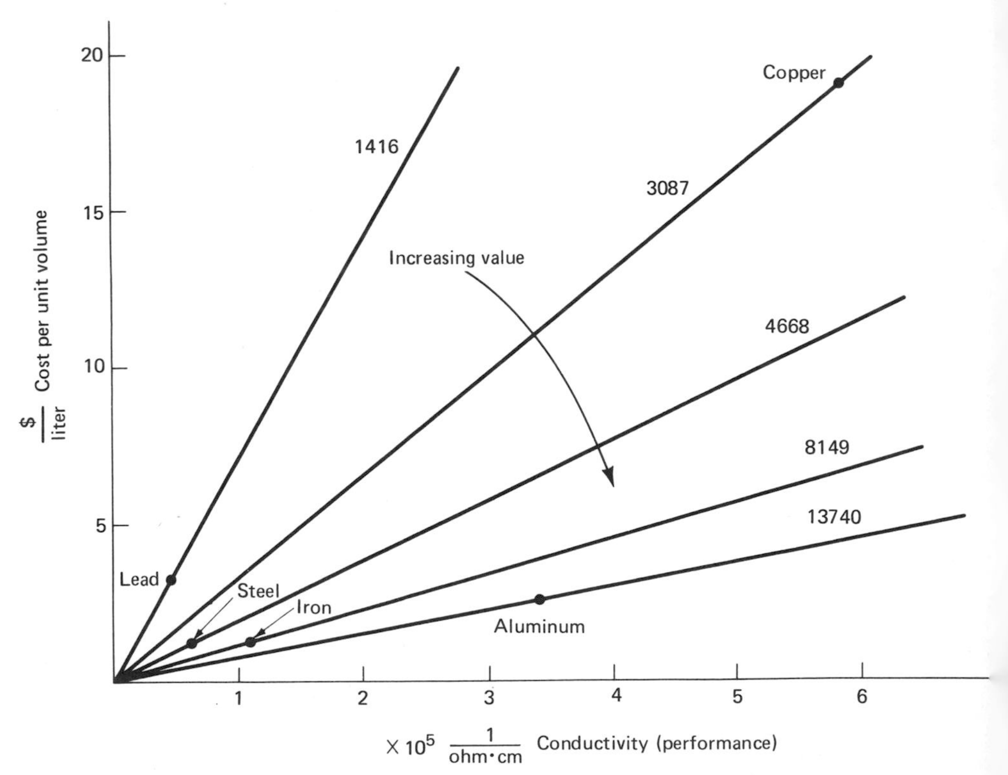

Figure 3.1. Cost versus conductivity of various metals. Straight lines radiating from the origin are isovalue lines. The lines making smaller angles with the horizontal axis have higher values.

eration (Table 3.2). Note that conductance depends on geometric dimensions (length and sectional area), whereas the cost is usually based on weight. Therefore, the cost has to be converted to dollars/volume

$$\text{conductive value} = \frac{\text{conductivity}}{\text{density} \cdot \text{cost}}$$

The engineer should also consider the cost of processing, joining, maintaining, and the scrap value of different materials. Design parameters have an important influence, too. If aluminum is used instead of copper in the windings of a motor, a larger rotor would be required because aluminum wires are bulkier than copper wires of the same electrical properties.

On the performance-cost plot, horizontal lines are constant-cost lines. Vertical lines are constant-performance lines. Straight lines from the origin are constant-value lines (isovalue lines). If two materials fall on the same isovalue line, they are equal in value. An isovalue line that makes a smaller angle with the performance axis has a higher value.

If we can decrease the cost and maintain the same performance, we have increased the value. This means removing unnecessary costs and unnecessary features.

Most features of a product cost money and they usually improve performance. But is the percentage increase in performance more significant than the percentage increase in cost? Note that the minimum cost does not imply the maximum value because performance does not remain constant. In the example of the typewriter (Section

TABLE 3.2. ELECTRIC CONDUCTIVE VALUE OF SIX COMMON METALS.

	Electric conductivity	Density	Cost		Conductive value = conductivity / cost
	$\frac{1}{\text{ohm} \cdot \text{cm}}$	$\frac{\text{gm}}{\text{cm}^3}$	$\frac{\$}{\text{kgm}}$	$\frac{\$}{\text{liter}}$	$\frac{\text{m}^2}{\text{ohm} \cdot \$}$
Copper	58.8×10^4	8.9	2.14	19	3087
Silver	61.4×10^4	10.4	52.3	544	113
Aluminum	34.5×10^4	2.7	0.93	2.5	13740
Iron	10.3×10^4	7.9	0.16	1.3	8149
Steel	5.9×10^4	7.9	0.16	1.3	4668
Lead	4.8×10^4	11.3	0.3	3.4	1416

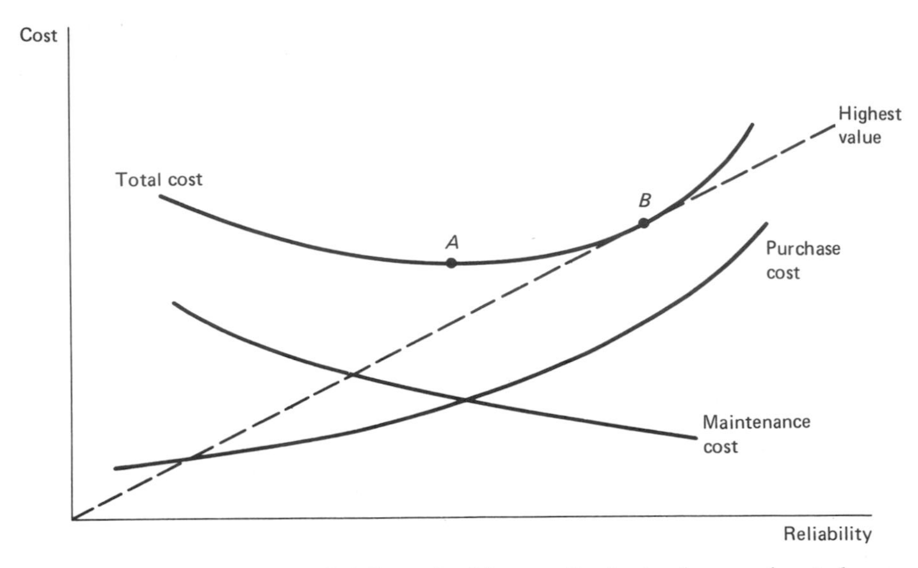

Figure 3.2. Cost versus reliability. In this case, the best values are located near the middle range of the performance (reliability). Point *A* represents the minimum total cost. Point *B* represents the maximum value.

3.1), the manual typewriter is the cheapest, but the electric typewriter has a better value. Two kinds of curves, shown in Figures 3.2 and 3.3, are particularly worth noting. Figure 3.2 shows a plot of reliability versus cost for a certain machine. A machine with a very low initial purchase cost ends up with a high maintenance bill. On the other extreme, a machine that has a very high initial purchase cost also offers limited value. The total-cost line (purchase cost plus maintenance cost) reveals the point of minimum cost, *A,* and the point of maximum value, *B.* The minimum-cost point is the tangent point between a horizontal line and the total-cost line. The maximum-value point is the tangent point of an isovalue line and the total-cost line. Figure 3.3 shows a plot of the weight of a glue versus its cost. As the size of the container gets larger, the cost per unit weight of glue decreases. Consumers are often seduced into buying more than they really need. A schoolboy quickly realizes that, regardless of how cheap it gets, he will not buy ten gallons of paper glue. Therefore, a cost ceiling sometimes exists to override the tendency to get maximum value.

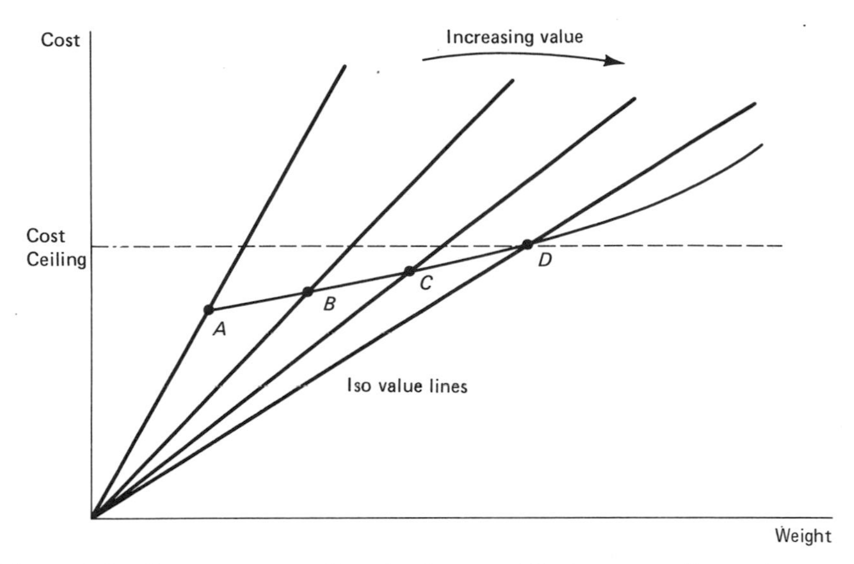

Figure 3.3. Cost versus weight. As the quantity increases, the cost per unit weight decreases. Points A, B, C, and D have increasing values. In the presence of a cost ceiling (as shown), point D would be the best choice.

3.1.3. Price Bracket and Target Price

In a society with abundant resources and no competition, prices are set according to needs. A product may be designed according to an arbitrary standard, and the cost is a result of the design. However, in businesses that have limited resources and tight competition, a target price is often decided for each product. In this case, the target price is a design parameter that determines the quality that can be afforded in the product.

Target prices are set through a careful study of consumer-buying patterns and their expendable income. The sales-expectation curve is found to have steps corresponding to price brackets: under $1, under $5, under $10, and under $20, for example. Assuming that the product can be made with four different sets of specifications, grades A, B, C, and D, the final choice of quality is made so that the prices rest comfortably within the suitable brackets. The quality has to be competitive with the qualities of similar products in the same price bracket. Before a person starts to design a calculator for over $100, he would check the sales of calculators within that price bracket from other manufacturers in order to get an indication of his potential market. The influence of price brackets is dramatically demonstrated in Polaroid cameras, stereo phonographs, and electronic pocket calculators.

Price brackets also influence price increases due to inflation. The psychological impact of a price increase from $9.90 to $10.10 is much larger than an increase from $9.40 to $9.60, although the percentage of increase is larger in the latter case. To ease the transition between price brackets, a 10-oz can of fruit can be reduced to 8- oz and sold at the same price. At a later date, the 10-oz cans will reappear at a 25% higher price. Target prices can also be set by market competition. Above the competitive price, the sales volume is small. As soon as price drops below the competitive price, volume starts to increase very substantially.

Let us clarify the terms *price* and *cost.* The price of a product is the amount of money it is sold for. Its cost is the amount of money needed to make or secure it. For a retailer, the difference between the cost and the price is the profit. For a customer, the selling price of a piece of merchandise is the cost. Sometimes, specifications have to be tailored and corners have to be cut to meet the target price. The following cost breakdown of an electronic pocket calculator illustrates such value analysis:

Component	Function	Cost
LED	output	$0.50
keyboard	input	$2.00
IC chip	computation	$1.00
circuit board	connection	$0.50
housing	protection	$0.20
battery	power source	$0.50
on–off switch	control	$0.10
labor	assembly	$0.80
	total	$5.60
	target cost	$4.00

First, the design engineer noticed that within the main function of input–compute–output, the cost of the keyboard seemed out of line. It might be that the technology of the keyboard was not as advanced as the chip and the LED (light emitting diode) display. With this finding, more engineering and purchasing effort was concentrated on lowering the cost of the keyboard. After a few months, the engineer came up with a $1.00 keyboard. There was still a discrepancy of 60¢ between the existing cost and the target cost. The following measures were suggested:

1. use smaller LEDs, and use a magnifying lens to make them more visible,
2. reduce the number of keys and extend their capabilities with control keys, e.g.: F key and 7 key = sine function; F key and 6 key = tangent function; IF key and 7 key = inverse sine; and IF key and 6 key = inverse tangent, where F = function and IF = inverse function,
3. sell the calculator without batteries and let the consumer buy his own,
4. cheapen the housing by using decals instead of hot-stamped letters and hot-stamped chrome.

A value study was conducted to identify the savings with the decrease in value of the calculator. Finally, the target cost was met by saving 10¢ through reducing the number of keys (suggestion 2), and 50 ¢ by not providing the batteries (suggestion 3).

In order for comparisons to yield helpful answers, the performance has to be suitably quantified and the cost has to be specific. Information collection is a very important part of decision making. It is important to know which features of a part make it costly. One of the specifications of the product may be on the verge of impossibility. The engineer should discuss the application with the supplier from the start so that costly qualities in the specifications can be identified.

3.2. SAFETY, RELIABILITY, AND ECOLOGY

Safety, reliability, and ecology are important considerations in design. These factors add extra cost to the product or the production process, and they put an additional burden on the engineer. Safe, reliable, and ecologically sound products will enrich society and win acceptance in the long run. Careful judgment is needed to balance the value of these rather abstract qualities with the cost and competitiveness of the product.

Accidents are costly, and somebody pays for them. Whether the people paying for the accidents are the consumers, the distributors, the manufacturers, or the insurance companies, society suffers. An unreliable machine may result in the mere termination of service or it may result in an accident. A machine that has a short useful life is uneconomical. Therefore, good reliability is also good economics.

Garbage is a by-product that has a negative value. If we were allowed to dump garbage in the street, we would save the fee paid to the garbage collector. Pollution of air, water, and soil are just different versions of saving money by abusing and exploiting the environment. A polluted environment is a source of discomfort and is often harmful to humans and animals. Yet, zero pollution is too heavy a burden for most industries. If the cost of a product has to include the cost of all the cleaning up (pollution control), probably the price would be beyond our reach. So we have to tolerate some damage to the environment for our short-term well being. Take cars for an example. With pollution-control devices, they have a higher initial cost and a lower gas mileage. Some cars even suffer in performance and reliability.

3.2.1. Safety is an Implied Warranty

An explicit warranty is a statement that the product will perform a described function for a stated period of time. Normally, if the consumer noticed a discrepancy between the claimed and actual performance, the manufacturer would refund, replace, or repair the product. A manufacturer can reduce his responsibility by making conservative claims or limiting the period covered by the warranty. However, this may impose a negative effect on sales.

Besides the explicit warranty, the manufacturer is also committed to an implied guarantee of safety to the user. If the user suffers from an accident during the normal usage of a product, he would expect compensation from the manufacturer.

How safe is safe? And what is the normal usage of a product? Many medicines, if taken carelessly, would produce adverse effects. Any bicycle, in the hands of an unskilled cyclist, is a potential hazard. Many chemicals would cause irritation or burns if they were swallowed or if they came into contact with the human eyes. Absolute safety is unrealistic for most products.

There are many conditional safety standards for products:

(1) as safe as they can possibly be,
(2) safe according to industry-wide standards,

(3) safe if used in the manner and situations specified in the instructions, and
(4) safe for a child who may behave irrationally.

Engineers and designers are concerned about the legal aspects of safety. Often decisions in court involve judgments not obvious to the designer. Moreover, the enforcement of laws and the interpretation of constitutional rights are different in different states, and may change with public movements. The information given here presents a general outline for designers. Specific details are available from government agencies such as the National Highway Traffic Safety Administration (NHTSA), Food and Drug Administration (FDA), Consumer Product Safety Commission (CPSC), and the Occupational Safety and Health Administration (OSHA), under the Department of Labor.

If an accident results in a court case, the engineer will have to prove that the consumer or user is totally responsible for the accident. For a child, the responsibility is on the parents. Using childproof containers and specifying "keep out of the reach of children" are common defensive practices. The ability to tinker differs from child to child. Nothing is really childproof, if a child is given enough time. The real function of childproofing is to reduce the chance of an accident and to increase the length of time needed for a child to succeed in harming himself. With sufficient time, the parents can notice the hazard and stop the child.

Reducing the responsibility of the manufacturer by specifying narrow situations and particular procedures does not always work. In court, the user is often not responsible for not following the fine print included in the instructions. It all depends on how reasonable and obvious the instructions are. Products are required to function in real-life situations, allowing for human error. Real-life conditions may include power failures, poor maintenance, accidental overheating, and a corrosive environment. Human errors include slight ignorance, negligence, and abuse. People are also known to panic and act irrationally in the face of danger. Though instructions do not exclude responsibility on the part of the manufacturer, clear and explicit warnings are often the best defenses. Whenever possible, safety warnings should be placed on the product. Crucial warnings are sometimes included as part of advertising. In the case of a toy, the safety standards are progressively more lenient with the higher age groups. By clearly labeling the age group, the designer has more flexibility. Toys are subjected to more stringent safety regulations than sporting goods. Sharp points and sharp edges are not allowed in toys. Darts, arrows, and knives are accepted as safe sporting goods. Therefore, a dart manufacturer has to label his product *not a toy* to prevent misuse.

Industry-wide standards are good guidelines for a designer, although such standards are not legally unchallengeable. There have been cases where an entire industry was challenged and ruled irresponsible. This is one way the standards improve and tighten. Years ago, electric power tools with just two wires (i.e., without the third ground wire) were supposely safe. Subsequent accidents proved the inadequacy of this practice, and grounding wires or double insulation were added. Tight standards cause inconvenience and increase the cost. Standards are usually set as low and least restrictive as possible.

In court, the victim usually has to prove only the existence of defects and hazards. Even if the defect is not caused by the negligence of the engineer, the manufacturer can be held liable for the injury. By now, the difficulty of maintaining superior product safety should be obvious. The remedy is just a careful, responsible, and alert attitude. An engineer should be responsive to prototype failures, field incidents, and complaints, and should not use excuses to disregard potential problems. Typical excuses are: "This is just a defect in the prototype, it is not inherent in the design."; "If only the user is more gentle with the controls. . ."; "There have been improvements made on the product, and the present design may not have the previous defects." In a relative sense, if a manufacturer is more careful than the others, he is less likely to be the first to run into trouble. Once an accident occurs, others in the industry can quickly review similar designs and initiate preventive measures. A good record of testing, consistent quality control, and prompt response to consumer feedback is useful evidence for a responsible attitude. Certification by independent private laboratories or consulting firms is also strong backing.

3.2.2. Safety Hazards and Regulations

There are many safety standards that apply to different categories of products. Table 3.3 lists the safety hazards for toys. This is a rather stringent set of standards because children are not presumed to know how to prevent accidents. Although the standards handicap design, experienced engineers soon get used to them. Sharp points and sharp edges can be overcome by using generous radii. It will also help if parting lines of molds are moved away from corners and edges (although this is undesirable in mold making) (Figure 3.4). Cuts from paper edges can be eliminated by serrating the edges. Small items could be excused if the toy were designed for children over 3 years old (see Table 3.3). The remedy for small items is obviously making them bulkier so that they cannot be swallowed by accident. A small item is defined as something that would fit completely inside the truncated cylinder shown in Figure 3.5. Articles weighing more than 10 lbs are exempt from drop tests. A toy may fail under the drop test, but the fragments should not contain small items, sharp edges, or sharp points. The sharp point and sharp edge regulation strongly encourages the use of high-impact and resilient materials. Pinch points can be avoided either by making the gaps so narrow that they cannot admit a finger or so wide that pinching cannot occur. Many materials, such as rubber and steel, can be highly inflammable if they are in the form of fine wires or thin sheets. The acceptability of an inflammable substance is based on the burning rate: less than 6 in./min for solids and 12 in./min for fabrics. Notice that by this definition, charcoal is not inflammable enough to be a hazard. Usually, by making a material thicker its burning rate can be lowered to the acceptable level. The hazard from a projectile can be decreased by lowering the kinetic energy ($\frac{1}{2}mV^2$) or by increasing the area of contact. The maximum allowable density of kinetic energy ($mV^2/2A$) is 2000 joules per m^2. The area A is taken as the projected area in the direction of motion.

Voluntary standards are issued by different trade associations. Independent testing companies such as the Underwriter's Laboratory also issue their own testing methods and standards.

In order for a safety standard to be accepted, it should be realistic and inexpensive to implement. In order for a standard to be meaningful, it has to eliminate or minimize the hazard. For this reason, standards usually take ten years or more to become mature and effective. Many new regulations have to be juggled and compromised under political and economic pressures. There is also personal freedom involved with government regulations. It is known that cigarettes, alcohol, coffee, and even bacon can result in health hazards, yet individuals are allowed to consume them at their own risk. There

TABLE 3.3. SOME COMMON HAZARDS OF TOYS [21-25].

Hazards	Comments
Heavy elements	To prevent poisoning, the maximum concentration of heavy elements allowed in paints is 0.06%. Examples of heavy elements are: lead, antimony, arsenic, cadmium, mercury, and selenium.
Dye extraction	If a fabric dye can dissolve significantly in saliva and gastric juices, it has to be tested for heavy elements.
Eye or skin irritation	Standards are set by experiments performed on rabbits and sometimes on people.
Dermal, inhalation, or oral toxicity	The material concerned is applied to skins of rabbits. Rabbits exposed to the suspected gas for 1 hr should not die within 2 wks. White rats are fed up to 1.6% of their weight of the material concerned.
Strings	The length and strength of strings are restricted to prevent strangling.
Plastic bags	The plastic film must be at least 0.0015 in. thick to prevent suffocation. Very small bags are excluded.
Projectiles	The energy-to-area ratio of a projectile has to be less than 2000 joules per m^2.
Pinch points	The gaps between moving components may crush fingers or toes. Examples of such "pinch points" are gaps between wheels and fenders, links of a chain, gaps in compression springs, hinge joints, etc.
Sharp points and edges	Toys for children under 8 years old should not have sharp points or edges that may puncture skin.
Small parts	Toys for children under 3 should not contain small parts that can be swallowed or inhaled.
Mechanical abuse	A toy may be first subjected to: (1) impact (drops), (2) compression, (3) torque-pull, (4) flexure, and (5) bite tests. If the piece is damaged, there should be no sharp points, sharp edges, or small parts.
Stability & overload	A ride-on toy or a seat should bear 3 times the weight of the child. With normal weight applied to the seat (6 in. above seat), the toy should be stable on a 15° slope.
Explosive substances	The material should not explode when subjected to (1) a spark, (2) a candle flame for 5 sec, and (3) a temperature of 130°F.
Flammability	Except strings, paper, and objects with area less than 1 in.^{2}, the burning rate (in./min) should be less than 6 for solids and less than 12 for fabrics.
Noise	The maximum continuous noise allowed is 100 db. The maximum impulsive noise allowed is 135 db.

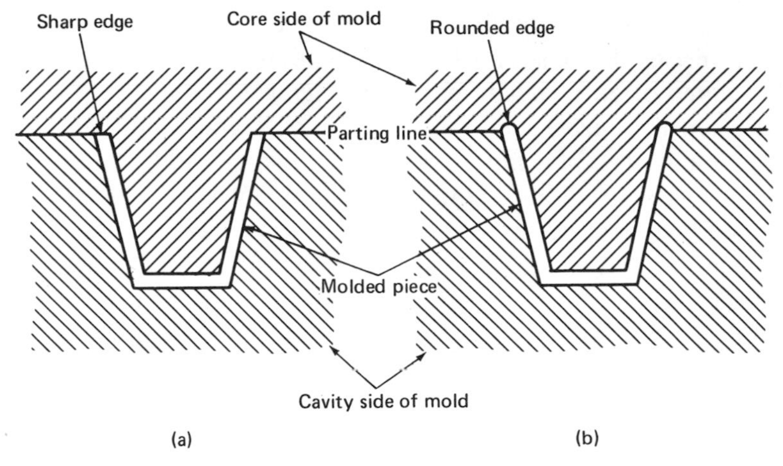

Figure 3.4. A safety hazard avoided by considering manufacturing features. (a) A sharp edge will be present if the parting line is placed at the corner of the piece. (b) To avoid sharp edges, move the parting line away from the edge.

seems to be a growing trend toward clearly stating the ingredients, preservatives, and colorants in packaged food and letting the consumer make individual choices.

In most companies, quality control and quality assurance are independent of the engineering design department and the manufacturing department. This independence is essential for unbiased judgment

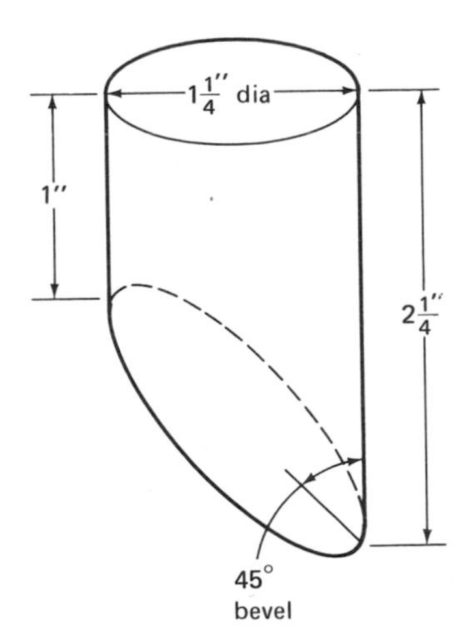

Figure 3.5. Definition of a small part is something that would fit into this truncated cylinder [21], [22]. The truncated cylinder represents the maximum size of a child's mouth.

and stronger action. Quality assurance is aimed at removing faulty designs, miscalculations, and inherent defects. The decision by quality assurance often necessitates design changes and drawing changes. Quality control curbs manufacturing defects and such variations in quality as misassembly, mislabeling, low quality molding, and faulty packaging. The methods of quality control include visual inspection, nondestructive testing, and random sampling. Date coding, batch coding, and serial numbering are widely used in high-volume consumer products. In case a defect is noticed after the product leaves the company, the date and batch codes will help to limit the size of the recall.

3.2.3. Fail-Safe Design

A fail-safe design is one that will not lead to accidents in case of functional failures. Since a functional repair is much cheaper than an accident, fail-safe designs are recognized as superior and necessary in many products. Take power steering as an example. Most power steering systems are designed so that the wheels can still be controlled without the power boost. If the fluid pump fails or the engine stalls, the car can still be steered.

Plastic bottles have been very successful in shatterproofing. Upon impact or abuse, the bottle may puncture or burst open but, unlike glass, the fragments are not dangerous. Some large, glass soft drink bottles have foam plastic coatings for increased safety. Uncoated bottles are known to explode upon impact, under the pressure of carbon dioxide. Photograhic flash bulbs also have plastic coatings to prevent shattering due to sudden heating. Automobile windshields are engineered to prevent shattering.

Safety brakes are needed in presses, cranes and hoists, which may fall under their own weights in the event of a power failure. Often the brake can also cease to function due to the power failure or an unknown fault in the circuit. To make a system fail-safe, the brakes have to be spring-loaded in the locking position. Thus, without any external help, the brakes will lock. Only when the power is on and the whole circuit is correct, will the brakes disengage. The same idea is applicable to electric, hydraulic, or pneumatic brakes. Devices that normally operate in one direction of rotation may need a reversing mechanism in case of accident. The clothing or hair of an employee may be caught between rollers and dragged into a machine. Slip clutches, shear pins, fuses, circuit breakers and pressure-relief valves are used to prevent overloads. If a bicycle is jammed under an electric garage door (or an elevator door), a slip clutch can prevent uncontrolled large forces that might damage both the bicycle and the garage door.

Interlocks are used to prevent human errors. As an example, the keys for a car with an automatic transmission cannot be removed unless the shift is in *park* position. In some cars with manual transmissions, the engine cannot be started unless the clutch is depressed. Safety switches are installed at the bottoms of some portable heating appliances. If an appliance, such as an electric room heater, is accidentally tipped over, it would automatically shut off. All of these are introduced because humans do not function perfectly, and human errors are essential design parameters.

Human errors can also be prevented by designing automatic systems and maintenance-free systems. A good example is the Polaroid SX-70 camera and film. The film is automatically ejected from the camera by a motor drive, as contrasted with the previous hand-pulling action. The film is self-fixing and requires no timing by the user. It is important to know that products designed to fit people appeal to more potential customers. People should not be required to fit products.

Emergency equipment in buildings has greatly improved during the last decade. Many buildings are equipped with emergency exits, emergency lights, which have self-contained battery packs, fire doors, smoke detectors, and automatic sprinkling systems. Emergency elevators are also installed in some tall buildings.

3.2.4. Factor of Safety and Reliability

The factor of safety is the ratio of strength to load. Strength is a characteristic of the machine component: the maximum force it can bear. The load is the actual force imparted to the component when the machine operates. The strength of a component may vary because of uncertainties in material properties and finishing processes. The load may exceed the estimated load when the machine is abused. A factor of safety larger than one leaves a margin for an understrength or an overload, and for the discrepancy between what is calculated and what actually happens. Sometimes a factor of safety as large as ten is used because the estimation of strength and load is very inaccurate. Since excess strength is wasteful, a better engineering practice is to obtain an accurate evaluation of the strength and the load and to use a small factor of safety.

Reliability is a statistical evaluation of the chance of success. Such an evaluation includes the statistical distribution of strength and load. It also includes the effect of time (aging and deterioration) and operating conditions (temperature, speed and environment). The concept of product reliability is largely associated with warranty and liability. In real life, nothing is 100% sure. Products can be expected to fail under extreme conditions. To design an absolutely indestructible and infallible product is unrealistic and uneconomical. A better practice is to make the product 99.9% reliable and to replace the one out of one thousand that is defective. The cost of providing a warranty is the product of the probability of failure and the cost of repairing a failure. The probability of failure is the complement of the reliability ($F = 1 - R$).

Though the reliability and the factor of safety are both indices describing the dependability of a product, they have quite different implications and applications.

A larger factor of safety means a higher reliability. The scale for factor of safety is between zero and infinity

$$0 \leqslant \text{factor of safety} < \infty.$$

The scale for reliability is between zero and one

$$0 \leqslant \text{reliability} < 1.$$

Reliability is a statistical evaluation to the real situation. The factor of safety is just a simple ratio to aid design. A product with a safety factor of 1.5, 2, or 4 presumably will not fail.

For components in series, the one with the smallest factor of safety determines the overall safety factor. As the load increases, the weakest component always fails first; thus, the weakest component acts like a fuse to protect the stronger components. Therefore, the safety factor of the system equals the smallest safety factor of the components. Reliability is a random occurence where the cause of failure may not be the load. The reliability R of a series components is the product of all component reliabilities:

$$R = R_1 \cdot R_2 \cdot R_3 \cdot R_4 \cdot R_5 \cdots R_n \qquad (3.1)$$

where R_1, R_2, R_3... are the reliabilities of the components, and n is the total number of components in series. Occasionally, a more dependable component will fail before a less dependable component. The addition of more components in a series will decrease the overall reliability. Figure 3.6 shows the schematic representation of the factor of safety and the reliability of a series system.

For compatible components in parallel, the system's safety factor is the summation of the component safety factors. When tension bars, which are compatible in stiffness, are put in parallel, their strength adds and so does the factor of safety. In actual design, the factor of safety of a parallel system is rarely used.

The evaluation of the reliability of components in parallel is rather different. A parallel system will fail only if all the redundant components fail. The probability of system failure F is the product of the individual components' probability of failure. Reliability is the complement of the chance of failure. A system that has 10% probability of failure has a reliability of 90%. Suppose we have a system of n components in parallel. Let F_i and R_i be the probability of failure and the reliability of the ith component, F and R be the probability of failure and reliability of the system.

$$F_i = 1 - R_i$$

$$R = 1 - F = 1 - F_1 \times F_2 \times F_3 \cdots \times F_n$$

$$R = 1 - (1 - R_1)(1 - R_2)(1 - R_3)(1 - R_4) \cdots (1 - R_n) \qquad (3.2)$$

To illustrate the dramatic difference in the overall reliabilities of a series system and a parallel system, let us connect four components of $R_1 = 80\%$ in two different ways (Figure 3.7):

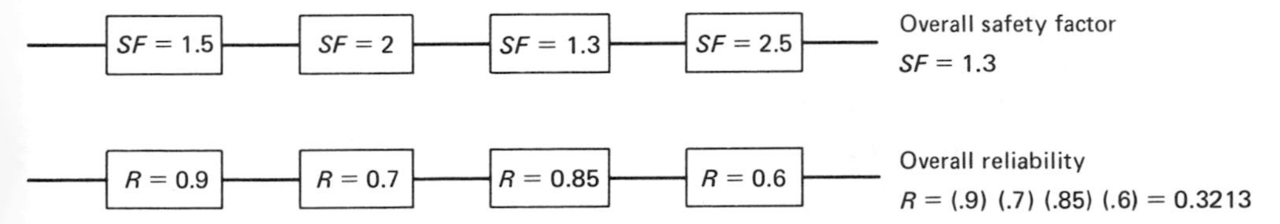

Figure 3.6. Reliability and the safety factor. The safety factor of the system equals the smallest safety factor of the components. The reliability of a series system is the product of the component reliabilities.

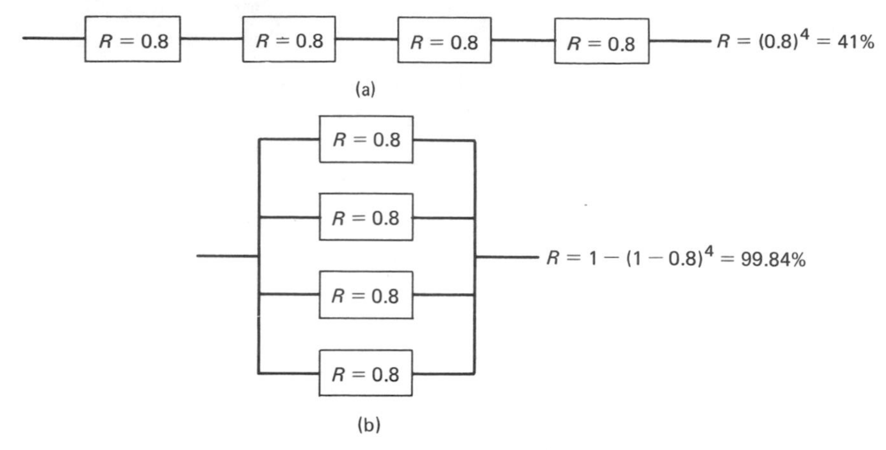

Figure 3.7. Reliability of series and parallel systems. Four identical components are connected in series (a) and in parallel (b).

(1) connected in series; the overall reliability is $R = (R_1)^4 = (0.8)^4 = 41\%$;

(2) connected in parallel; the overall reliability is $R = 1 - (1 - R_1)^4 = 1 - (0.2)^4 = 99.84\%$.

3.2.5. Reliability Design

The safety factor and the reliability can be converted from one to another if statistical data are available about the load and the strength. For simplicity, we will limit our discussion to the special case where the load and the strength are both normal distributions.

A tensile bar with a mean strength μ_s would have a frequency distribution as shown in Figure 3.8. If the load is very precise, the probability of failure is represented by the area to the left of the load line. The difference between the strength and the load, the safety margin $\mu_s - \mu_L$, can be expressed as x times the standard deviation of the strength σ_s. If $x = 0$, the load line coincides with the mean strength μ_s, and there would be 50% failure. If x is very large, the load line would be at the extreme left of the strength distribution. Thus, the failure rate would be very small. Table 3.4 shows x versus the reliability, this relationship being a characteristic of the assumed normal distribution. The safety factor can be related to x and σ_s/μ_s (the ratio of standard deviation of strength to mean strength) by the following equation:

TABLE 3.4. TABULATION OF RELIABILITIES AND SAFETY FACTORS FOR EQUATION 3.3.

$R = P(z > \mu_L)$	x	Safety factor (based on $\sigma_S/\mu_S = 0.08$)
50%	0	1
80%	0.84	1.07
90%	1.28	1.114
95%	1.645	1.15
99%	2.33	1.23
99.9%	3.08	1.327
99.99%	3.7	1.42

$$\text{safety factor SF} = \frac{\text{strength}}{\text{load}} = \frac{1}{1 - x\left(\dfrac{\sigma_s}{\mu_s}\right)} \tag{3.3}$$

Using the widely accepted value of 8% for (σ_s/μ_s) in fatigue, the safety factors are tabulated in Table 3.4.

The above method assumes that the load is very precise. In actual machines the load may vary, and there should be a wider safety margin. The combined effect of load and strength distributions is shown in Figure 3.9. The shaded area, the interference, itself has a distribution with:

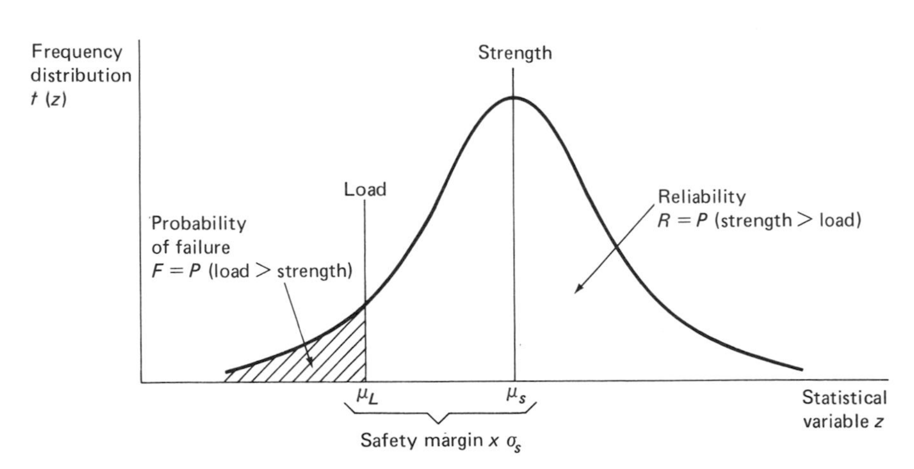

Figure 3.8. Evaluation of the safety factor when the load has no deviation. x = the ratio of safety margin to standard deviation; the safety margin = mean strength - mean load = $x \cdot \sigma_S$. See Equation 3.3.

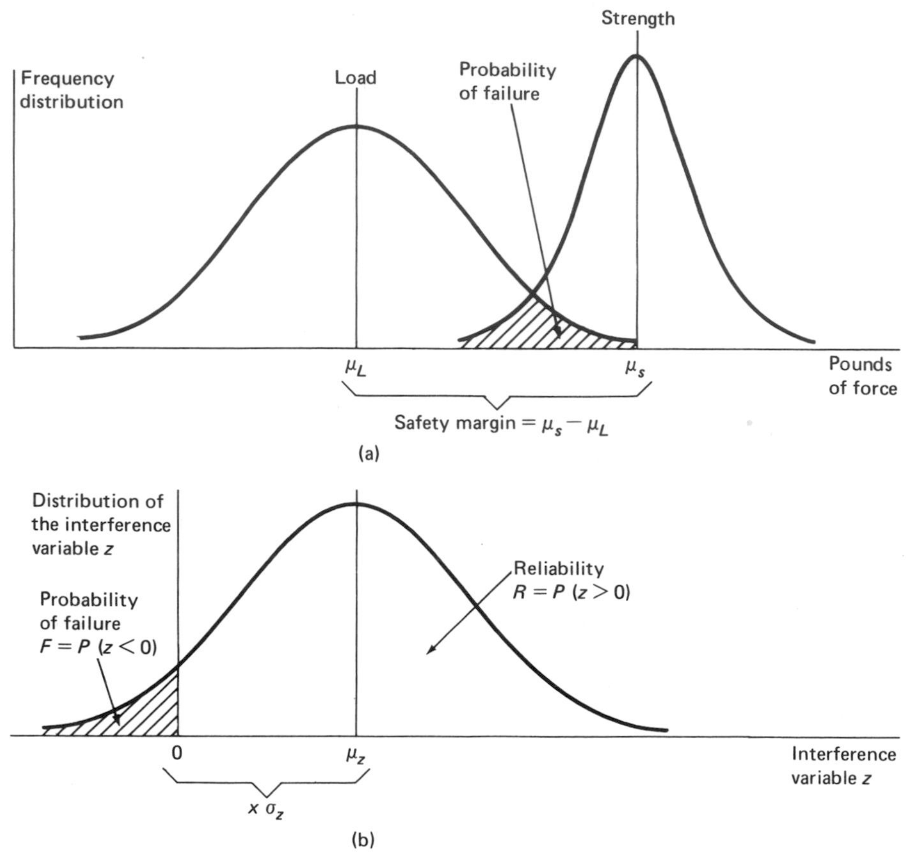

Figure 3.9. Illustration of the interference theory for normal distributions. (a) shows statistical distributions of load and strength, where μ_s = mean strength, μ_L = mean load, σ_s = standard deviation of strength, and σ_L = standard deviation of load. (b) the interference variable has a normal distribution with

$$\mu_z = \mu_s - \mu_L$$
$$\sigma_z = \sqrt{\sigma_s^2 + \sigma_L^2}$$

$$\text{mean } \mu_z = \mu_s - \mu_L$$
$$\text{standard deviation } \sigma_z = \sqrt{\sigma_s^2 + \sigma_L^2}$$

where z is the interference variable. The conversion from reliabilities to safety factors depends on which factors are known, where:

σ_s = the standard deviation of strength,
σ_L = the standard deviation of load,

μ_s = mean strength, and
μ_L = mean load.

Given σ_s/μ_s and σ_L/μ_L, the following formula gives the safety factor:

$$(SF)^2\left[1 - x^2\left(\frac{\sigma_s}{\mu_s}\right)^2\right] - 2(SF) + \left[1 - x^2\left(\frac{\sigma_L}{\mu_L}\right)^2\right] = 0 \qquad (3.4)$$

Given σ_s/μ_L and σ_L/μ_L, the following formula gives the safety factor:

$$SF = x\sqrt{\left(\frac{\sigma_s}{\mu_s}\right)^2 + \left(\frac{\sigma_L}{\mu_L}\right)^2} + 1 \qquad (3.5)$$

Given σ_s/μ_s and σ_L/μ_s, the following formula gives the safety factor:

TABLE 3.5. CONVERTING FROM RELIABILITIES TO SAFETY FACTORS.

$R = 90\%$ ($x = 1.28$)

Load σ_L/μ_L	Strength σ_s/μ_s 1%	2%	5%	10%
0%	1.01	1.03	1.07	1.15
5%	1.07	1.07	1.09	1.16
10%	1.13	1.13	1.15	1.20
20%	1.26	1.26	1.27	1.31
50%	1.64	1.64	1.65	1.67
100%	2.28	2.28	2.29	2.31

$R = 95\%$ ($x = 1.645$)

Load σ_L/μ_L	Strength σ_s/μ_s 1%	2%	5%	10%
0%	1.02	1.03	1.09	1.20
5%	1.08	1.09	1.12	1.22
10%	1.17	1.17	1.19	1.27
20%	1.33	1.33	1.35	1.40
50%	1.82	1.82	1.84	1.88
100%	2.65	2.65	2.66	2.70

$R = 99\%$ ($x = 2.33$)

Load σ_L/μ_L	Strength σ_s/μ_s 1%	2%	5%	10%
0%	1.02	1.05	1.13	1.30
5%	1.12	1.13	1.18	1.33
10%	1.23	1.24	1.28	1.40
20%	1.47	1.47	1.50	1.60
50%	2.17	2.17	2.19	2.28
100%	3.33	3.34	3.36	3.47

$R = 99.9\%$ ($x = 3.08$)

Load σ_L/μ_L	Strength σ_s/μ_s 1%	2%	5%	10%
0%	1.03	1.07	1.18	1.45
5%	1.16	1.17	1.25	1.48
10%	1.31	1.32	1.37	1.57
20%	1.62	1.62	1.67	1.84
50%	2.54	2.55	2.59	2.76
100%	4.08	4.09	4.15	4.36

Values in the table are safety factors obtained from Equation 3.4.

Example: $R = 90\%, \frac{\sigma_s}{\mu_s} = 1\%$ and $\frac{\sigma_L}{\mu_L} = 0\%$

Safety factor SF = 1.01

$$SF = \frac{1}{1 - x\sqrt{\left(\frac{\sigma_s}{\mu_s}\right)^2 + \left(\frac{\sigma_L}{\mu_L}\right)^2}} \qquad (3.6)$$

Since the first case is the most frequently encountered, it is also tabulated in Table 3.5. Equation 3.6 reduces to Equation 3.3 if the load has no deviation.

Example. A type of cable is known to have a standard deviation in strength of 5% of the mean strength. In a particular crane, the dynamic load during hoisting has a 20% standard deviation from the dead load. How large a safety factor is needed for 99.9% reliability? What should the mean strength of the cable be in order to hoist a 1000-pound load?

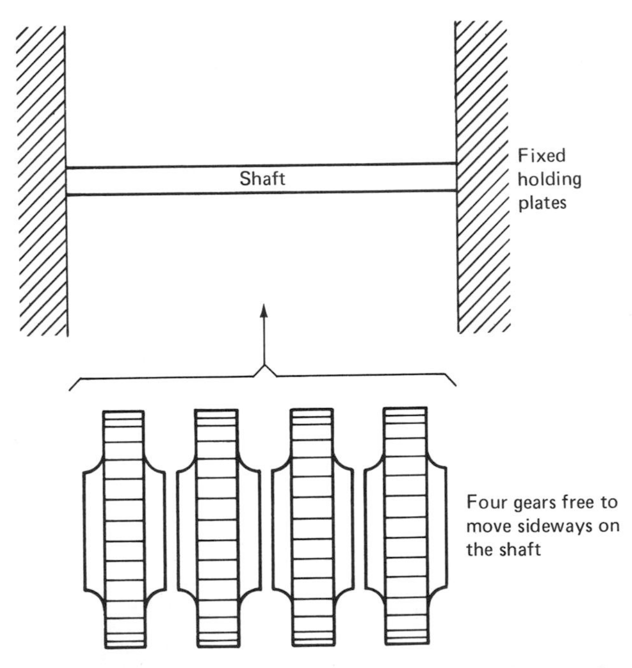

Figure 3.10. Cumulative tolerances for the thickness of the gears has to be controlled within limits. If the overall thickness exceeds the width between the holding plates, the gears may jam.

Given

$$\frac{\sigma_S}{\mu_S} = 5\%,$$

$$\frac{\sigma_L}{\mu_L} = 20\%.$$

$$R = 99.9\%,$$

$$\text{safety factor} = 1.67 \text{ (from Table 3.5).}$$

The cable should have a mean strength of 1670 lbs.

An example of maximizing the reliability of a series system is given in Section 2.4.2. Figure 3.7 shows how redundant (parallel) components can be added to increase reliability. Problems with mixed series and parallel components can be optimized by dynamic programming. Figure 3.2 shows an example of maximizing value by choosing the proper reliability. The cost of increasing reliability and the cost of repairing a failure should be balanced to minimize the overall cost.

Cumulative tolerances can really make a part expensive. If four gears can float sideways on a shaft (Figure 3.10), the cumulative tolerance is the sum of the individual tolerances. Therefore, a ±0.010″ cumulative tolerance would require an individual tolerance of, at most, ±0.0025″. If we allow individual tolerances to be larger than 0.0025″, there is a possibility that the gears may jam if they are all at their upper limits. The chance of reject is 0.3% if the individual tolerance is doubled, i.e., increased from 0.0025″ to 0.005″ (Figure 3.11). Some of the rejects can be saved by selective assembly. It seems economical in many production situations to gain larger tolerances by allowing rejects. The suitable reject rate depends on the control and the policy of the company. If the product is assembled and serviced by factory technicians only, a higher percentage of rejects and rework can be tolerated. Consumer-assembled kits have to be more reliable. If the product is a military item, a very high reliability is required since field failure is too costly.

3.2.6. Recyclable Design

Pollution control and resource conservation can both be achieved by recycling, i.e., making usable goods out of things that would otherwise become garbage. The level and extent of recycling are determined by economic factors. As the quality of ores dwindles and the

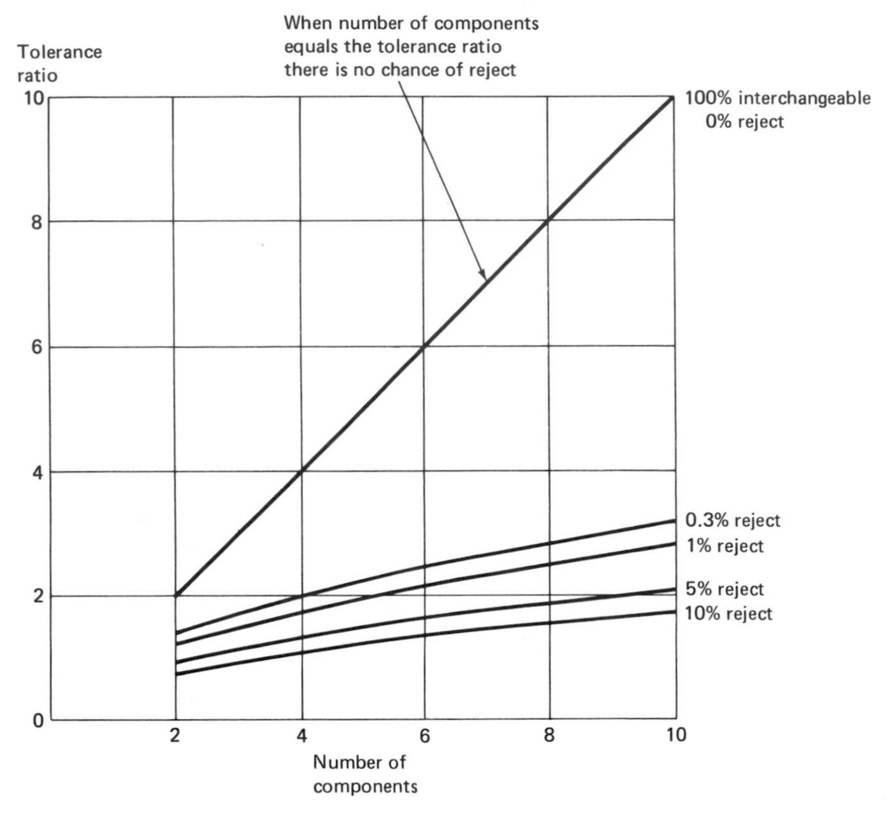

Figure 3.11. Allowing for rejects can cut costs for cumulative tolerances [28].

$$\text{Tolerance ratio} = \frac{\text{cumulative tolerance}}{\text{individual tolerance } T}$$

For parts with equal tolerances T, the overall tolerance is:

$$\text{cumulative tolerance} = \text{tolerance ratio} \times T$$

For n parts with individual tolerance T_i not equal, the tolerance T is replaced by the root-mean-square of the tolerances:

$$\text{cumulative tolerance} = \text{tolerance ratio} \times \sqrt{\frac{T_1^2 + T_2^2 + T_3^2 + \cdots + T_n^2}{n}}$$

cost of land filling increases, recycling is likely to increase and become the ultimate solution. Other factors include tax credits, consumer attitude, cost of pollution control, cost of collection, and transportation. Paper wastes may be used as fuel for power generation or as raw material for recycled paper; the scraps from metal stamping are remelted; the chips from lumber processing are made into reconstituted materials; the sprues and runners of thermoplastic parts are usually ground and mixed with virgin material; old ships can be converted into steel bars for use in reinforced concrete. Two other examples are the recovery of lead from automobile batteries and the recovery of silver from photographic processing. Design for recycling encompasses several ideas: (1) recycling by repair, (2) recycling by conversion, (3) design for easy separation, and (4) allowing for contamination.

Products can be recycled as a whole or as functional subassemblies. Automobile components can be recycled as rebuilt parts. The module and the snap-fit concepts can be utilized to make a product easily disassembled. Automobile components, such as seats and radiators, can be designed with standard mounts so that they can be used in different models of cars. The use of inserts at critical places can facilitate repair and replacement. Such examples are found in the brushes and bearings of electric motors and in the bearings of connecting rods for engines.

Many products can be recycled by conversion. At the end of their intended functions, they are used for another purpose. The leftover blanks from sheet-metal stamping can be used as perforated metal screens. Some jelly containers are designed to serve ultimately as drinking glasses. It has been suggested that glass bottles can be shaped like bricks so that empty bottles can be used to build houses. It has also been suggested that radioactive waste from nuclear plants can be used to sterilize food and sanitize waste.

The usefulness of garbage can be increased by sorting. One could separate garbage into glass, plastic, paper, tin cans, and aluminum cans. (Trash cans could be marked with "This can is for waste paper only.") The ease of separation can be quantified by entropy. Garbage that can be easily grouped into high-concentration components has a low entropy. Garbage with various materials finely dispersed has a high entropy. There is a natural tendency for materials to get mixed up, that is, for entropy to increase. To decrease entropy, external work or energy is required. The effort toward easy separation can start with the product designer. Copper wiring in an automobile can be centralized into one cable, which can be easily demounted. Again, snap-fits and modules can help. Ferrous materials might be separated by magnetic processes. Nonferrous materials might be separated by heat, solvent, density, etc. Markers or tracers can be embedded in the wanted materials to activate sensors in the sorting processes.

The tolerance of contamination is essential for the use of recycled material. If a plastic part is clear-transparent, it has to be made of 100% virgin material. A black plastic part can accommodate more regenerated material because color degrading can easily be concealed. Plastic parts that are to be vacuum-metalized or painted can be made of reclaimed material of any color. Some contamination is more damaging than others. For example, copper impurities make steel brittle. Aluminum, which does not hurt steel as much, may be used to replace copper in car wiring and radiators, in order to eliminate copper impurities in the reclaimed steel.

The following problem illustrates the degrading commonly associated with recycled material (Figure 3.12). Degrading can be any kind of loss of usefulness. Assume we can measure degrading as a percentage loss of strength. Let d_1 and d_2 be the pre- and postprocessing degrees of degrading. The processing degrades the material by $D\%$ $(d_2 - d_1 = (1 - d_1) \times D)$. When the machine is first turned on, 100% virgin material is fed in ($d_1 = 0$) and $d_2 = D$. In the second cycle, $R\%$ of regrind is introduced to the preprocess material ($d_1 = D \cdot R$) and $d_2 = D + D(R - RD)$. In the third cycle $d_2 = D + D(R - RD) + D(R - RD)^2$. It is easily observed that the amount of degrading is a geometric progression with the number of cycles of operation. In a steady state, the amount of degrading is the sum to infinity of the geometric series. After many cycles,

$$d_2 = \frac{D}{1 - R(1 - D)}$$

The amount of regrind R that can be used is limited by the allowable degrading T in the final product and the degrading during processing D.

$$d_2 = \frac{D}{1 - R(1 - D)} \leqslant T \tag{3.7}$$

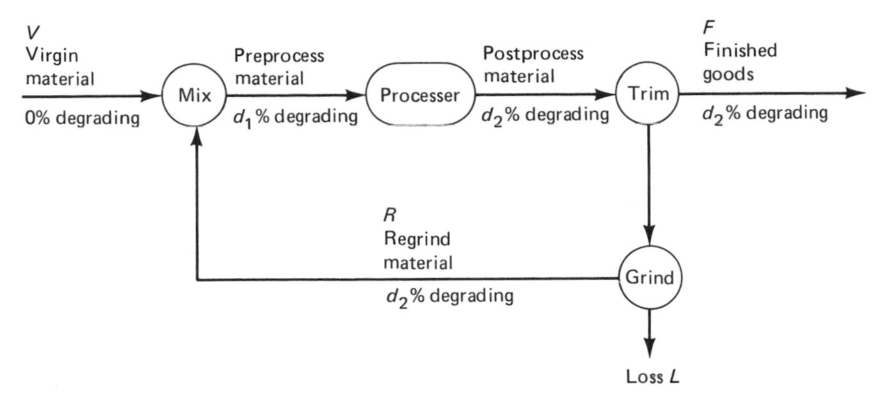

Figure 3.12. Degrading of material during processing.
System input = system output: $V = F + L$
Preprocessing material = postprocessing material: $V + R = F + L + R$
Amount of degrading during processing $= D = \dfrac{d_2 - d_1}{1 - d_1}$

Design criteria:

$$R \leqslant \left(1 - \frac{D}{T}\right)\left(\frac{1}{1 - D}\right) \tag{3.8}$$

where

R = % of regrind,
D = process degrading, and
T = allowable degrading in the final product.

Suppose a paper material degrades 25% during each processing. If the allowable degrading in the final product is 36%, how much recycled paper can be used? $D = 25\%$, $T = 36\%$; therefore, $R \leqslant 41\%$.

In this example, the product would have 25% degrading even if it were made from 100% virgin material. The incremental degrading due to recycling is only 11%.

Degraded materials may be channeled to a less demanding use. There are also chemical treatments that can rejuvenate materials.

3.3. MANUFACTURING OPERATIONS IN RELATION TO DESIGN

Suppose every family is self-sufficient. People grow their own vegetables, hunt their own animals, make their own tools, and build their own houses. Nobody is really proficient at any particular job. That is why this way of living is associated only with primitive cultures. When people specialize in individual trades, they naturally become more efficient. Through exchange (buying and selling), society as a whole benefits from improved productivity.

Figure 3.13 shows the price structure of a modern manufacturing company. The selling price can be divided into the indirect cost, the direct cost, the tax, and the profit. The direct cost includes the costs of material, labor, and equipment. As a way to simplify cost estimation, nominal ratios are usually established between the material cost, the direct cost, and the selling price. A typical ratio is

$$\text{material cost} : \text{direct cost} : \text{selling price} = 1:3:9 \tag{3.9}$$

In this case, a product containing $1 of raw material costs $3 to make, and has to be sold at $9 in order to make a reasonable profit. This nominal cost formula works well as long as we have a narrow group of similar products. Using the above ratios, plastic products made of engineering plastics that cost about 80¢/lb would have a selling price of $7.20/lb [32]. Because of this cascading effect, a 10¢ saving in the material cost could lead to 90¢ reduction in the selling price. The nominal ratio may be quite different for a completely different industry. The food packaging industry may have a ratio of 1:2:5. The ratio between the selling price and the direct cost is sometimes called the *operational overhead ratio*. This ratio describes

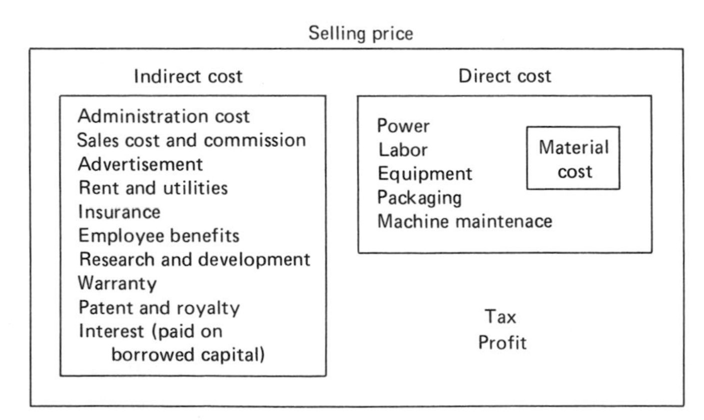

Figure 3.13. The price structure of a modern manufacturing company. Processing overhead ratio = direct cost/material cost = 3. Operational overhead ratio = selling price/direct cost = 3.

the efficiency and the risks in the operation. Typically, for companies specialized in advanced technology or speculative products, the operational overhead ratios are high.

Since we are mostly interested in the design of a product, we will concentrate on the direct cost of a product

$$\text{direct cost} = \text{material} + \text{power} + \text{labor} + \text{equipment} \tag{3.10}$$

Usually, the costs of material, power, and labor are directly proportional to the production volume. The equipment cost is independent of the production volume. The cost component that is proportional to the production volume is a variable cost.

$$\text{total cost} = \text{variable cost} \times \text{quantity} + \text{fixed cost} \tag{3.11}$$

$$\text{unit cost} = \text{variable cost} + \frac{\text{fixed cost}}{\text{quantity}} \tag{3.12}$$

The two equations above are illustrated in Figure 3.14. When the quantity of the product is limited, the cost of equipment has to be amortized over the production period or the production quantity. A set of dies used to produce common coins can be considered as contributing nothing to the cost of each coin because of a huge quantity. A set of dies used to produce a limited collector's mintage contributes significantly to the cost of each coin. This equipment cost per product decreases with the production quantity. If a product is not profitable because the fixed cost is high, it is possible to "make it up in volume." However, a product that is not profitable because of a high variable cost cannot be saved by increasing the sales volume. In fact,

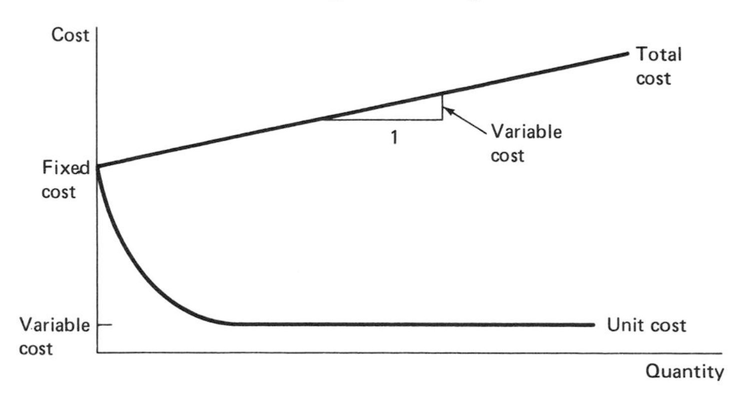

Figure 3.14. Total cost and unit cost plotted against quantity of production.

the more you sell, the more you lose. The distinction between the fixed and the variable costs is critical in the profit-volume relationship.

Sometimes it is difficult to distinguish between the variable cost and the fixed cost. Take labor for example. If workers can be hired and laid off from day to day to match the production need, the labor cost would be a variable cost. If recruiting efforts or labor unions prevent the change of personnel with the fluctuation of production need, the labor cost would be a fixed cost. Some cost engineers prefer to use accounting data from the history of the company to determine whether the cost of one constituent varies with the company's sales.

A question that frequently arises is whether a company should purchase components or make them in-house. Figure 3.15 shows a simplified model, which assumes that the unit purchase price is constant (independent of quantity). There is a break-even quantity at which a manufactured item is comparable in cost to a purchased item. For quantities larger than the break-even quantity, it is more economical to manufacture the parts, and vice versa. If the unit purchase price decreases with quantity, the break-even point will be at a larger quantity. The break-even point in Figure 3.15 will shift to the right.

The question of make or buy goes beyond just figures. There is more control over the quality of a component if it is manufactured in-house. There is less shipping, receiving, and inspecting. The company would also be building up its strength instead of possibly help-

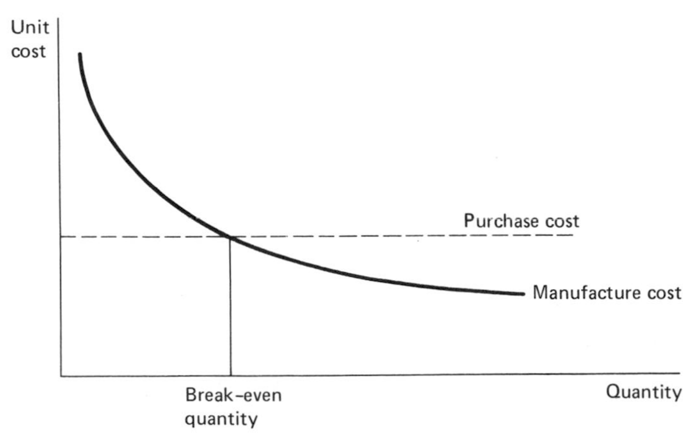

Figure 3.15. Costs of purchasing and manufacturing cross at the break-even point.

ing a potential competitor. On the other hand, a company cannot just get into a new business and expect to make money. It might be better to stay in one's specialty and leave unrelated jobs to others. Purchased items are cheaper for small quantities because there is no capital tied up. Purchasing parts from the outside also has the advantage of flexibility. Assume that a component of a car is purchased. The drop in sales of one motor company and an increase in the sales of another would not influence this component supplier as long as the total automobile market is constant. The component supplier can make the part cheaper than an individual motor company because of a larger volume. The motor company is hurt less by the fluctuation in sales because of less capital investment.

The following list includes ten aspects of manufacturing operations that will lower the cost of products if they are considered in the design stage.

3.3.1. Standardization and Interchangeability

Standardization involves the reduction of numerous options to the most popular ones, or the ones corresponding to "round numbers" in dimensions. Standardization can be applied to components, products, materials, processes, and even organizational procedures. Standardization increases the volumes in the standardized items and allows components to be interchanged. However, a standard item may not do a job as effectively as a custom-made item. Standard-sized garments often cannot fit individuals very well, but they are much cheaper than tailor-made ones.

The advantage of standardization is best seen in the fastener industry. If standardization is not applied, the number of varieties of fasteners will be astronomical. With proper standardization, manufacturers and users can deal with a manageable number of varieties. In some appliance companies, rivets, eyelets, and screws are limited to about five varieties each. Though the narrow selection imposes some inconvenience on the designers and engineers, it simplifies bookkeeping and secures better bargains for the larger quantities that result.

If components are standardized, a company may be able to use a certain component in different products. Interchangeability is desirable in terms of inventory control and product line flexibility. There would be one blanket order for a family of products, which decreases the burden of forecasting sales for individual products. Standardization can also lead to enough volume to justify automation.

3.3.2. Standard Components

A stock item is always cheaper than a custom-made item. In a design process, an engineer should always consider using a standard component first.

Figure 3.16 shows a standard rubber tube and a custom-made polyethylene tube. At first, it was believed that a standard tube could not give enough flexibility and could not stand the high temperature and pressure. Therefore, the custom-made tube was used. After further investigation, it was found that a surgical rubber tube can perform the job just as well. Pound for pound, surgical rubber is more expensive than polyethylene. Nevertheless, the surgical rubber tube is 20% cheaper than the custom-made tube because the surgical rubber tube is a stock item.

Besides being cheaper, standard items require little or no lead time. They are either in stock or can be readily produced. Custom-made items require a lead time because they usually need special tooling and handling. Standard components are likely to be more reliable

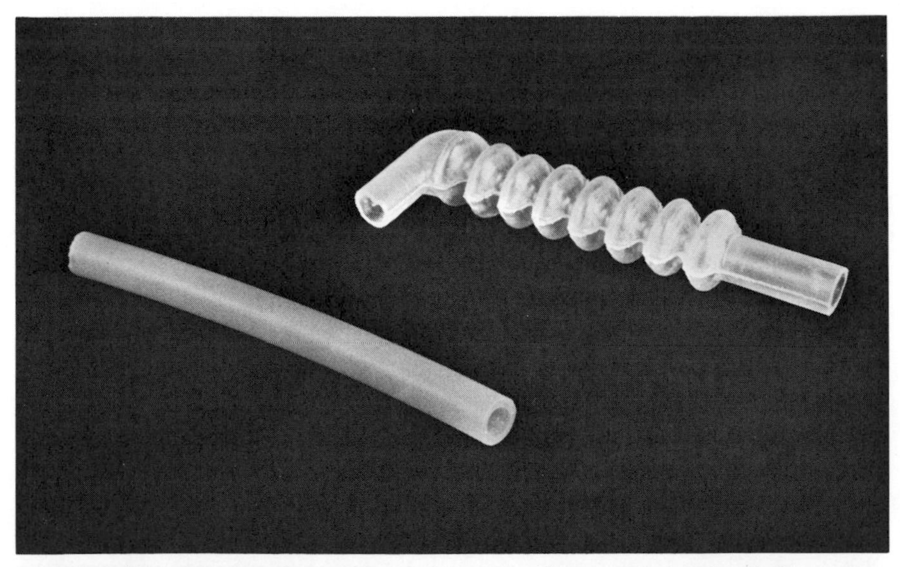

Figure 3.16. An extruded rubber tube and a blow-molded polyethylene tube. The standard tube (left) is made of a more expensive material, but is still cheaper than the custom-made tube.

because their characteristics and weaknesses are well known through years of field experience. Flexibility in volume is another advantage for standard items, since they can be ordered in any quantity at any time. Custom-made parts usually have to satisfy a minimum volume requirement.

When repair is needed, replacements for a standard component are easier to find. If a standard component is not suitable for a certain design, modification of standard components should be considered.

3.3.3. Module and Subassembly

If a component can be used in a wide variety of products, it is more economical to make it a subassembly. The tuner, the amplifier, and the speaker of a radio can be separated into distinct components. Calculator keyboards, IC chips, and digital displays are made into standard subassemblies, too. The same keyboard can be used for a variety of calculators. The fluctuation in sales of individual models will not influence these subassemblies as long as the total sales volume remains steady. The modular concept may be a great help to a continued program of product improvement. The new generation product can utilize most of the old modules except a few new, improved modules. Therefore, modular design can resist obsolesence. Of course, a certain sacrifice has to be made. For example, standard IC chips might have 14, 16, or 24 pins. If a particular circuit uses only 13 pins, there will be one unused pin on the chip. This is still worthwhile because the chips are made by high-speed automatic machines.

A module is a self-contained component with a standard interface to other components of a system. A module is sometimes more costly than an integral design because of the extra cost of fittings and interconnections. However, a modular construction decreases the tooling cost and increases the production quantity. It also offers flexibility in sizes and functions, ease of maintainance, and ease of custom-tailoring. Since a module can be detached and isolated from the rest of the system, malfunction can be easily located, and a broken component can be replaced easily. A component can also be fully checked and tested before it is inserted into, or combined with, the rest of the product. *Product Engineering* [33] described an interesting application of modular construction in 14th-century armor: "One suit made in Innsbruck had 87 pieces that could be fitted together, puzzle-like, into 12 different suits, for horse, foot, jousting, combat, and so on."

Repeatability within a product can be utilized with modular design Figure 3.17(a) shows a book cabinet whose height can be increased easily by stacking more units together. Note that a larger system is replaced by identical smaller units—a way to decrease tooling costs and increase production volume. The modules may have different height and different functions as long as the fittings are standard. Some modules may have small drawers, others can have glass doors, partitions, etc. (Figure 3.17(b)). Figure 3.18(a) shows a plastic box that has identical top and bottom halves. Figure 3.18(b) shows a novel gear design utilizing left and right symmetry.

The modular concept is advantageously used in systems of interchangeable lenses for single-lens reflex cameras. As an example, a 2X magnifying unit when used with a 50-mm lens produces the equivalent of a 100-mm lens. Close-up lenses, bellows, extension tubes, and macro lenses can be arranged in many different combinations to get the desired magnification and effects. Similar concepts allow the Zeiss Co. to custom-build microscopes to meet the needs of

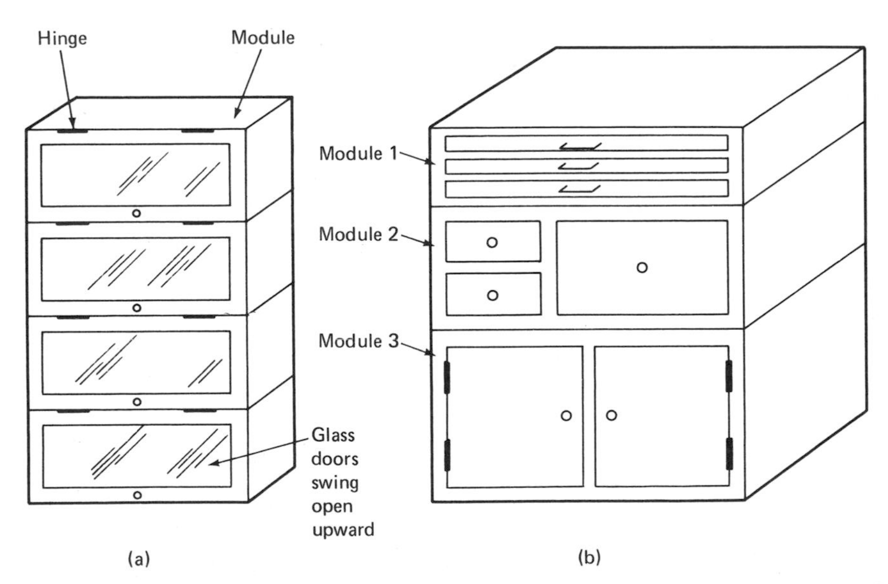

Figure 3.17. Modular cabinets. The bookcase (a) is made of four identical modules stacked together. The storage cabinet (b) uses three different modules.

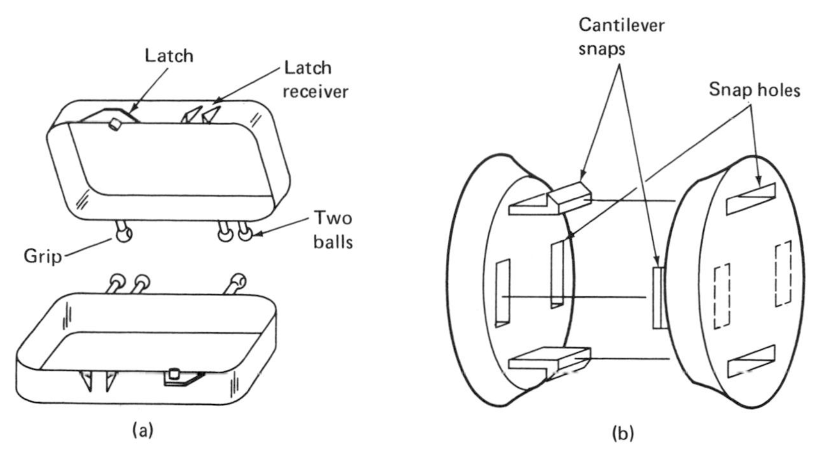

Figure 3.18. Modular design utilizing symmetry. The plastic box (a) is made by snapping together two identical pieces. An enveloping worm gear (b) is difficult to mold. DuPont's designers created a novel design consisting of two identical halves. Additional key-studs and holes (not shown) are used to insure alignment.

metallography, geology, petrography, chemistry, biology, and many other disciplines [34]. The system allows (1) upright or inverted viewing, (2) transmitted or reflected light, (3) visible, ultraviolet, or infrared light. Despite such diversity, there are only twelve different modules, permitting economic production of large quantities.

Electric fuses, light bulbs, and batteries are designed in modules. Certain tool manufacturers make modular garden tools: the tools, which may be trimmers, scissors, drills, spotlights, or saws are separate modules snapped onto the battery power packs. Electric kilns, ovens, and stoves are also moving toward modular designs.

3.3.4. Group Technology

As an extension of standardization and subassembly, group technology takes advantage of similarities among products to improve productivity. Many machine components or product components may be classified into functional groups, geometrically similar groups, weight groups, material groups, manufacturing groups, assembly groups, and so forth. Classification codes can be set up so that design engineers, manufacturing engineers, purchasing agents, and accountants can quickly identify the nature and cost of a component. Such codes impart information and work like driver license numbers, library cataloging numbers, and the universal product code. A typical code may consist of ten numbers and letters.

A coding system helps engineers to eliminate duplicate design effort and facilitate information retrieval. An engineer setting out to design a valve can check all the previously designed valves and adopt a suitable one. Without such a functional grouping code, an engineering department may have redesigned a similar component several times simply because it is too difficult to uncover records of earlier designs. Good coding also helps to bring to attention similar parts that can be made into a standard subassembly. Figure 3.19 shows five shafts that can practically be represented in one single drawing. A fixture is set up to include all the features of such a "composite design." Then, individual versions can be obtained by skipping some steps and features in the manufacturing process. There may not be

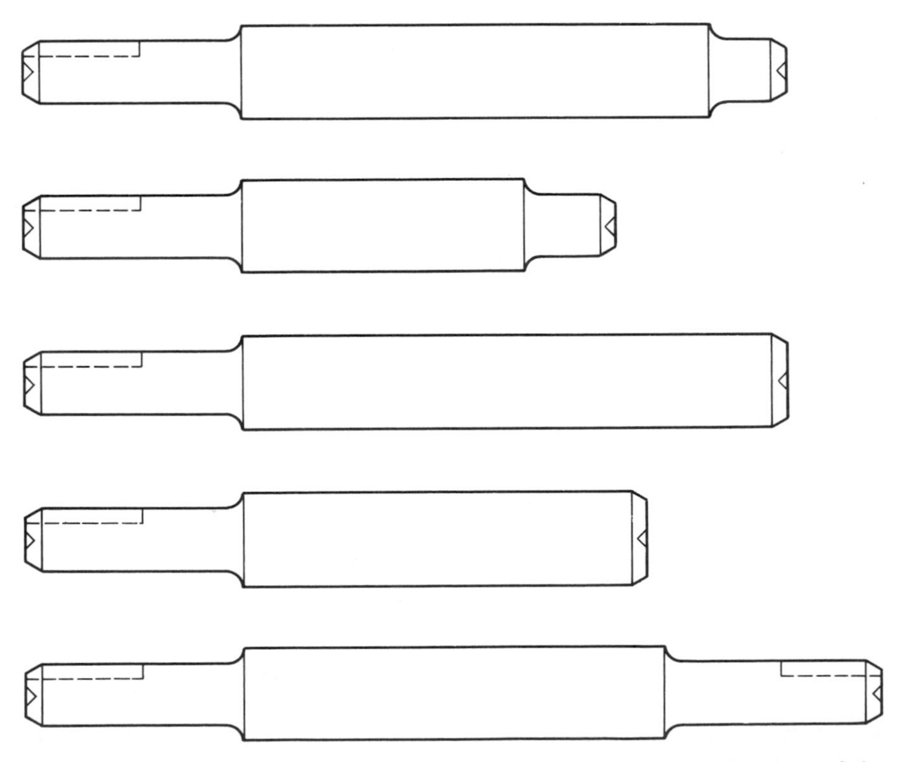

Figure 3.19. Five shafts with identical fittings on the left ends. Because of the similarities of these shafts, they can be grouped together for manufacturing in one batch.

an actual component that has all the features—the composite component may be hypothetical. The cost of design proliferation is more than just wasting design time. It means more tooling cost, more warehouse storage, more separate purchasing, and more assembly variations. Cost-conscious engineers should think in terms of design rationing, which means limiting the number of new designs and tightening the requirements for justifying a new design. The coding system can be blended into the design procedure such that it becomes a systematic guideline for standardizing design features.

Group technology is most useful in manufacturing, scheduling, motion study, and quality control. Products with the same operation can be grouped together and run in one batch. Since many small orders become a large group production, set-up time and training time are reduced. A shop supervisor can quickly estimate the labor required in a new product by drawing on his experience from similar products represented by the same manufacturing code. The shop's machine tool capacity can be scheduled by adding the required machine times of all the products requiring the same machine. The purchasing agents and the accountants may be able to estimate the cost of a product from its code without an elaborate study of the product's function. The grouping code serves as a common language that ties together marketing, engineering, manufacturing, and purchasing. Discrepancies and inconsistencies can be identified and eliminated.

3.3.5. Continuous Processing and Assembly Line

There are basically two types of layouts for the machinery. One is the *batch-job* layout, the other is a *continuous processing* set-up. They are sometimes called the *workshop* and the *assembly line*, respectively. We can consider a restaurant as a batch-job arrangement because the waiters serve different customers one group at a time. The order goes through different stages of maturity with time. On the contrary, in a cafeteria, people come in continuously and are served in different stations along a line. Then they pay and leave the line. As far as the service line is concerned, all its functions are performed simultaneously, and there is no change in activity with time. The cafeteria style is very efficient for a large number of customers because the workers on the line do not walk around; all they do is dish out the food or collect money continuously. No order is taken; only minimum bookkeeping is needed. On the other hand, a restaurant can provide individual attention and variations not provided in a cafeteria. A restaurant usually serves a greater variety of food, over longer periods of time because the food is cooked to order.

The economic choice between the two manufacturing philosophies depends on the production volume and the variety of products being manufactured. A company may manufacture many different products, yet, if all the products contain a standard component, an assembly line may be set up for that component. When group technology was first introduced, many production engineers discovered groups of components with quantities large enough to justify a continuous process. Such grouping and utilization of continuous processing resulted in increases in productivity and substantial savings.

A typical workshop will have machines grouped according to functions: turning, milling, drilling, and cutting. Different batch jobs would be routed through different paths in the shop depending on the processing sequence (Figure 3.20). Because of the functional grouping, the breakdown of one mill causes no problem. The work can be shared among the remaining mills. Because of the variety of jobs, some jobs have to be moved several rounds across the factory before they are finished. The major disadvantages of a plant layout by machine functions are the large amount of effort in handling, receiving, and scheduling at each station, and the required storage

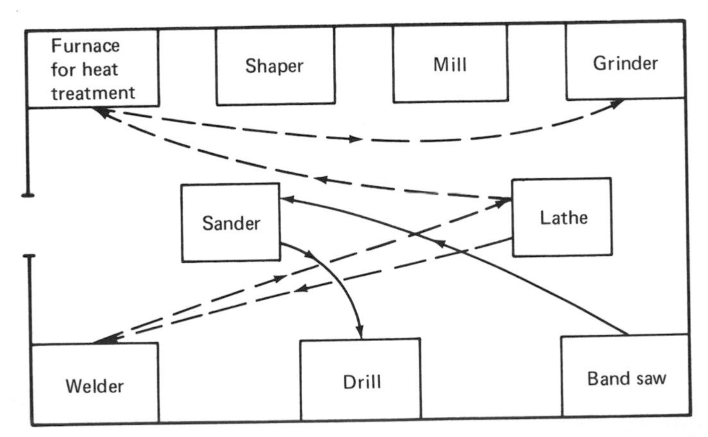

Figure 3.20. A plant layout with machines grouped by function. Job no. 1, saw to sander to drill. Job no. 2, lathe to welder to lathe to furnace to grinder.

capacity at each station for the workpieces awaiting to be processed. Since each station has its workload independently scheduled, a piece that is given low priority at each station would be seriously delayed. Since all components have to be finished before a product can be assembled, the delay of one component will hold up a whole product. A workshop is characterized by a long fabrication time and a large amount of in-process inventory. The in-process inventory can increase the cost of the operation by increasing the investment in inventory.

An assembly line is efficient partly because each worker is confined to a small portion of the job and becomes extremely efficient at it. The machinery arrangement is tailored to the sequence of the product fabrication so that the product travels a minimum distance through the plant (Figure 3.21). The flow rate of the product is large and there is little storage capacity at the stations. The amount of floor space required is proportional to the duration of the operation. Reduction of waiting time (aging) physically decreases the floor space. Therefore, fast-curing glues and fast-acting enzymes are

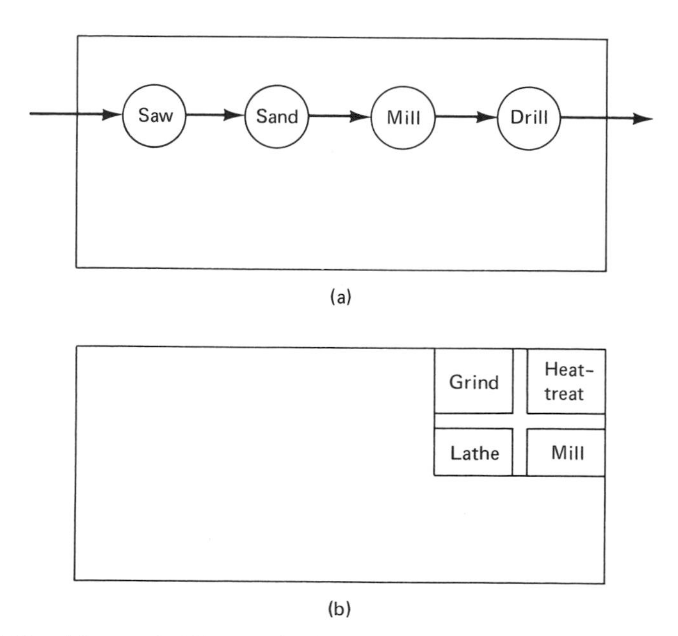

Figure 3.21. Line and cell organization. The line layout (a) is in sequence of product fabrication. (b) shows the cell organization in a machine shop for high-volume components.

valuable for reducing costs. An even flow rate in the assembly line is important: bottlenecks or idle segments in the line should be avoided. To help balance the line, the speeds of all fabrication and assembly steps should be comparable. One way to achieve balance is to make the production speed the lowest common multiple of individual machine capacities (or worker assembly speeds). Even if a drill press is utilized to only 10% capacity, one has to be placed in the assembly line. Five assembly lines, each utilizing only a small percentage of the drill press capacity, still require five separate drills. A breakdown in one machine will hold up the complete assembly line. Notice that if machines are grouped by their functions, there would be very little idle capacity, and equipment failure would be less damaging. One version of continuous processing that applies to a machine shop is the *cell organization* (Figure 3.21). A lathe, a grinder, a heat-treatment furnace, and a mill would be grouped into a "shaft division." Efficiency is improved because once the stock material enters the cell, there is minimum moving and scheduling. A cell can be controlled by two operators running four machines. Note that a cell is like an assembly line in that the production rate has to be quite high to make it all worthwhile.

The arrangement of machines and workers in the product fabrication sequence is the first step toward automation. Individual workers can be replaced by automatic machines at appropriate times. When all the workers are replaced by automatic machines, tuned to function as a unit, the system is completely automated.

3.3.6. Automation

Machines have proved to be very useful in extending man's muscle power. Modern automatic machines can replace people's brain power to inspect, coordinate, and make decisions. The completely automated factory would have raw materials dumped in at one side and boxes of finished product delivered on the other side. Such a system would be very productive because there is no human error, no delay, no waste, and no labor cost. Once the production passes the break-even quantity, the manufacturing cost is down to a minimum. The production speed and product reliability would also be much better. Few companies are so advanced because of economic reasons. The huge capital investment in such a system would require extremely high production volume and no change in the product over many years. Only processing companies, packaging companies, and some high-volume, consumer-product companies can come close to meeting these requirements. For most manufacturing companies, automation may be mixed with manual operations in different segments of production. The assembly of a standard component (or module) may be automated, while the assembly of the product may be manual. The central problem with automatic machines is the lack of flexibility. In order to get enough earnings out of the investment, designs have to be frozen over five to ten years. Economic pressures sometimes counter engineering progress. Once a design is changed, the old equipment is practically useless. Therefore, new designs are not accepted easily. To overcome this, engineers always strive for more versatile forms of automation.

One of the first versatile machines is the indexing machine tool. Such machines may have several stations of operations controlled by mechanical cams and relay logic. A workpiece can be automatically registered, drilled, tapped, milled, etc.

The difficulty of mechanical control is overcome by numerical control—software programming. The controlling commands are words and numbers recorded on paper tapes or magnetic tapes or cards. Numerical control is used with great success in machining that involves several tool changes, different machining angles, etc. Production quantities of several pieces to several thousand pieces are easily handled by such machines.

The most versatile form of automation is the industrial robot. A robot can be programmed to do many simple jobs such as loading presses and quenching workpieces. Modern industrial robots have six axes of movement and many gripping actions, though they lack the visual perception and creativity of human workers. Visual pattern recognition is very useful in the performance of work, and has drawn much research attention. Creativity is not needed in repetitive work. In fact, the lack of creativity and the inability to be bored are factors in favor of the robot. Utilization of robots is also a safety precaution in harsh environments where noise, harmful radiation, and chemicals abound. Some robots, called remote manipulators, are directly controlled by an operator, such that the visual feedback loop is completed by a person.

3.3.7. Specifications and Tolerances

Specifications are a set of conditions that qualify or disqualify a product. Specifications that are easy to meet reduce the number of rejects and lower the cost. However, specifications should not be lowered so far that quality is jeopardized. What is the minimum quality required? What is the highest quality of a component that would yield significant improvement in the overall product? What is the rate of increase of cost with quality? Figure 3.22 shows a plot of cost versus quality, such as surface finish or strength. We would specify a quality to the right of the minimum, but to the left of the point where the cost turns sharply upward. As an example, pure gold in jewelry usually means 98% to 99% gold. To ask for 99.99% gold jewelry would double the price, but would add insignificant quality to the jewelry. The quality of an individual component should also be balanced with the overall quality of the product. Ordinarily, one would not fit plastic crystals into pure gold rings.

Tolerance is the allowable variation in the design specifications. A shaft specified to have a 1.000″ diameter might have an actual diameter in the range 0.999″ to 1.001″. Note that the last digit implies the required accuracy. Figure 3.23 shows the statistical distribution of a manufacturing process. The parts that fall outside the tolerance range are rejected. If the rejection rate is too high, the manufacturing

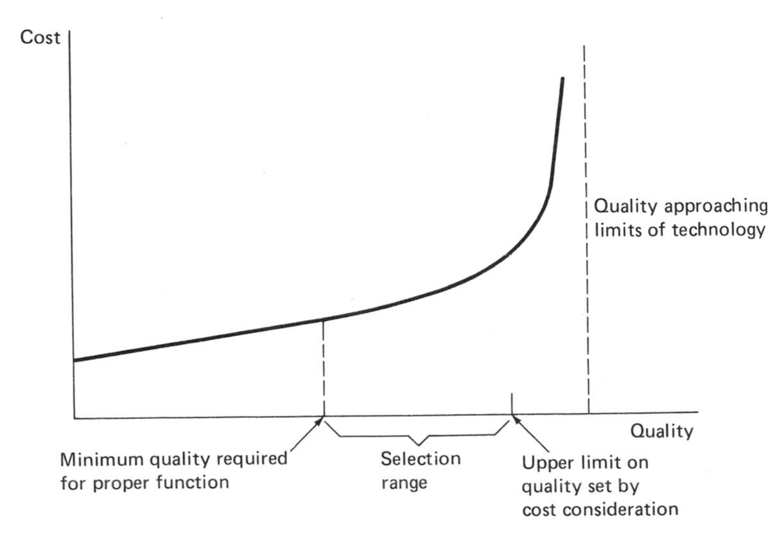

Figure 3.22. Cost versus quality. The selection range is set by considerations of function and cost.

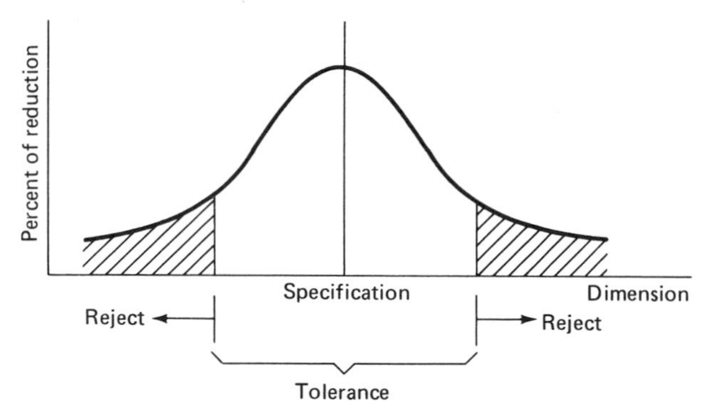

Figure 3.23. Tolerance and reject rate for a hole of specified diameter.

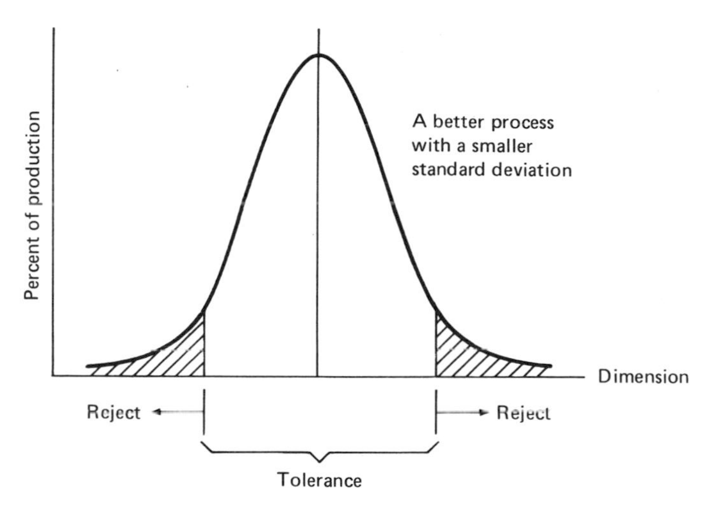

Figure 3.24. An improved manufacturing process reduces the percentage of rejects.

$\frac{\text{Tolerance}}{\text{Standard deviation}}$	Reject rate
2.56	20%
3.29	10%
3.92	5%
5.15	1%
5.6	0.5%
6.6	0.1%

The required ratio of tolerance to standard deviation for a given rate of reject is shown above. The reject rate can be decreased by larger tolerance or better processing (smaller standard deviation).

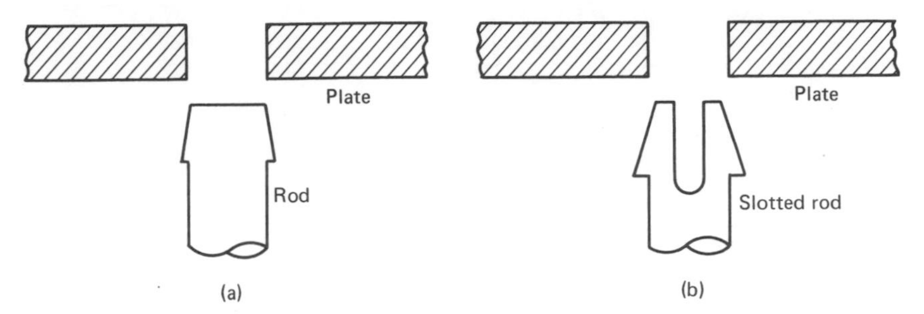

Figure 3.25. Component stiffness, interference, and tolerance. Two stiff components (a) require a very tight tolerance. Softer members (b) allow much wider tolerance.

process must be improved so that there is less scattering around the specification (Figure 3.24). In any event, a tight tolerance costs more.

Figure 3.25(a) shows a solid rod that snaps through a hole. Because both the rod and the plate are very stiff, the interference at the barb has to be very small and very precise. By cutting a slot in the rod (Figure 3.25(b)), the interference can be increased and the tolerance on the barb will be much larger. In many cases, the tolerance can be increased by altering the interacting parts.

3.3.8. Compact Design

Compact design should be considered for products whose performance is independent of their size. Portable equipment would gain value from miniaturization. Compact designs conserve material, reduce shipping costs, and reduce storage costs. Modern television sets have large picture tubes but small overall dimensions (Figure 3.26). Modern pocket calculators and cassette tape recorders show what miniaturization can do. Microfilm technology helps to store huge volume of data and documents.

Up to the early 1960s, nails of the same size were just dumped into boxes and shipped. Because such handling resulted in the tangling of nails, a box bigger than the actual volume of nails was needed. The invention of a special packaging machine enabled the alignment and stacking of nails in a small package. The savings in the cost of paper, labor, and shipping could pay for the machinery in one or two years. Such machines are now used worldwide for packaging ferrous nails.

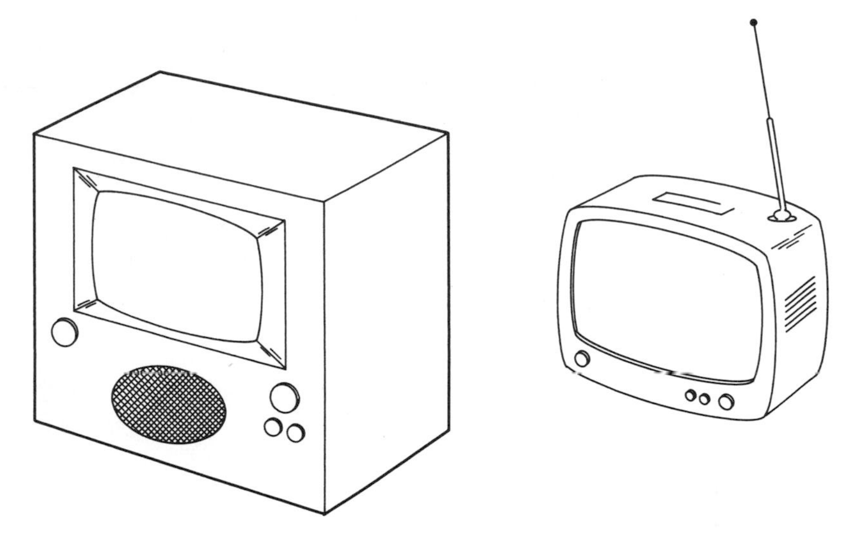

Figure 3.26. Old versus new television designs. The performance of a television set is related to the size and quality of the picture tube, but is independent of the overall size. The screen area should be increased and the ancillary equipment miniaturized.

Garbage is often compressed before it is moved. Junk cars are compacted into dense blocks for easy handling and recycling. Gaseous fuels and chemicals are shipped in pressurized cylinders or in liquid form. Many fluids such as fruit juices, shampoo, antifreeze, and paints are sold in concentrated forms. All of these illustrate the unquestionable value of compacting.

During the sixties and seventies, compact fruit trees have revolutionized commercial and backyard fruit orchards. By grafting dwarfing rootstock or inter-stem, fruit trees can be made to bear more fruits at an earlier age. Some genetic dwarf trees are produced by mutation or chance seedling. The size of fruits from such a dwarf tree is the same as that of fruits from a standard tree. Dwarf trees are easier to care for. The space for a standard apple tree can be used for 16 dwarf apple trees. Fruit production per unit area of land can be increased by 5 to 10 times.

3.3.9. Foldable, Stackable, and Consumer-Assembled Goods

Some furniture comes in several pieces that are to be assembled by the consumer; baby furniture is often made foldable; plastic

(a)

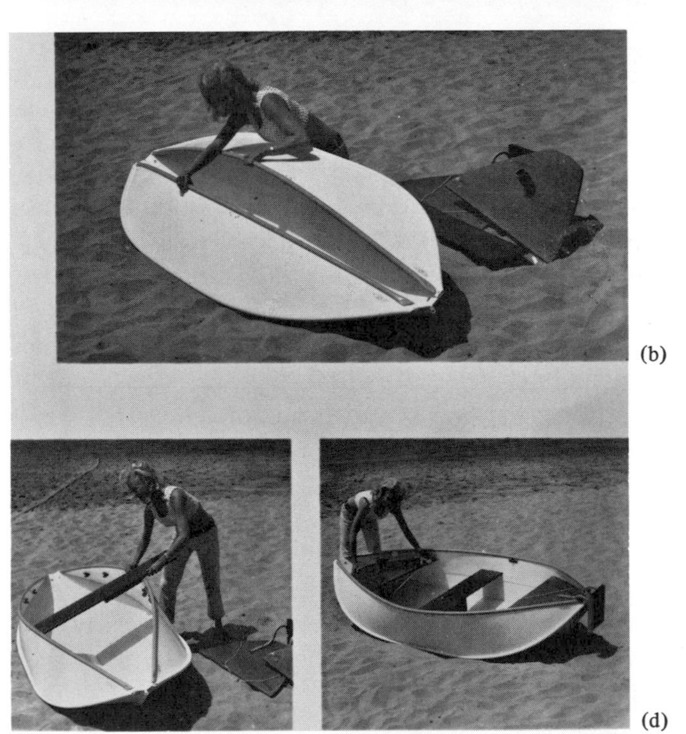

(b)

(c) (d)

Figure 3.27. An ingeniously-designed, foldable boat. The boat is made from sheet propylene material with propylene hinges. Because propylene plastic has a specific gravity less than one, this boat is truly unsinkable. (a) shows the boat folded for convenient handling; (b), (c) and (d) show several of the steps to unfold the boat. (Courtesy K Enterprises, Menlo Park, Calif.)

and paper cups are sold in stacks; a set of three suitcases can be made to nest inside one another—these are techniques to save packaging and shipping costs.

Foldable and stackable furniture definitely provides convenience. The ease of shipping and storing it adds to its value. Figure 3.27 shows a foldable boat made of polypropylene plastic. The advantage of folding is a major attraction.

3.3.10. Complementary Products and Differential Pricing

A company that manufactures winter goods will have its factory idle for a large part of the year. A complementary product line might include summer goods that can use the manufacturing facilities when winter goods are not being produced.

For shipping purposes, low-density objects are usually charged by volume. Heavy objects are charged by weight. The idea is that a truck has certain volume and weight limitations. It makes no difference if the truck is moving a chunk of steel with excess volume capacity or a shipment of pillows with unused weight capacity. We can make use of the excess volume and excess weight capacities by shipping combinations of steel and pillows. Such complementary product shipment would greatly increase productivity. The problem gets quite complicated when there are several products to be moved and each product has a different value, a different weight, and a different volume. To minimize the number of truckloads would be a good problem for linear programming. The solution scheme is similar to the mixing problem and the diet problem described in Section 2.3.1 and Section 2.3.2.

Time variations in demand may be smoothed out by a differential pricing system. Power companies have to build a whole plant based on the peak load. If the difference between the peak load and the slack load is small, the plant would be very productive. On the other hand, if the slack load is frequent and is much smaller than the peak load, the plant would be idle most of the time. The ratio of the average load to the peak load is the load factor for that consumer, which may be a manufacturing company. If a consumer has a poor load factor, he will be charged extra for the irregular demand of power. Often, it is worthwhile for a factory to schedule its activities so as to increase the load factor and decrease the cost of power. Telephone companies provide lower rates for long distance calls during weekends as a way to stimulate usage during slack time. The use of differential pricing to smooth peak demands is a common practice in computer services, hotels, car rentals, and movie theaters.

Products can also be complementary in raw materials, financing, and marketing. If we could design products to utilize a variety of different grades of wood, the price of the wood would be lower. Then lumber companies could sell all the yield from a whole forest instead of cutting and selling selectively.

3.4. ECONOMIC ASPECTS OF PRODUCTION EQUIPMENT AND SYSTEMS

The following discussion covers four basic aspects of economic analysis: break-even analysis, compound-interest analysis, selection of alternatives, and determination of economic life. Three aspects of system control are presented here: system balance, system soothing, and inventory control.

3.4.1. Break-Even Analysis

Break-even analysis uses a simple linear model that relates cost, revenue, saving, and profit to the quantity of a product. The method is very effective for short-term problems.

Figure 3.28 shows a typical break-even analysis diagram with the y-axis representing the cost and the x-axis representing the quantity. The line OR, the purchase cost, increases in direct proportion to the quantity. The manufacturing cost (line FG) consists of a fixed cost F and a variable cost V. The break-even point is where the two lines intersect. For quantities less than the break-even quantity, it is cheaper to buy. For quantities larger than the break-even quantity, it is cheaper to manufacture. The difference between the manufacturing cost and purchase cost is the savings, which is positive for quantities larger than the break-even quantity.

The same diagram can be used to analyze the sales of a product. The line OR describes the revenue, which is the product of the unit price and the quantity sold. The line FG represents the cost of manufacturing. To the right of the break-even point, revenue is greater than the cost of manufacturing, and the company makes a profit.

The mathematical relations for these two situations are listed in Figure 3.28. We will now apply break-even analysis to three different problems.

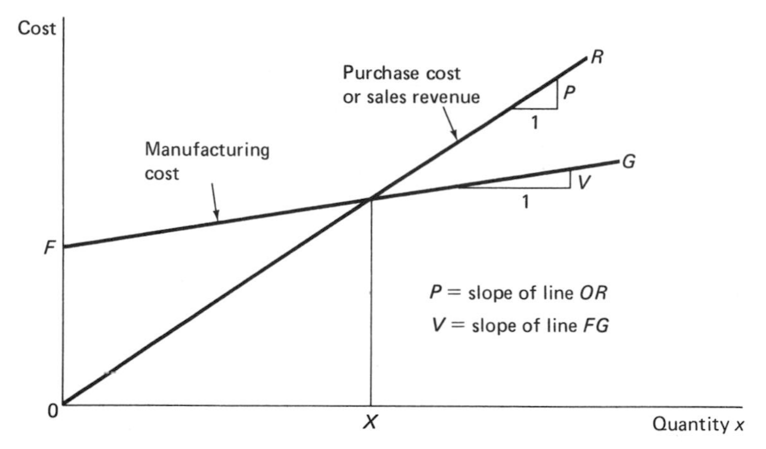

Figure 3.28. Break-even analysis for savings and profits.

Savings from make-or-buy	Profit from sales	Eq.
let P = unit purchase price	let P = unit selling price	
F = manufacturing fixed cost	F = manufacturing fixed cost	
V = manufacturing variable cost	V = manufacturing variable cost	
x = quantity	x = quantity	
purchase cost = $P \cdot x$	revenue = $P \cdot x$	(3.13)
manufacturing cost = $F + V \cdot x$	manufacturing cost = $F + V \cdot x$	(3.14)
savings = purchase cost − manufacturing cost	profit = revenue − cost	
$= (P - V)x - F$	$= (P - V)x - F$	(3.15)
break-even quantity	break-even quantity	
$X = \dfrac{F}{(P - V)}$	$X = \dfrac{F}{(P - V)}$	(3.16)

Example 1. Photocopies cost 5¢ each if they are purchased from a commercial duplicating shop. Suppose the rent of a copier is \$400/month and the cost of chemicals and paper is 1¢ per copy, how many copies/month will you need to justify a rented copier? From the statement of the problem, $P = 5¢$, $F = \$400$, and $V = 1¢$. The break-even quantity X can be obtained by Equation 3.16.

$$x = \frac{F}{(P - V)} = \frac{\$400}{(5 - 1)¢} = 10{,}000 \text{ copies.}$$

Example 2. In a battery factory, the manufacturing engineer notices that if he adds a piece of automatic machinery, he could save some labor and spoilage cost. The company sells 20,000 C size batteries each month at 15¢ each. The manufacturing fixed cost is $1700/month and the variable cost is 5¢ per battery. The proposed automatic machine would increase the fixed cost by $250/month and decrease the variable cost by 2¢ per battery. Should this extra machinery be added?

We could apply Equation 3.15 to check the profit with and without the additional automatic equipment: without the additional machinery, profit = $300/month; with the additional machinery, profit = $450/month. We would conclude that it is advantageous to add the automatic machine.

Example 3. A small photographic processing company is considering an advertising campaign about discount processing. The manager thinks that the increase in quantity would offset the loss of revenue due to the discount. The company develops and prints 1000 rolls of film monthly at an average price of $3 per roll. The fixed cost of the operation is $1700/month, and the variable cost is $1 per roll. It is estimated that a $200-campaign, coupled with a 50¢ discount would increase the business by 50% for a month. How much more profit would this campaign bring if everything works out as expected?

Equation 3.15 can be used again to find the profit with and without the campaign: without the campaign, profit = ($3 - $1) × 1000 - $1700 = $300; with the campaign, profit = ($2.5 - $1) × 1000 × 150% - $1900 = $350. The advertising campaign would bring in $50 extra profit. Besides, the publicity involved with the campaign would probably increase future business.

To use the break-even analysis properly, extreme care is needed in determining the fixed cost. Some elements of the fixed cost are shown in Figure 3.13. Sometimes an apparently fixed cost might change due to large changes in sales. Very often, administration costs, rental costs, and utility costs cannot be broken down product by product. Such difficulties limit the accuracy of the break-even analysis.

After a product has exceeded the break-even quantity and the fixed cost is paid for, a company can sell the product at a lower price and still make a profit. This marketing method is called *dumping*. It is usually carried out by: (1) selling to a different consumer sector, such as students, elderly citizens, a foreign market, trial offers, or low-price charity projects, or (2) by using a different brand name with an inferior image. The important point is to protect the original market from the influence of dumping.

3.4.2. Compound Interest and Investment Decisions

At 10% rate of annual interest, \$100 will become \$110 in one year, \$121 in two years, \$133 in three years, and \$146.41 in four years. In this case, we say a present worth of \$100 has a future worth of \$146.41 after four years. The same \$100 borrowed over four years can be repaid in many different ways. We can choose to repay \$10 the first, second, and third years and repay \$110 after the fourth year. Another common type of payment is by a uniform annual payment A. We would pay \$31.55 each year for four consecutive years. The relationship between the present worth P, the future worth F, and the uniform annual payment A is described in the following equations:

$$F = P(1 + i)^n \qquad \text{or} \quad P = F\frac{1}{(1 + i)^n} \tag{3.17}$$

$$A = P\frac{i(1 + i)^n}{(1 + i)^n - 1} \qquad \text{or} \quad P = A\frac{(1 + i)^n - 1}{i(1 + i)^n} \tag{3.18}$$

where

i = the effective annual interest rate,
n = the number of years,
P = the present worth,
F = the future worth after n years, and
A = the uniform annual payment.

The effective interest rate is slightly larger than the nominal interest rate if the interest is compounded more frequently than yearly, such as semiannually or quarterly. Let the interest be compounded

m times per year and nominal interest rate be i'. The effective interest rate $i = (1 + i'/m)^m - 1$.

The interest rate is the speed at which a sum of money grows. To double a sum of money, it takes 70 years at 1% interest rate, 13 years at 5.5%, and 9 years at 8%. Suppose a company has some money on hand, the objective is to invest the money and get the highest earnings possible. In other words, we would like to maximize the interest rate i (the rate of return). The purchase of a piece of productive equipment involves a large initial investment and subsequent earnings. Such cash flow is analogous to making a deposit in a savings account and making periodic withdrawals. Figure 3.29 shows several project proposals presented to the top management of a company. If the project is accepted, extra capital has to be allocated to the project. The proposals are judged by two factors: (1) the amount of initial investment, and (2) the interest rate i, which is the index of attractiveness. The top management may make one of the following decisions.

(1) If the company has only $200,000 available for investment, it may choose proposed projects 2 and 6 because of the high interest rates.
(2) If the company has access to abundant capital, or could borrow money at 10% interest, it may decide to take all the projects with expected interest of 10% or higher. Thus, proposed projects 1, 2, 3(a), 4, and 6 will be adopted.
(3) If all the internal investments are not attractive enough, the company may channel the available capital to external investments such as bonds, saving accounts, or the like.

Note if the interest rate for borrowed money is 10%, and the interest rate from the investment is 10%, the two would just cancel each other. With the investment just enough to pay for itself, the company would make no profit on the project, but would bear the risk of business losses due to prediction errors. Normally, the interest from investment must be higher than that for the borrowed capital because of risk, tax, handling charges in bank loans, and other hidden expenses. If the risk of a project is greater, a larger interest rate is required to justify the project. There are other intangible factors that cannot be expressed in money terms. The company may be operating at a low-profit level to gain control of the market or to establish a

Proposed projects	Initial investment	Interest rate
Project 1	$ 32,000	17%
Project 2	$150,000	32%
Project 3 (a)	$ 20,000	18.6%
(b) increment	$ 12,000	9.4%
Project 4	$ 10,000	15%
Project 5	$ 25,000	8%
Project 6	$ 50,000	22%

Figure 3.29. Selection of projects based on interest on investment.

favorable image with the customers. Sometimes product A is sold at a low profit to stimulate sales of product B. An example is the sales of low-price cameras to stimulate the sale of film. The trend of price changes, the flexibility of the system with sales fluctuations, personnel problems, and labor problems may all enter into the decision.

In the remaining part of this section, we will concentrate on the mathematics needed to determine the unknown interest rates (rate of return) on investments. The known factors are: the number of years n, the cash flow at date zero and at the end of each year (P, F, and A). The solution involves cash-flow diagrams, present worth calculations, and interpolation methods. Interest-rate tables are provided in Appendix 2.

A cash-flow diagram is a schematic representation of all the cash transactions (receipts and disbursements) on a time scale (Figure 3.30). The horizontal axis is the time scale, starting on the left with date zero (present). Usually, each increment on the horizontal axis represents one year, and time progresses toward the right. Receipts are positive incomes indicated by arrows pointing upward. Disbursements or investments are indicated by arrows pointing downward. Figure 3.30(b), (c), and (d) show the three plans to repay $100 borrowed for 4 yrs at 10% interest rate.

A present worth comparison is one in which all the cash transactions on the cash-flow diagram are converted to equivalent values at date zero.

initial investment (arrow down) =
present worth of all receipts (arrow up)

net cash flow =
present worth of receipts − initial investment = 0 (3.19)

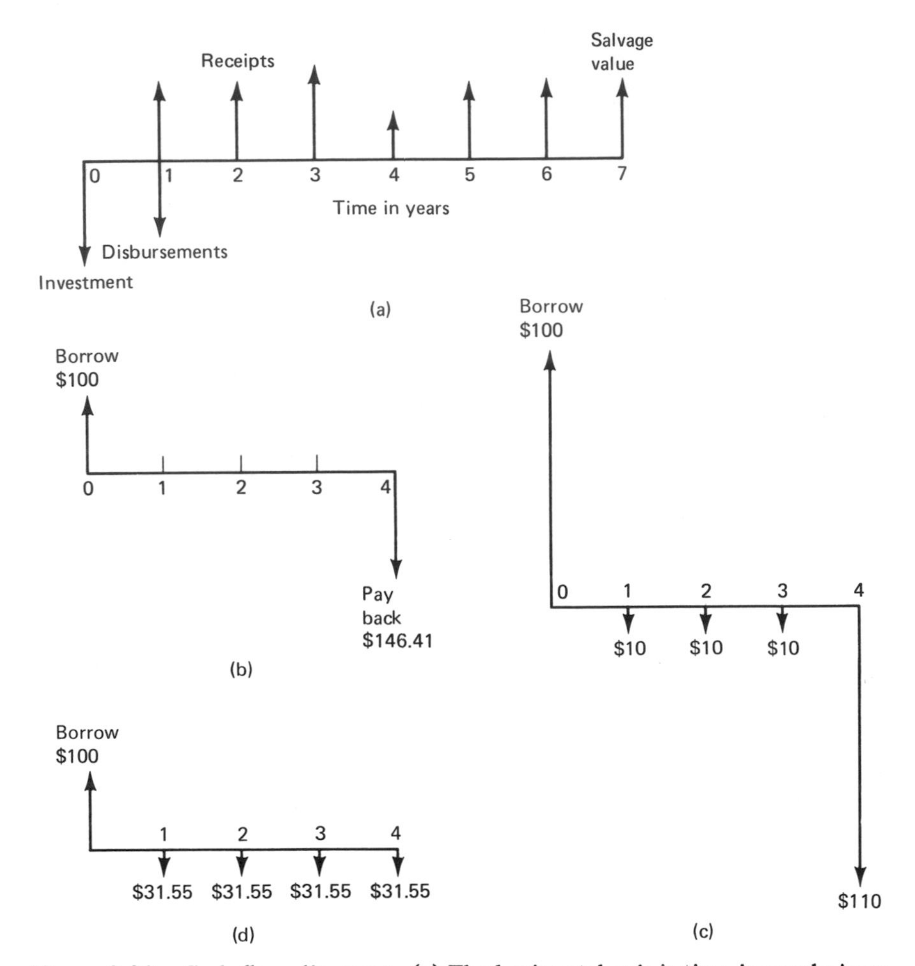

Figure 3.30. Cash-flow diagrams. (a) The horizontal axis is time in yearly increments. Vertical arrows represents cash transactions. Income is shown as upward arrows. (b) Present worth and future worth equivalence at 10% interest rate. (c) Payment of interest for three years. The sum of money borrowed remains constant. The last payment includes the interest and the capital. (d) Four uniform annual payments.

Note that with the correct interest rate, the present worths of up and down arrows should cancel each other: the net cash flow is zero. We will study three examples to get a clear understanding of how the interest rates shown in Figure 3.29 are obtained.

Example 1. An appliance company is considering buying the tools and machines to manufacture a new coffee brewer. The initial investment is estimated to be $32,000. The coffee brewer would have

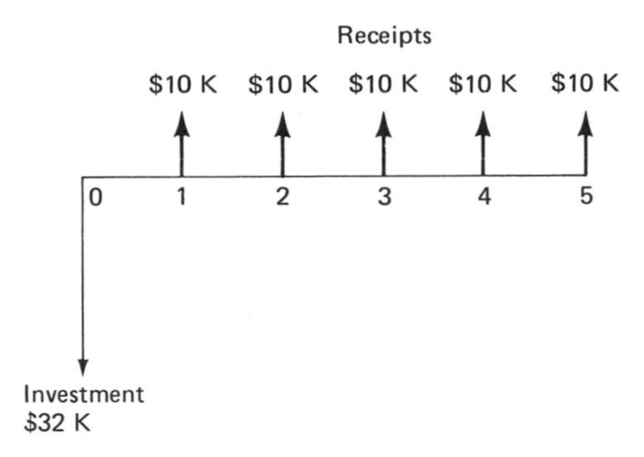

Figure 3.31. Cash-flow diagram for Project 1 (Example 1).

strong sales for 5 years; it is expected to be obsolete after the fifth year and the tools will have no salvage value. The net earnings from sales, after paying for labor, materials, operation costs, and tax is $10,000 each year. (The cash-flow diagram is shown in Figure 3.31.)

The problem is reduced to these data:

Project 1: Investment in a new coffee brewer
  Alternative 1–do nothing; $P = 0$, $A = 0$
  Alternative 2–accept the proposal; $P = -\$32{,}000$, $A = \$10{,}000$, $n = 5$
Question: What is the interest rate?

Because this is a new product, rejection of the proposal means do nothing ($P = 0$, $A = 0$). We can apply Equation 3.19 with the help of a P/A interest table.

$i = 10\%$ $P/A = 3.791$ net cash flow = $37,910 - $32,000 = $5910
$i = 12\%$ $P/A = 3.605$ net cash flow = $36,050 - $32,000 = $4050
$i = 15\%$ $P/A = 3.352$ net cash flow = $33,520 - $32,000 = $1520
$i = 20\%$ $P/A = 2.991$ net cash flow = $29,910 - $32,000 = -$2090
$i = 25\%$ $P/A = 2.689$ net cash flow = $26,890 - $32,000 = -$5110

The correct interest rate, which would make net cash flow zero, is somewhere between 15% and 20%. By plotting net cash flow against the interest rate (Figure 3.32), we can interpolate to get $i = 17\%$, where the net cash flow = 0. A mathematical formula can be used to do the interpolation:

$$\text{the unknown } i = i_1 + (i_2 - i_1) \times \frac{|NCF_1|}{|NCF_1| + |NCF_2|} \qquad (3.20)$$

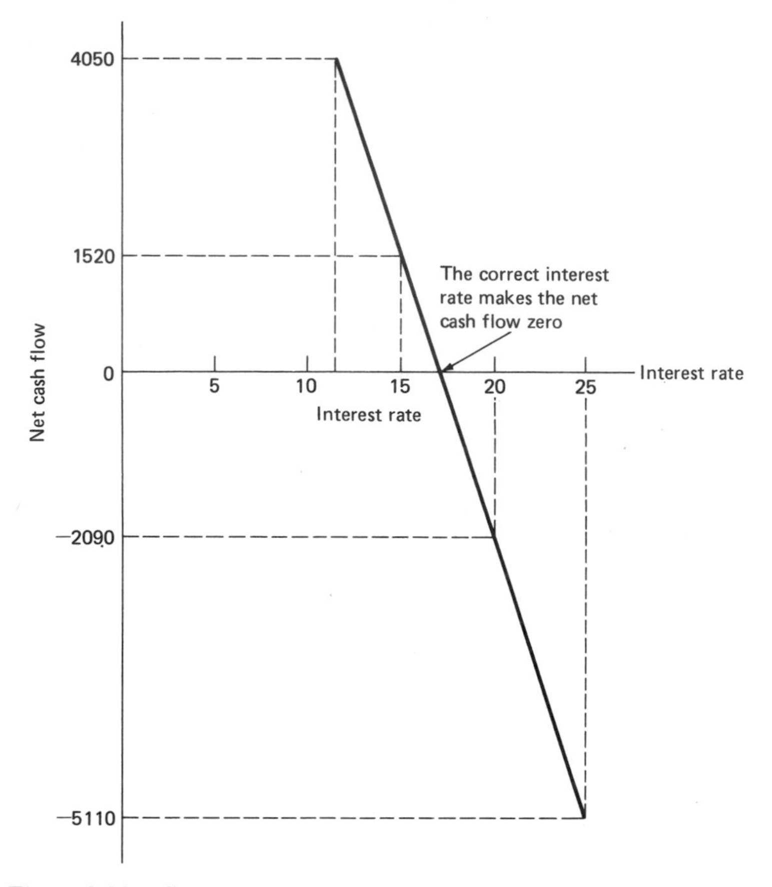

Figure 3.32. Graphical interpolation to find the unknown interest rate.

where i_1 and i_2 are two approximations on the interest rate. i_1 is lower than the actual interest so that the net cash flow NCF_1 at i_1 is positive. i_2 is higher than the actual interest, and the net cash flow NCF_2 at i_2 is negative.

Figure 3.33 shows an alternative way to find the unknown interest rate i if P/A and n are given. For problems involving nonuniform annual earnings, Figure 3.33 is useful as a rough guide to get an estimate of i to be used in the interpolation method.

Example 2. An engineering firm is considering installing a computerized drafting system. The present staff consists of 25 persons. Fifteen persons will be trained to work with the computer. The other

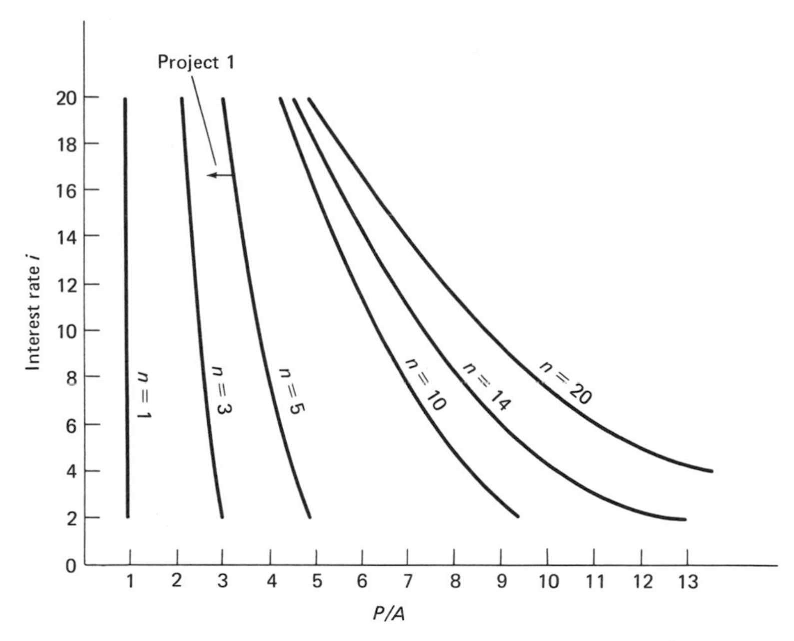

Figure 3.33. Finding unknown interest rate i with n and P/A given.

10 will be transferred to the manufacturing division. The computer system would require a 3-yr start-up period. The first-year, the system would be completely nonproductive. In the second year, the system can replace 3 draftsmen. In the third year, the system can replace 6 persons. From the fourth year on, the system would replace 10 persons. The computer system is expected to be obsolete in 15 yrs with a salvage value of ⅙ the initial cost. The salary is $11,500 per person. There is an additional cost of $500 for the fringe benefits per person per year. The computer system has an initial cost of $150,000 and a yearly cost of $20,000 for operation and maintenance.

Because this project involves an essential function of the company, rejection of the project means keeping the existing system. Keeping the existing system means no immediate investment but a higher disbursement later on. The unknown interest rate can be obtained by first taking the difference between the two alternatives (Table 3.6) and then converting all cash flow to the present values (Table 3.7). The uniform annual payment formulas cannot be used because the receipts are irregular. In Table 3.7, the two guessed interest rates

TABLE 3.6. CASH FLOW FOR THE TWO ALTERNATIVES IN EXAMPLE 2.

Project 2: Computerized drafting system

	Alternative 1	Alternative 2			Alt. 2–Alt. 1
Year	Present manual method (×1000)	Computer systems (×1000)			Difference between the two alternatives
0	0	-150			-150
1	-300	-20	-300		-20
2	-300	-20	-264		16
3	-300	-20	-228		52
4	-300	-20	-180		100
5	-300	-20	-180		100
6	-300	-20	-180		100
7	-300	-20	-180		100
8	-300	-20	-180		100
9	-300	-20	-180		100
10	-300	-20	-180		100
11	-300	-20	-180		100
12	-300	-20	-180		100
13	-300	-20	-180		100
14	-300	-20	-180		100
15	-300	-20	-180	25	125

Question: What is the interest rate that would make the net cash flow for (Alt. 2–Alt. 1) equals zero?

(30% and 35%) yeild net cash flows of \$13,450 and -\$21,670. The correct interest rate would make the net cash flow zero (Equation 3.19). We can find the correct interest rate by interpolation (Equation 3.20)

$$\text{unknown interest rate } i = 30\% + (35\% - 30\%)\frac{13.45}{13.45 + 21.67} = 31.9\%.$$

Example 3. This is an example to illustrate the comparison of several alternatives and the importance of analyzing incremental investiments. In a company, a segment of the manufacturing is now performed manually. The alternatives are: buy a semiautomatic machine, buy a fully automatic machine, or rent a fully automatic machine. The alternatives and the statement of the problem are given in Table 3.8. We can start out by comparing Alternative 2 to Alternative 1;

TABLE 3.7.*

Alt. 2-Alt. 1	i_1 = 30% interest P/F	PW	i_2 = 35% interest P/F	PW
-150	$n = 0, P/F = 1$	-150.00	$P/F = 1$	-150.00
-20	$n = 1, P/F = 0.7692$	-15.38	$P/F = 0.7407$	-14.81
16	$n = 2, P/F = 0.5917$	9.47	$P/F = 0.5487$	8.78
52	$n = 3, P/F = 0.4552$	23.67	$P/F = 0.4064$	21.13
100	$n = 4, P/F = 0.3501$	35.01	$P/F = 0.3011$	30.11
100	$n = 5, P/F = 0.2693$	26.93	$P/F = 0.2230$	22.30
100	$n = 6, P/F = 0.2072$	20.72	$P/F = 0.1652$	16.52
100	$n = 7, P/F = 0.1594$	15.94	$P/F = 0.1224$	12.24
100	$n = 8, P/F = 0.1226$	12.26	$P/F = 0.0906$	9.06
100	$n = 9, P/F = 0.0943$	9.43	$P/F = 0.0671$	6.71
100	$n = 10, P/F = 0.0725$	7.25	$P/F = 0.0497$	4.97
100	$n = 11, P/F = 0.0558$	5.58	$P/F = 0.0368$	3.68
100	$n = 12, P/F = 0.0429$	4.29	$P/F = 0.0273$	2.73
100	$n = 13, P/F = 0.0330$	3.30	$P/F = 0.0202$	2.02
100	$n = 14, P/F = 0.0254$	2.54	$P/F = 0.0150$	1.50
125	$n = 15, P/F = 0.0195$	2.44	$P/F = 0.0111$	1.39
	NCF_1	13.45	NCF_2	-21.67

*By assuming interest rates of 30% and 35%, the net cash flow at date zero can be calculated. The correct interest rate, somewhere between 30% and 35%, would produce zero net cash flow.

TABLE 3.8. STATEMENT OF PROBLEM FOR EXAMPLE 3.

Project 3: An investment involving 4 alternatives

Alternative 1–retain the present machine

$P = 0, A = -\$10{,}000$

Alternative 2- buy semiautomatic machine

$P = -\$20{,}000, A = -\$4000$

Alternative 3- buy fully automatic machine

$P = -\$32{,}000, A = -\$1800$

Alternative 4–rent fully automatic machine

$P = 0, A = -\$9000$

number of years $n = 8$

Question: How should these alternatives be compared? What is the highest interest rate from these investments?

the interest rate i is 24.95%. When Alternative 3 is compared to Alternative 1, the interest is 19.5%. When Alternative 4 is compared to Alternative 1, there is a saving of $1000 every year without any initial investment. Since Alternative 1 is definitely unattractive, we would reject Alternative 1.

We can compare Alternative 2 to Alternative 4 and get an interest rate of 18.62%. Comparing Alternative 3 to Alternative 4 gives an interest rate of 15.33%. These values are lower than those we get by using Alternative 1 as a reference, but they are more realistic. Even poor investments may look favorable when compared to extremely bad alternatives. We know that Alternative 2 is better than Alternative 3. However, the 15.33% interest rate from Alternative 3 is not all that low. An engineer can argue that 15.33% is higher than the 10% interest rate of borrowed money.

To resolve the difficulty of whether or not to reject Alternative 3, we have to find the incremental interest rate by comparing Alternative 3 to Alternative 2. The incremental investment is $12,000, and the incremental saving is $2200 per year. The incremental interest rate is found to be 9.42%. This rather low incremental interest rate is the hidden uneconomical portion of a large decision. As a rule, if a decision can be broken down into smaller segments, each segment should be considered as a separate decision. The combination of 18.62% and 9.42% gives a 15.33% for Alternative 3 when compared to Alternative 4. If the incremental investment is made by borrowing money at 10% interest, the company would suffer a loss of 0.58% for the borrowed money.

The above comparisons are summarized in Table 3.9. The two interest rates with asterisks are the correct comparisons to be entered in Figure 3.29.

Several concepts are very important in economic studies:

(1) All the alternatives should be identified to make sure that the best solution is not left out.
(2) The system being compared economically must be equal in all other respects. The system must be equally good in engineering performance, safety, and ecology, as well as other intangible factors. If other factors are not equal, the economic benefits have to be weighed with the other factors.
(3) Only the differences between the alternatives are important.

TABLE 3.9.

Comparison of alternatives				Unknown i
Alt. 2–Alt. 1	$P = -\$20,000$	$A = \$6000$	$n = 8$	24.95%
Alt. 3–Alt. 1	$P = -\$32,000$	$A = \$8200$	$n = 8$	19.5%
Alt. 4–Alt. 1	$P = 0$	$A = \$1000$		–
Alt. 2–Alt. 4	$P = -\$20,000$	$A = \$5000$	$n = 8$	18.62%*
Alt. 3–Alt. 4	$P = -\$32,000$	$A = \$7200$	$n = 8$	15.33%
Alt. 3–Alt. 2	$P = -\$12,000$	$A = \$2200$	$n = 8$	9.42%*

*Only these two comparisons are valid.

Since we always subtract one from another, costs common to both can be neglected.

(4) The comparison should be made between the best solutions. Exaggerated interest rates obtained by comparing to extremely poor alternatives are not representative of the true situation.

(5) If a large project can be broken down into smaller segments, each segment should be considered a separate increment. Lumped projects and package deals usually contain uneconomical portions that are covered up by the overall favorable impressions.

3.4.3. The Economic Life of Machines

It is obvious that if a piece of machinery is obsolete or is broken beyond repair, it should be replaced. Such functional deficiency is only one of the reasons that production equipment is replaced. Often a perfectly functional machine is replaced because of economic reasons. This means the machine is not productive enough to be attractive in a business competition. The business might have changed in size so that the original machines are inefficient, or a technological breakthrough may introduce cheaper methods.

The economic life of a machine is shorter than the functional life usually because the costs of operation and maintenance increase as the components age and deteriorate. Antique cars and trucks can be made to function well if the owner is willing to pay excessive maintenance costs.

Let us look at the economic life of a taxi:

Example 1. A \$3500 car is to be equipped to be a taxi. The finished taxi costs \$5000 at date zero and has a resale value as shown in the first column of Table 3.10. The taxi consumes $3000 worth of gas every year, and the salary of the driver is $8000. Because of the manufacturer's warranty, the maintenance cost is zero the first year. After the first year, the maintenance cost increases by $300 per year, as shown in column 2 of Table 3.10. The earning from the taxi is $13,000 every year.

Based on the data in Table 3.10, the interest rate on the initial investment is shown as: -20% for the first year, 8.3% for a 2-yr operation, 17.9% for a 3-yr operation, 19.1% for a 4-yr operation, 19.1% for a 5-yr operation, etc. The maximum interest rate can be obtained by using each new taxi for 4 yrs and reselling it at the end of the fourth year for $1500. Notice also that the interest rate remains rather high if the economic life of the taxi is within 3 to 6 yrs. If the taxi is used longer than 6 yrs, the soaring maintenance bill will make the old taxi less economical than buying a new one. The interest that can be obtained from such an investment would also depend on the resale value of used cars. If the resale value remains high up to a certain age and takes a large plunge, it would be desirable to terminate the economic life just before the plunge to gain a high salvage value. Terminating a machine's economic life just before major maintenance may also be considered. A common characteris-

TABLE 3.10*

Year	Market price or resale value	Operation and maintenance	Salary	Receipt	Net year's income	Interest rate
0	5000					
1	2000	-3000	-8000	13,000	2000	-20%
2	2000	-3300	-8000	13,000	1700	8.3%
3	2000	-3600	-8000	13,000	1400	17.9%
4	1500	-3900	-8000	13,000	1100	19.1%
5	1000	-4200	-8000	13,000	800	19.1%
6	500	-4500	-8000	13,000	500	18.5%
7	0	-4800	-8000	13,000	200	17.4%
8	0	-5100	-8000	13,000	-100	17.1%
9	0	-5400	-8000	13,000	-400	16.1%
10	0	-5700	-8000	13,000	-700	14.4%
11	0	-6000	-8000	13,000	-1000	11.1%

*The economic life of a taxi is determined by looking at the interest rate versus years of use. The most economical life is 4 to 5 yrs, with an interest rate of 19.1%.

tic of economic life is a rather large range of peak performance. In this case, the choice between 3 to 6 yrs can be based on budgetary considerations. The anticipation of future price drops on new products can also be a good reason for putting off replacement.

Example 2. The use of interest rates comparison is usually more difficult than the annual cost comparison in the study of economic life. Moreover, if a car is used as a company-owned car or a privately-owned car, there will be no receipts. Calculation of interest rates is not possible. Table 3.11 shows basically the same data as Table 3.10, except there is no driver's salary and no receipts. To make a judgment on the economic life, we convert all the costs into an overall average annual cost. The number of years that will minimize the overall average yearly cost will be the most economic life of the machine. The computation method for the overall average yearly cost is shown in Table 3.11. The method requires an assumption about the interest i ($i = 15\%$ for this example). Typically, this will be the

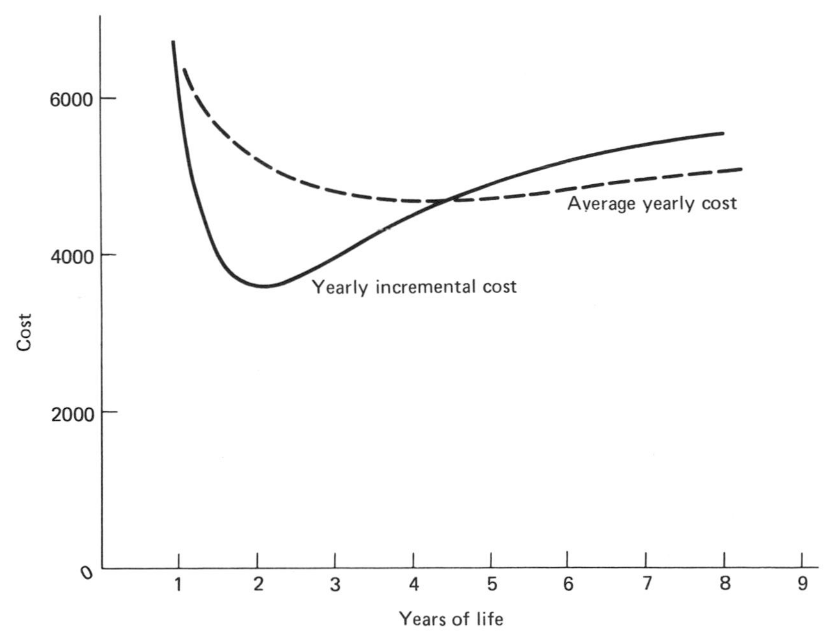

Figure 3.34. Increment and average costs. The increment year's cost (column 5 of Table 3.11) bottoms earlier than the average yearly cost (column 8). In general the behavior of the average lags behind that of the increment.

TABLE 3.11.*

			Yearly incremental cost			Overall average yearly cost		
Year	Resale value	O & M	Capital cost	O & M cost	Total cost	Capital cost	O & M cost	Total cost
0	5000							
1	2000	-3000	3750	3000	6750	3750	3000	6750
2	2000	-3300	300	3300	3600	2145	3140	5285
3	2000	-3600	300	3600	3900	1614	3272	4886
4	1500	-3900	800	3900	4700	1451	3398	4849
5	1000	-4200	725	4200	4925	1343	3517	4860
6	500	-4500	650	4500	5150	1264	3629	4893
7	0	-4800	575	4800	5375	1202	3735	4937
8	0	-5100	0	5100	5100	1114	3834	4948
9	0	-5400	0	5400	5400	1048	3927	4975
10	0	-5700	0	5700	5700	996	4014	5010
11	0	-6000	0	6000	6000	955	4095	5050
	Col. 1	Col. 2	Col. 3	Col. 4	Col. 5	Col. 6	Col. 7	Col. 8

*The economic life of production equipment with no receipts can be determined by minimizing the annual cost (column 8). The most economic life is 4 yrs with an annual cost of $4849/yr. The interest rate is assumed to be 15%.

Columns 1 and 2 are the same as those in Table 3.10.

Column 3 is obtained by considering the interest paid on the borrowed capital for the year and the yearly decrease in the resale value; e.g., for year 1, the sum of interest ($5000 \times 0.15 = 750$) and loss of value (3000) is $3750.

Column 4 is the same as column 2.

Column 5 is the sum of columns 3 and 4.

Column 6 is obtained by transforming values in column 3 into uniform annual costs. First, all the costs in column 3 up to the date are transformed to an overall cost at date zero:

$$\text{overall cost at date zero} = \sum_{i=1}^{n} \text{cost}_i \cdot (P/F)_i$$

Then, the overall costs at date zero are converted to annual costs by $(P/A)_n$. For example, for year 3,

$$A = [\text{cost}_1 (P/F)_1 + \text{cost}_2 (P/F)_2 + \text{cost}_3 (P/F)_3] / (P/A)_3$$
$$= [3750 \times 0.8696 + 300 \times 0.7561 + 300 \times 0.6575] / 2.283$$

Column 7 is obtained from column 4 by the procedure used with column 6.

Column 8 may be obtained by adding columns 6 and 7. Column 8 can also be obtained from column 5 by the procedure used with column 6.

minimum interest rate acceptable to the company. The results from the annual cost method and the interest rate method are comparable.

Another important quantity in Table 3.11 is the yearly incremental cost (column 5). From year 2 to year 4, the incremental cost is less than the average cost (column 8). This causes the average to go down year by year. At the optimum economic life, the incremental cost equals the average cost: the two curves cross over (Figure 3.34). After year 5, the incremental cost is higher, which pulls the average up. In general, the behavior of the average lags behind that of the incremental.

3.4.4 Law of Diminishing Returns

In any system, there is an optimum combination of inputs that will produce output at the lowest cost. Consider farming, where a good balance of labor, fertilizer, machinery, and land is needed. In a factory, a balance of manual labor, supervising labor, equipment, and floor space is essential.

Assume that a particular factory is operating inefficiently because of an insufficient labor force. After hiring a few more workers, the output per worker increases. Further hiring will increase the gross output further, but the output per worker will soon start to decrease. At this point, the factory has overhired for the amount of machinery and floor space. The law of dimishing returns says that if we increase only one input, the output per unit of this input will decrease after a certain point. Figure 3.35 shows a simplified cafeteria food service line. If 1 worker does the cooking, moving, and serving, the efficiency is very low. With 4 workers, the output per worker is 1 customer/min. With 5 workers, the output per worker is 1.2 customers/min. As the number of workers increases beyond 5, the output continues to increase, but very slowly. Ten workers can produce a gross output of 8 customers/min, but the output per worker is only 0.8 customer/min.

After arriving at an optimum combination of inputs (e.g., man-machine-space), all inputs should be adjusted appropriately when one input is increased.

The optimum combination is often described in the form of student-teacher ratio, worker-foreman ratio, hydrocarbon-protein ratio, etc. A deviation from the optimum ratio means a higher cost and a lower efficiency.

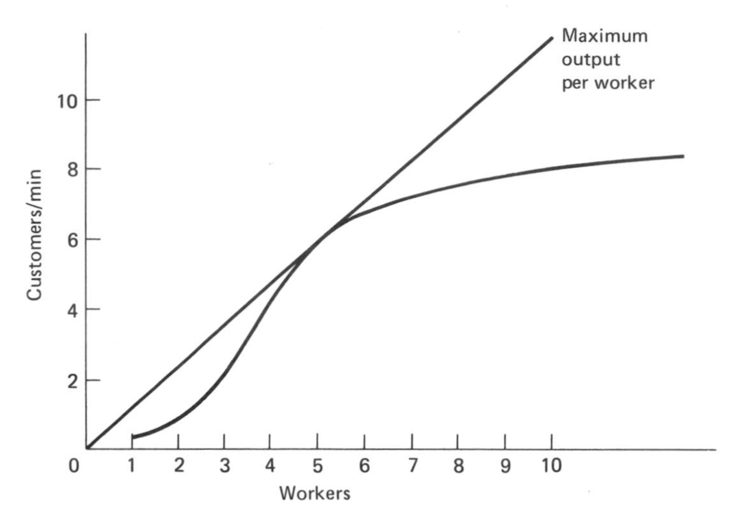

Figure 3.35. Plot of the output against one of the inputs. The output per worker is maximum at 5 workers. When the number of workers exceeds 5, the output continues to increase, but the output per worker decreases.

3.4.5. Soothing Fluctuations in Operations

It is expected that as the need for the output fluctuates, idle capacities will appear in a production system. To match the capacity to the need could reduce the cost. However, it costs to change capacity, and it costs more to change it suddenly. The cost of changing may result from recruiting efforts, relocation efforts, training, physical movement of equipment, and additional administration costs. The soothing action is essentially a trade-off between the cost of idle capacity and the cost of adjusting the system.

Figure 3.36 shows three plots of capacity and need against time. In Figure 3.36(a), since the need falls off very briefly, it may be more economical to hold the full capacity and save all the adjustment and shuffling. In Figure 3.36(b), the need fluctuates and settles to a lower level. In Figure 3.36(c), the need fluctuates periodically. There are many possible ways of changing the capacity. Zero idle capacity and no adjustment would not be the most economical alternative. The problem here is to minimize the combined cost of idle capacity and adjusting the system. One typical set of relations may be:

$$\text{cost of idle capacity} = \$1000 \times (\text{units of idle capacity})$$

$$= \$1000\,(X_i - \text{need}_i) \qquad (3.21)$$

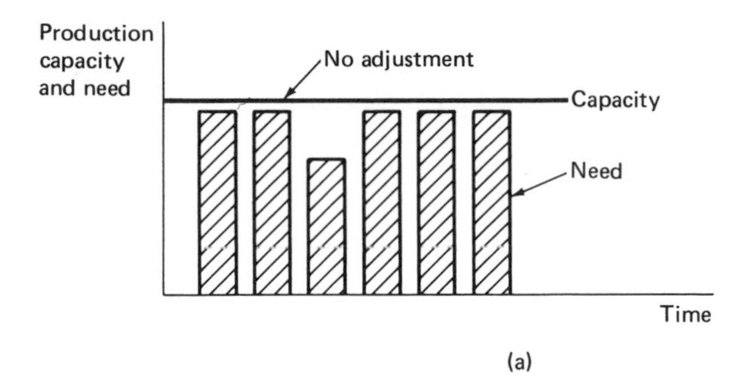

(a)

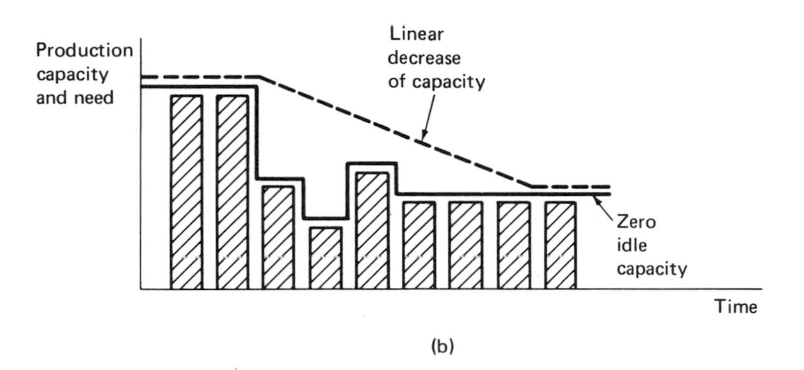

(b)

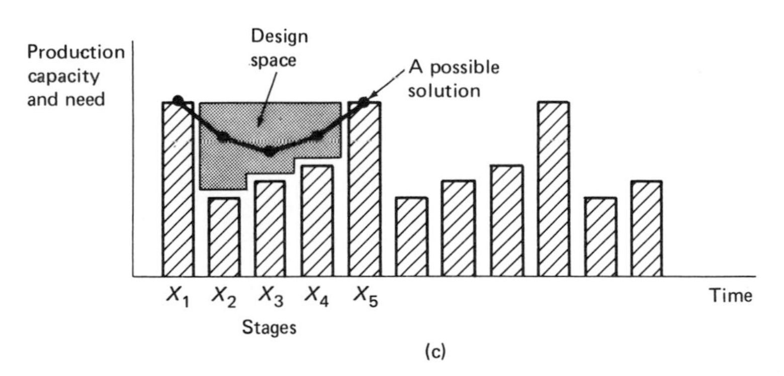

(c)

Figure 3.36. Soothing operation fluctuation. For a brief decrease in need (a), the best policy might be to maintain the full capacity. (b) shows many alternatives to tackle fluctuation of need. In (c), the design space is between the peak level and the need. The needs for stages 1, 2, 3, 4, and 5 are 100, 40, 50, 70, and 100, respectively. The objective is to pick a set of X_2, X_3, X_4 such that the cumulative cost cost_5 is minimum.

cost of adjusting the system = \$50 × (units of change)²

$$= \$50\,(X_i - X_{i-1})^2 \quad (3.22)$$

where

X_i = the capacity at a stage i, and
need_i = the need at a stage i.

The problem carries the following assumptions. (1) The capacity is not allowed to fall below the need. Idle capacity is zero or positive. (2) The capacity at the beginning and end of the study period is known or established. Only transient capacities need to be determined. (3) At the peak of the need, the idle capacity is zero.

The dynamic-programming method will be used in which each stage is a time segment within the study period. Figure 3.36(c) shows the feasible design space for the variables X_2, X_3, and X_4. The upper limit of all three variables is 100 units while the lower limits are the corresponding needs: $\text{need}_2 = 40$, $\text{need}_3 = 50$, and $\text{need}_4 = 70$. Note that X_1 and X_5 are known: $X_1 = \text{need}_1 = 100$ and $X_5 = \text{need}_5 = 100$. Compare Figure 3.36 and Figure 2.29. In the soothing problem, the design variables can vary in a continuous range of values, whereas in the machine-allocation problem, the variables are limited to discrete values. The cumulative cost of soothing in each stage:

$$\text{cost}_i = \text{(adjusting cost)} + \text{(idle capacity cost)} + \text{(previous cumulative cost)}$$

$$= 50(X_i - X_{i-1})^2 + 1000(X_i - \text{need}_i) + \text{cost}_{i-1}$$

$$\text{cost}_2 = 50(X_2 - X_1)^2 + 1000(X_2 - \text{need}_2) \quad (3.23)$$

$$\text{cost}_3 = 50(X_3 - X_2)^2 + 1000(X_3 - \text{need}_3) + \text{cost}_2 \quad (3.24)$$

$$\text{cost}_4 = 50(X_4 - X_3)^2 + 1000(X_4 - \text{need}_4) + \text{cost}_3 \quad (3.25)$$

$$\text{cost}_5 = 50(X_5 - X_4)^2 + \text{cost}_4 \quad (3.26)$$

Problem: minimize cost_5.

Cost_3 at stage 3 is a function of X_2 and X_3. To solve the problem stage by stage, we need to minimize cost_3 with respect to X_2 and express cost_3 as a function of X_3 only. This will progress to cost_4 and cost_5.

Put 3.23 in 3.24:

$$\text{cost}_3 = [50(X_2 - 100)^2 + 1000(X_2 - 40)] + [50(X_3 - X_2)^2 + 1000(X_3 - 50)] \qquad (3.27)$$

Differentiate Equation 3.27 to minimize cost_3 with respect to X_2

$$\frac{\partial \text{cost}_3}{\partial X_2} = 0, \quad X_2 = \frac{X_3}{2} + 45 \qquad (3.28)$$

Putting 3.28 in 3.27 will give cost_3 as a function of X_3 only

$$\text{cost}_3 = 25X_3^2 - 3500X_3 + 207{,}500 \qquad (3.29)$$

Put 3.29 in 3.25. Progress to the next stage.

$$\text{cost}_4 = [50(X_4 - X_3)^2 + 1000(X_4 - 70)] + [25X_3^2 - 3500X_3 + 207{,}500] \qquad (3.30)$$

Differentiate 3.30 to minimize cost_4 with respect to X_3

$$\frac{\partial}{\partial X_3}(\text{cost}_4) = 0, \quad X_3 = \frac{2}{3}X_4 + 23.3 \qquad (3.31)$$

Putting 3.31 in 3.30 will give cost_4 as a function of X_4 only.

$$\text{cost}_4 = 16.67X_4^2 - 1333.3X_4 + 96{,}666.75 \qquad (3.32)$$

Put 3.32 in 3.26. Progress to the final stage.

$$\text{cost}_5 = [50(100 - X_4)^2] + 16.67X_4^2 - 1333.3X_4 + 96{,}666.75 \qquad (3.33)$$

Differentiate Equation 3.33 to minimize the overall cost.

$$\frac{\partial}{\partial X_4}(\text{cost}_5) = 0, \quad X_4 = 85 \qquad (3.34)$$

We can substitute back and find X_3 and X_2.
Put 3.34 in 3.31. Solve for X_3

$$X_3 = 80 \qquad (3.35)$$

Put 3.35 in 3.28. Solve for X_2

$$X_2 = 85.$$

The minimum overall cost is obtained by substituting Equation 3.34 in Equation 3.33. The minimum overall cost_5 is \$115,027.

The method can be applied to a variety of situations with constant, linear, and higher-power terms in the cost of adjustment (Equation 3.22). The characteristics of the solution are strongly influenced by the square term in Equation 3.22.

3.4.6. Inventory Control and Batch Quantity

In order to maximize the return from investments, it is necessary to keep down the working capital and keep up the profit. A major element of the working capital is the cost of holding inventory. By decreasing inventory, without hurting sales or operations, a company can increase the percentage of profit (not the absolute dollar profit).

The time factor also enters the picture. It is said that regardless of whether a product is useful or useless, the difference is just how much can be sold and how long it takes to sell it. In a supermarket, an average loaf of bread may be on the shelf 24 hrs before it is sold. In an antique shop, a clock may be on display for 10 yrs before it is sold. If a sum of money generates profit every 24 hrs, a smaller profit can be accepted. In other words, the rate of return (interest rate) on a sum of money is equal to the turnover rate times the percentage of profit per transaction. The manufacturer's suggested retail prices assume an average turnover rate. A discount is possible if the turnover rate is faster than the average. In a discount chain store, the percentage of discount has to be related to the turnover rate of each item. Not every kind of merchandise is discounted the same; slow-moving items cannot be discounted at all. In a hardware store, odd-size nuts and bolts and large wrenches (unpopular sizes) are money-losing items because the invested capital will not generate profit for long periods of time. Such idle and nonproductive investments have to be tolerated for the completeness of supply.

There are several interesting effects of holding inventory. A company can save inventory cost by filling orders quickly. If all the production capacity is concentrated on one job and each job is completed before another is started, there will be only one unit of in-process inventory. If, however, several jobs are undertaken concurrently and manpower is diverted, there would be many units of in-process inventory. Some mail order companies can operate on zero inventory by purchasing the goods after an order has arrived. This is inefficient in several ways: (1) there is more purchasing and bookkeeping effort,

(2) a bulk discount cannot be obtained, and (3) it usually takes long periods of time to complete the delivery. Inventory is susceptible to inflation and depreciation. During 1974 and 1975, when electronic calculators were on a constant downward price change, it was very difficult for retailers to make a profit on old inventory. On the other hand, if a material is in short supply, the inventory may earn its profit on just the price rise. As an example, aluminum requires a lot of electricity to smelt. Holding aluminum inventory may be a way to bank electric power, which is getting ever more expensive.

The batch quantity is the production or purchase quantity with one order. A large batch has less "startup," bookkeeping and transaction. The disadvantages are the increased requirements for warehousing, insurance, and capital investment. There may also be problems of obsolescence, deterioration, and pilferage. Inventory control in a factory is the control of the stock level so that the right amount is on hand to meet the production objective and sales level. During the 1970s, many minicomputers were introduced to help control inventory [40]. Savings in in-process inventory cost and in speeding up production by efficient coordination has been effected. In the sheet metal industry, there may be several jobs that require similar raw materials. It is economical to combine the sheet materials into one order. In other cases, materials stocked up for a job may be used for another job temporarily if the need for the latter is more urgent.

The following is a model to optimize the batch quantity Q so that the cost of the inventory is minimum. This optimization is a trade-off between the cost of ordering (or start-up cost in production) and the cost of holding inventory. Figure 3.37 shows several sawtooth plots of inventory versus time. On the upward side of the curve, the inventory is building up. On the downward side of the curve, it is draining due to sales or production activities. The bottom of the saw-teeth in Figure 3.37(a) represents the safeguard inventory when a re-order is issued. Figure 3.37(b) shows one large batch versus three small batches. Figure 3.37(c) shows the inventory pattern if the production batch is instantly delivered. This is the case if the purchased material is delivered when ordered.

Let the production rate for generating the inventory be R_a and the sales rate for draining the inventory be R_d. The upward slope in Figure 3.37 represents $R_a - R_d$, whereas the downward slope represents R_d only.

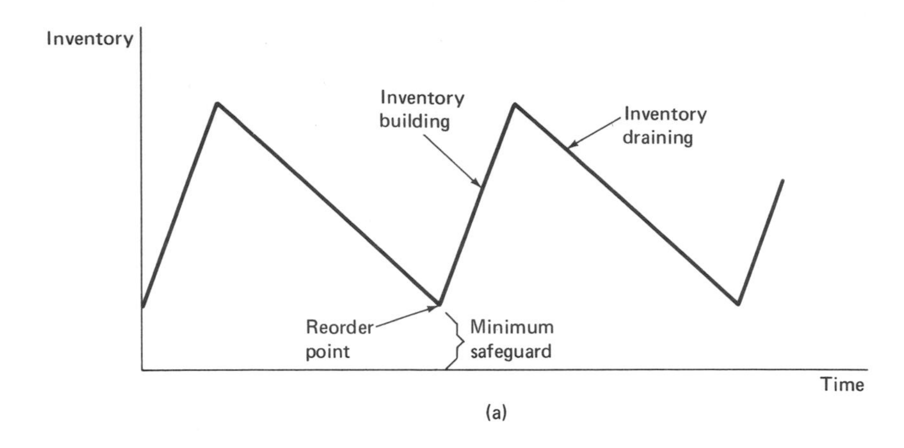

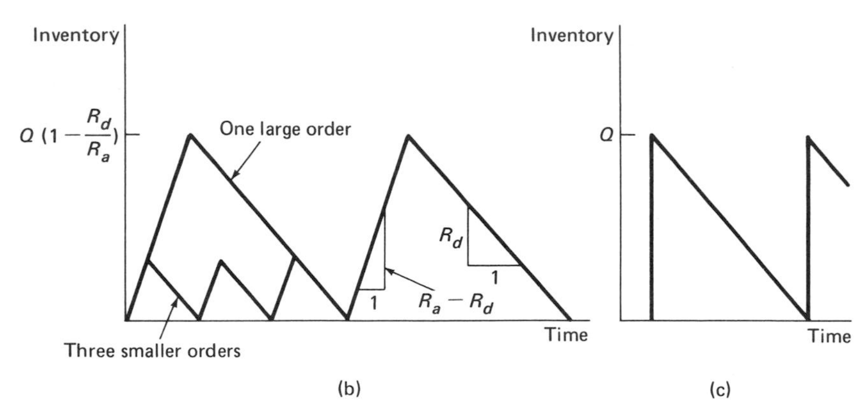

Figure 3.37. Inventory level versus time. (a), general; (b), large batch versus small batch; (c), instantly delivered production or purchased inventory $R_a = \infty$.

$$\text{the cost of ordering} = C_1 \times (\text{the number of orders}), \tag{3.36}$$

$$\text{the cost of inventory holding} = C_2 \times (\text{average inventory}), \tag{3.37}$$

where C_1 and C_2 are cost constants.

$$\begin{aligned}\text{the yearly cost of inventory control} &= \text{cost of ordering} \\ &\quad + \text{cost of holding} \qquad (3.38) \\ &= C_1\,(\text{number of orders}) \\ &\quad + C_2\,(\text{average inventory})\end{aligned}$$

Note that the cost of the inventory itself and the cost of the minimum safeguard inventory are not included. This is because such costs are independent of the batch quantity Q.

$$\text{the number of orders per year} = \frac{\text{quantity used per year}}{\text{quantity per order}} = \frac{R_d}{Q} \tag{3.39}$$

For a given number of orders per year, the cycle time t is also fixed (Figure 3.38). If the order is instantly delivered, $R_a = \infty$, the peak inventory equals the batch quantity Q (Figure 3.37(c)). If the production rate R_a is less than infinity, the peak inventory is $Q(1 - R_d/R_a)$. The peak inventory is lowered because part of the batch quantity is used during the time period it is generated (Figure 3.38). Note that if the production rate equals the use rate ($R_a = R_d$), the inventory would not accumulate at all. The average inventory is half the peak inventory:

$$\text{average inventory} = \frac{Q}{2}\left(1 - \frac{R_d}{R_a}\right) \tag{3.40}$$

Using 3.39 and 3.40 in 3.38

$$\text{yearly inventory control cost} = C_1 \frac{R_d}{Q} + C_2 \left(\frac{Q}{2}\right)\left(1 - \frac{R_d}{R_a}\right) \tag{3.41}$$

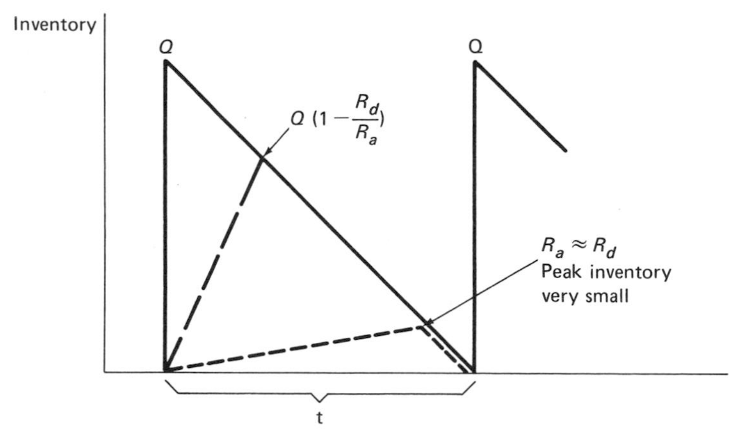

Figure 3.38. The peak inventory and production rate R_a. If $R_a = \infty$, the peak inventory is Q. The peak inventory is lower if R_a is smaller.

Equation 3.41 can be optimized by $d(\text{cost})/dQ = 0$

$$\text{optimum batch quantity } Q = \sqrt{\frac{C_1}{C_2} \frac{2R_d}{(1 - R_d/R_a)}} \tag{3.42}$$

The minimum inventory cost can be obtained by substituting 3.42 into 3.41.

A special case arises when $R_a = R_d$. The optimum batch quantity is undefined because the incoming inventory is used immediately. There is no accumulation of inventory, no storage needed, and no reordering required.

4. DESIGN FOR STRENGTH WITH MINIMUM MATERIAL

Several concepts for efficient design were discussed in Chapter 1. In this chapter, we will introduce some practical ways to obtain greater strength while using less material.

4.1. LIGHTWEIGHT TRUSSES

Pin-jointed structures are quite efficient and simple to design. First, we will discuss structures under different types of load. The last two subsections will give a generalized view of different trusses and will relate trusses to frames.

4.1.1. All-Tension and All-Compression Systems

A structure composed of only tension members, or only compression members is more efficient than one containing both. The best designs among the all-tension or all-compression systems are those that conform to the guidelines dictated by the force-flow concept.

In Figure 4.1, all members have the same sectional area and can bear the same axial load (100 lbs). In order to compare the efficiencies of structures, let us define a strength index (SI) for a structure, to be its strength-to-weight ratio relative to a reference structure. Here, the straight bar, with an ideal force flow, is used as the reference: its SI is 100%. Among the four symmetrical structures (a, b, c, and d) that have uniform stress, the SI indicates that efficiency decreases as the configuration deviates from the configuration with the

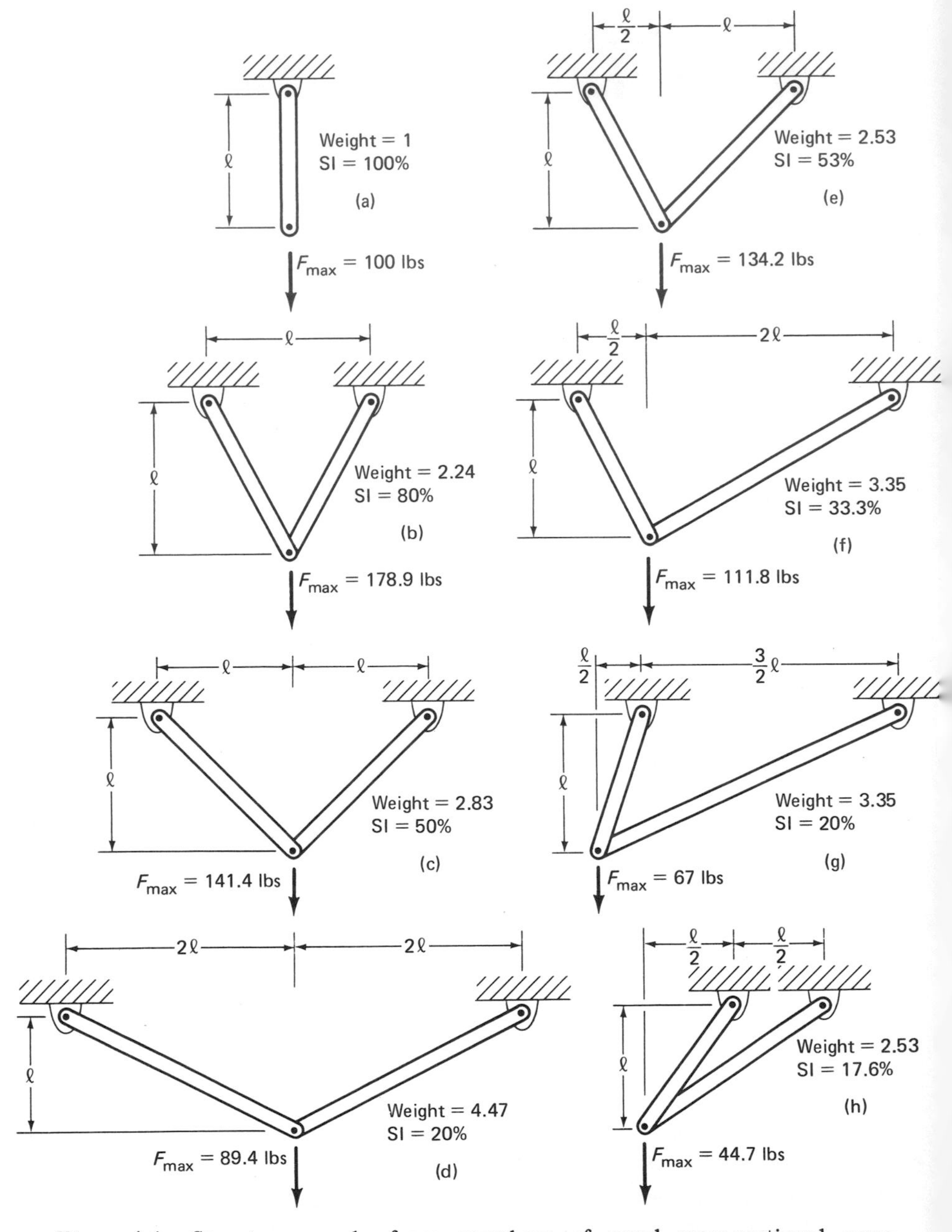

Figure 4.1. Structures made from members of equal cross-sectional areas. Structure (a) is used as a reference. Its SI is 100%. All other structures are rated with respect to (a).

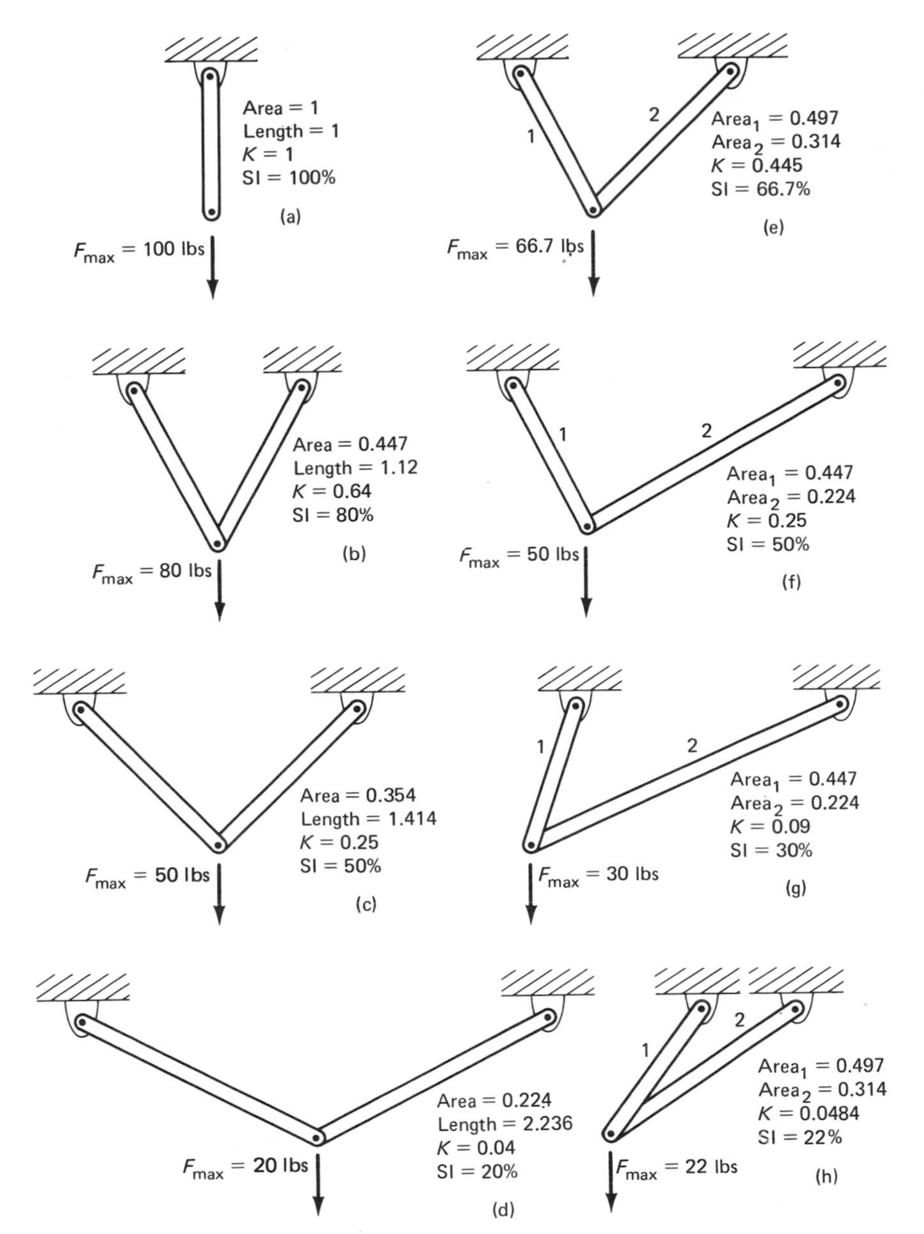

Figure 4.2. Structures with uniform stress distribution made from equal amounts of material. The structures are similar to the corresponding structures in Figure 4.1 with the sectional area scaled down by the formula: area = F/S_y. After the sectional areas are scaled, the stress is equal in all members. The SI is proportional to the square root of the spring rate, K.

ideal force-flow pattern (a). The asymmetrical structures (e, f, g, and h) may be inefficient because of three reasons: (1) nonuniform stress distribution, (2) deviation from the ideal force-flow pattern, and (3) presence of both tension and compression members in one structure (g, h).

The stresses in asymmetrical structures can be made uniform by scaling the sectional areas so that they are proportional to the forces in the members. Figure 4.2 shows such a system where all the structures have the same weight and all stresses are uniform. The asymmetrical structures in Figure 4.2 show improvement over the similar structures in Figure 4.1. Since the stress is uniform, the material's elastic energy capacity (modulus of resilience R) is fully utilized.

$$R = \tfrac{1}{2} S_y \cdot \epsilon_y \tag{4.1}$$

$$\text{energy capacity of the structure} = \text{volume} \cdot R.$$

The energy capacity is the same for all structures because they all have the same volume. The energy capacity is also half the product of the maximum load and the maximum deflection.

$$\text{energy capacity of the structure} = \frac{1}{2} F_{max} \cdot \delta_{max} = \frac{(F_{max})^2}{2K} \tag{1.7}$$

where

δ_{max} is the deflection, and
K is the spring rate.

Therefore, the spring rate, K, of a structure is proportional to the square of the maximum load (strength) of the structure. For a given volume of material and for a uniform stress, maximizing the strength in terms of the geometry of a structure is the same as maximizing its stiffness K:

$$\text{volume} \cdot R = \text{constant} = \frac{(F_{max})^2}{2K} \tag{4.2}$$

$$(F_{max})^2 \propto K \tag{4.3}$$

The tension bars in Figure 4.3 all have a uniform stress and the same volume of material. They all have the ideal force-flow pattern. The

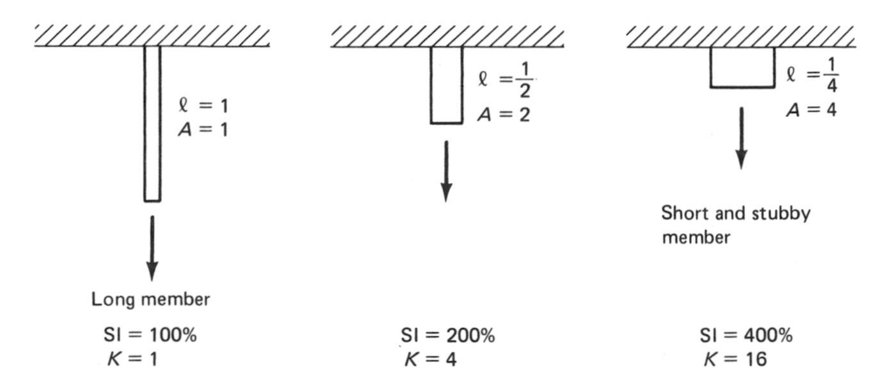

Figure 4.3. Strength and stiffness. If the stress is uniform and the volume of material is the same, the strength is proportional to the square root of the stiffness. Note that perfect force flow exists in all members.

difference in strength is represented by the square root of their stiffnesses. Careful judgment is needed in such a comparison, because tension bars of different lengths are not equal in function.

There are two kinds of redundancy in a structure: redundant members and redundant supports. The two kinds of redundancy have to be treated separately. Redundant members share loads in proportion to their stiffnesses. A weak, stiff member will greatly degrade the overall strength of the structure. If the redundant members have a stiffness ratio equal to their strength ratio, the stress is uniform, and the design is balanced. However, uniformity of stress is a necessary but not sufficient condition for maximum efficiency (see Figure 4.2). Structures with redundant members will not be more efficient than the best determinate structure. A structure with redundant members can be converted into a determinate structure simply by taking out some members. This usually yields several alternative determinate configurations, one of which may be more efficient than the others. The redundant structure, then, is just a mixture of the more efficient and the less efficient. If the determinate structures derived from the redundant structure are equally efficient, the redundant structure may be equally efficient. Structures with redundant supports can be more efficient than those with determinate supports. Redundant supports can stiffen a structure without increasing the volume of material or causing nonuniform stress distribution. If the stress is uniform, a stiffer structure is stronger (see Figures 4.2 and 4.14).

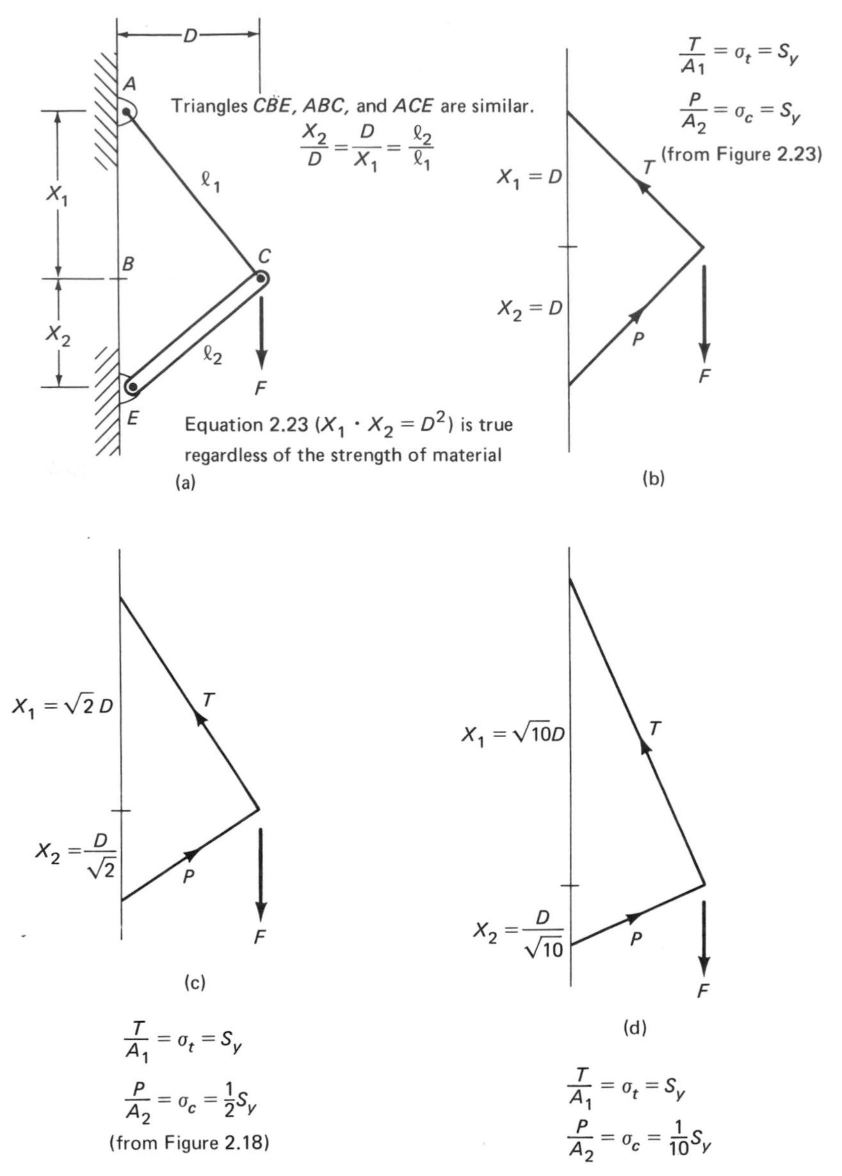

Figure 4.4. The orthogonal relationship between tension and compression members. In (c) and (d) the allowable compressive stress is less than S_y because of buckling considerations.

4.1.2. Structures Involving both Tension and Compression Members

We have optimized two such structures in Section 2.2.3, Figure 2.18, and Section 2.3.3, Figure 2.23. These two examples are shown in Figure 4.4. The dominant characteristic is that the tension and compression members are perpendicular to each other ($X_1 \cdot X_2 = D^2$). The length ratio X_1/X_2 is controlled by the ratio of the allowable tensile stress to the allowable compressive stress. Typically, $X_1/X_2 > 1$ because compression members may buckle, and the allowable compressive stress is lower than the allowable tensile stress.

When tensile members are orthogonal (perpendicular) to compressive members, the forces in these members will not cancel each other. The structures (g) and (h) in Figures 4.1 and 4.2 are inefficient because portions of the member forces are cancelled. If the angle be-

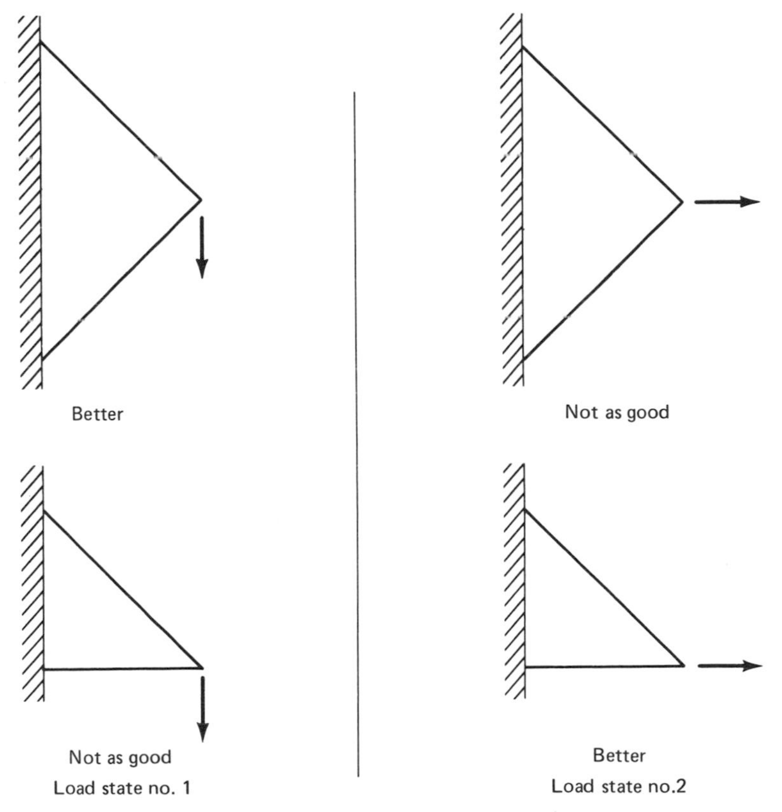

Figure 4.5. An optimum structure is not applicable to all load conditions.

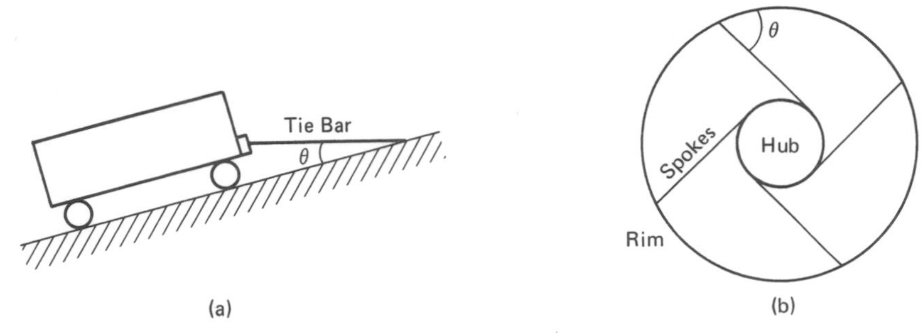

Figure 4.6. Trusses to resist shear motion. A tie bar (a) stops the relative motion between a cart and a track. Spokes (b) transmit torque from the hub to the rim of a wheel.

tween a tension and a compression member is 0, the two members together would have no net strength because of complete cancellation of forces.

As the loading conditions change, the optimum configuration changes. Figure 4.5 shows two structures that are both superior under a specific load. Sensitivity analysis can be used to determine the range of loads where an optimum structure remains best.

4.1.3. A Truss to Resist Shear Motion

Figure 4.6 shows two cases where a tie bar is used between bodies that can move parallel to each other. If the objective is to minimize the force in the tie bar, the angle θ should be 0. In such case, the length of the tie bar is infinite. The minimum length of the tie bar is achieved when the angle is 90°, but the force in the bar becomes infinite. The proper objective is to minimize the volume of material in the tie bar. The solution as shown in the tripod problem in Section 2.2.3 (Figures 2.16 and 2.17) is a bar at an angle of 45°.

The 45° angle can be explained by the elasticity theory: the principal tension or compression stresses make a 45° angle with the maximum shear stress.

4.1.4. Michell Strain Field

All the observations in the three subsections above can be generalized in one theory: the Michell strain field. A Michell strain field is a map of the directions of principal strains over the design space for a

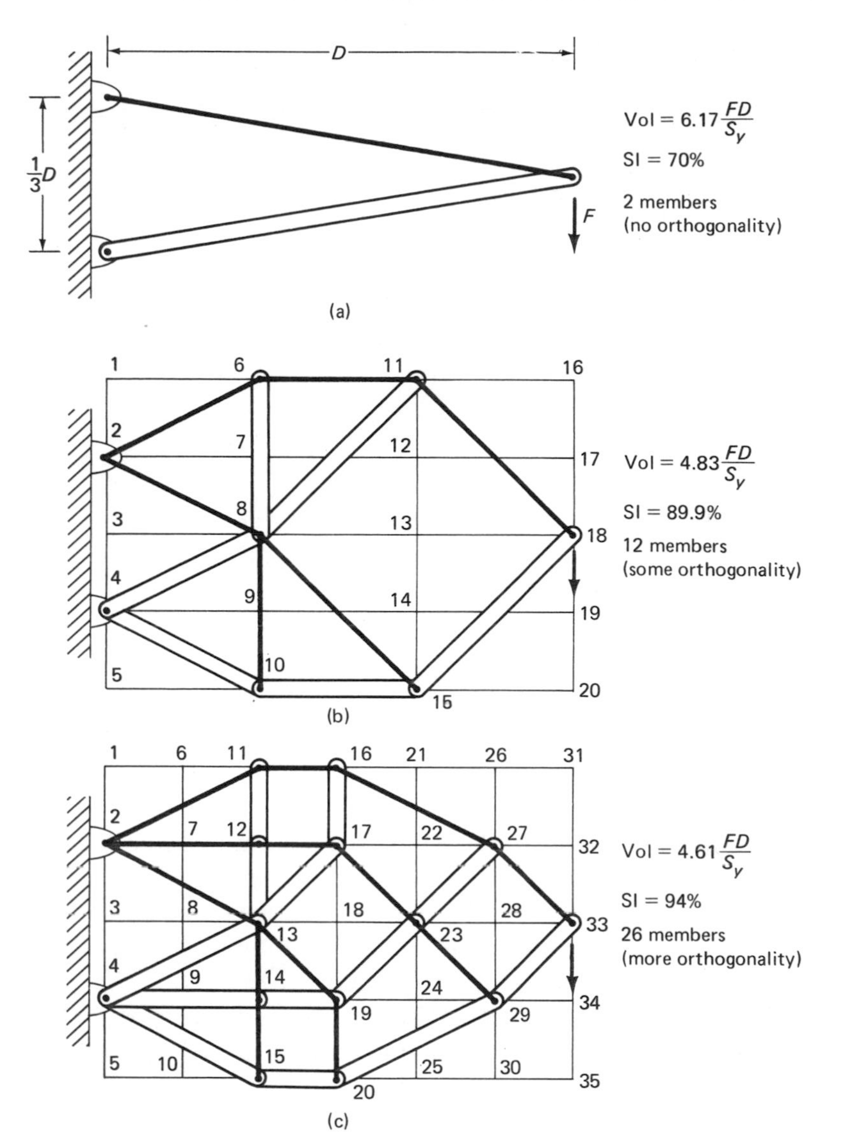

Figure 4.7. Comparison of structures with varying degrees of orthogonality. Hemp [42].

(b) reproduced by permission of the Controller of Her Britannic Majesty's Stationery Office.

(c) from the Proceedings of the International Congress of Applied Mechanics, 1964. Springer Verlag, Berlin, Germany.

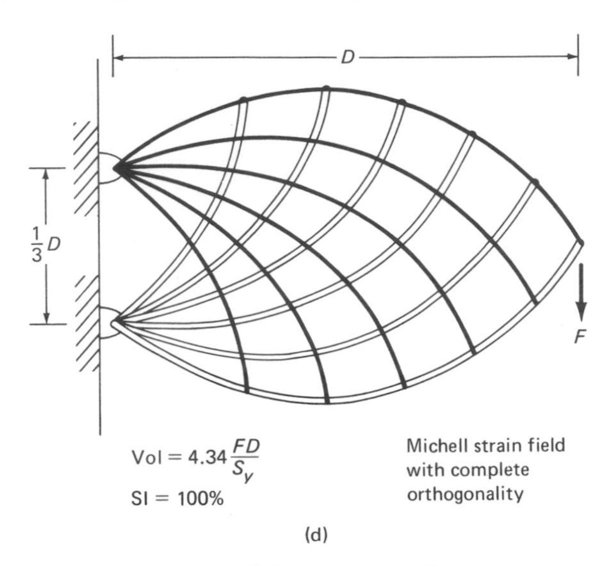

Figure 4.7. (*continued*).
(d) with permission from Oxford University Press.

set of forces. A structure with all members in the direction of the strain field is potentially the most efficient structure. Intuitively, when the member is in the direction of the strain field, there is straight and direct force flow.

The strain field concept and application can best be illustrated by an example from Hemp [42]. Figure 4.7 shows a force F to be supported by some kind of truss. The simple 2-bar truss (a) is inefficient (SI = 70%) because tension and compression members are not orthogonal to one another. By a linear-programming technique, structures can be synthesized with shorter members (b) and (c). The improved structures have many orthogonal members and much better SI's. A grid of nodes is initially laid in the design space. The number of nodes in Figure 4.7 (b) and (c) are 20 and 35, respectively. At the end of the linear-programming analysis, only the most desirable nodes will be utilized. The more nodal positions allowed initially, the better the choice that can be made, and the more orthogonality that will be incorporated. The exact solution for this Michell field is shown in Figure 4.7(d), where tension and compression members are orthogonal everywhere. The volume of this ideal structure is 4.34 FD/S_y, and this structure is used as a reference for the strength indices of other structures. Its SI is 100%.

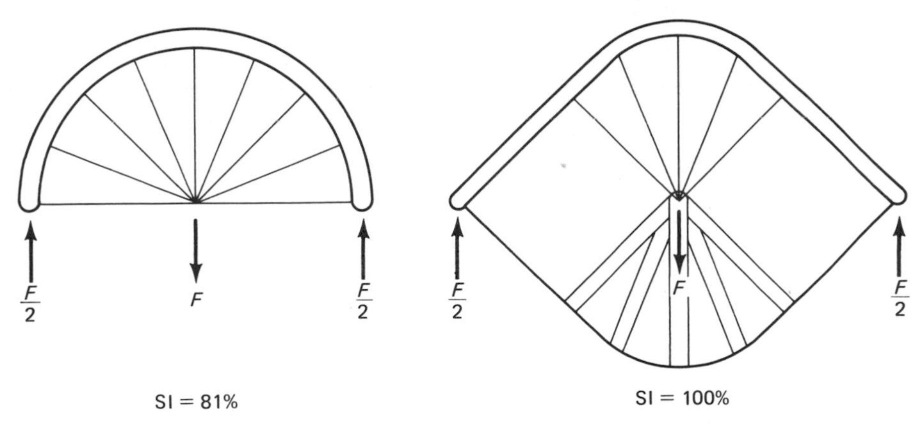

Figure 4.8. Two strain fields with complete orthogonality [41].

One shortcoming of the strain field is that it is not unique. Figure 4.8 shows two strain fields for 3-point bending (one center load and two end supports). Both fields have perfect orthogonality, but they have different efficiencies. Another feature of the strain-field method

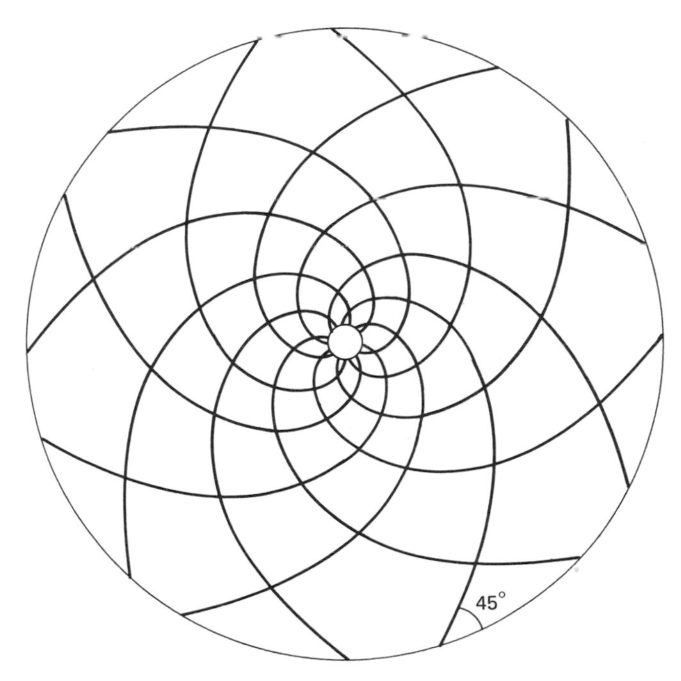

Figure 4.9. Michell strain field for a wheel under torsion.

Figure 4.10. A plastic gear designed so that the spokes are under pure compression.

is that the resultant structure may not be stable. In general, the strain-field method gives very useful clues to a good design. It also establishes the reference line for evaluating the less efficient structures.

The Michell strain field for a wheel under torsion between the hub and the rim is shown in Figure 4.9. This concept is easily applied to a plastic gear (Figure 4.10). For an idler gear, however, the Michell strain field is shown in Figure 4.8(a).

4.1.5. From a Truss to a Frame

A truss is a good starting point for structural analysis and optimization. Each member experiences only axial forces and the force-deflection relationships are linear. Because of its mathematical simplicity, the geometry of a truss can be synthesized by linear programming.

Real structures are actually not pin-jointed. Most truss members are connected by welding, bolt-joints, or rivet-joints, which can transmit moments. Even if a pin is used, friction at the hinge can transmit some moment.

A pin-jointed structure has the advantage of being able to articulate into configurations that have no bending moments in the joints. Fig-

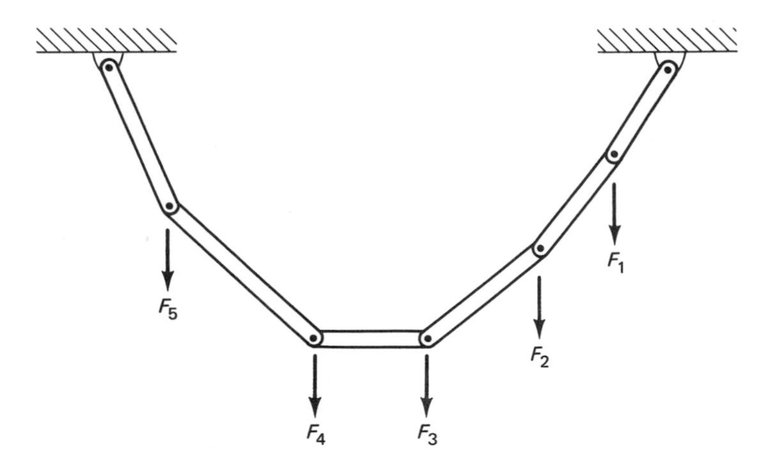

Figure 4.11. A suspension structure.

ure 4.11 shows a suspension structure in which all members are in pure tension. If this catenary is inverted to form an arch, the arch will be very efficient. The St. Louis Gateway Arch is designed on this catenary principle (Figure 4.12).

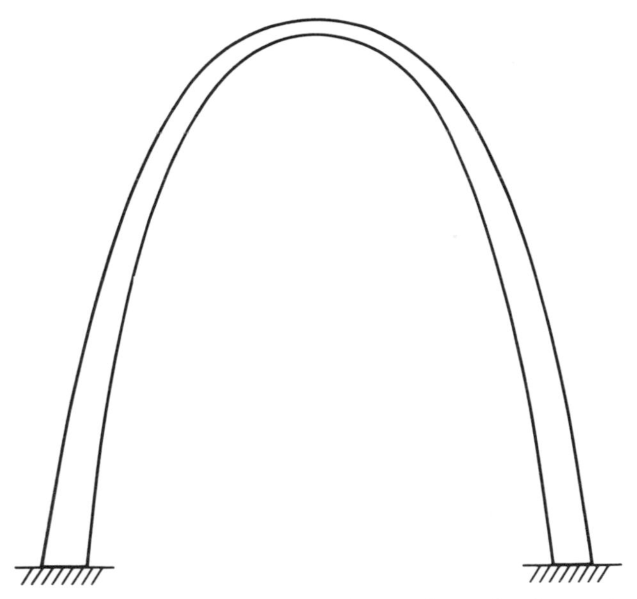

Figure 4.12. The St. Louis Gateway Arch is built to the shape of a catenary so that the whole arch is under pure compression.

If the joints are frozen by replacing the pins with welds, the structure as a whole becomes stiffer. The stiffness may be good or bad depending on whether the moment transmitted through the joints is constructive (e.g., stabilizing) or destructive (causing nonuniform stress distribution). The moment developed after a joint is frozen is $EI\theta/l$, where θ is the angular change that would have taken place if the joint were pinned. In Figure 4.13, the buckling strength of the main compression member is doubled after the lower hinge is frozen. This additional strength is a result of the stabilizing moment transmitted through the weld joint. In Figure 4.14, the beam becomes stronger and stronger as pins are replaced by welds. In this case, the support reaction moments, obtained after the welds are used, reduce the bending in the beam without making the stress less uniform. In Figure 4.15, the horizontal member is loaded close to yielding by F_x. When F_y is applied, point A would deflect downward. If a pin is used at point B, the horizontal member can deflect downward with no change in stress. If joint B is a weld-joint, the horizontal member would experience bending stresses due to F_y. The stress distribution becomes less uniform, and the additional stress due to the bending may cause a failure.

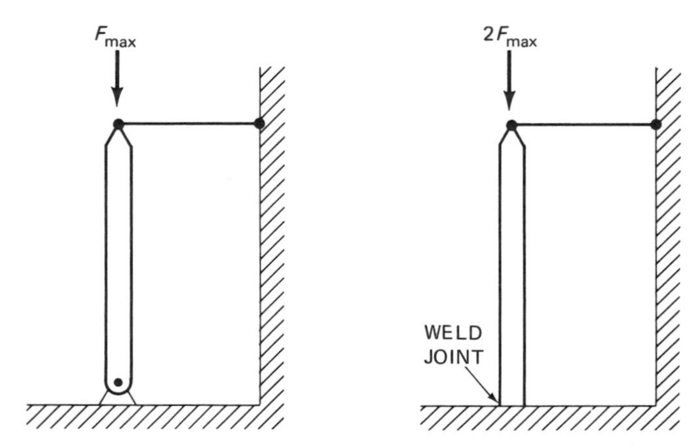

Figure 4.13. The strength of the vertical member is doubled after the pin-joint is replaced by a weld joint.

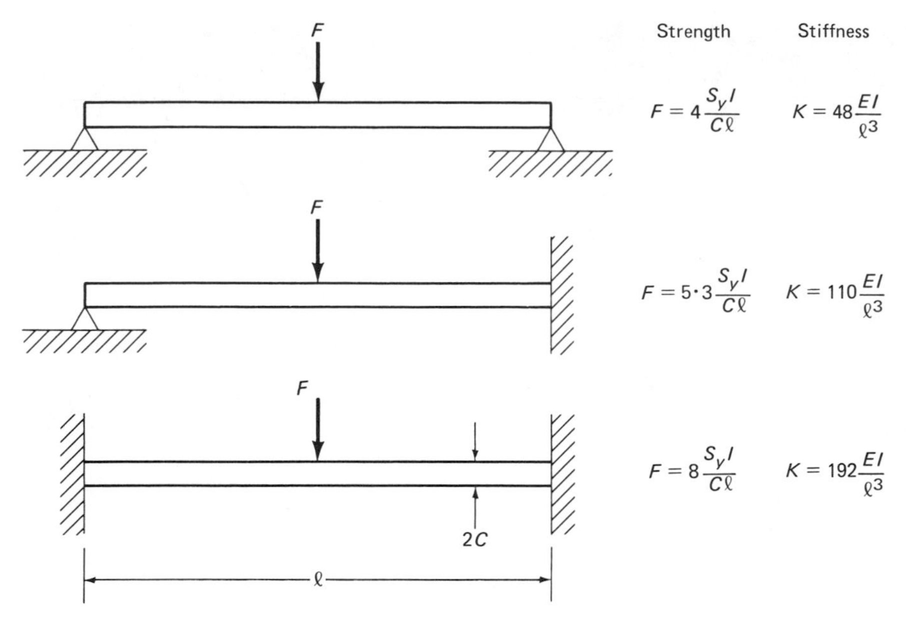

Figure 4.14. Adding redundant supports increases strength and stiffness if the uniformity of stress distribution is maintained or improved.

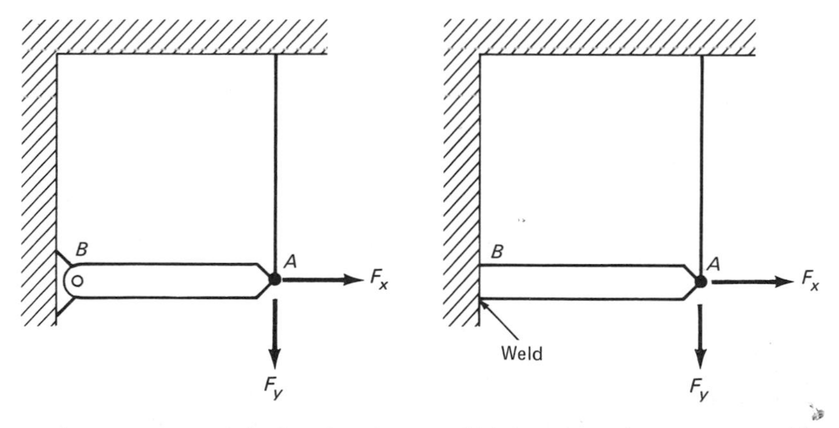

Figure 4.15. An example showing that a weld-joint can make stress nonuniform and decrease the strength of the structure.

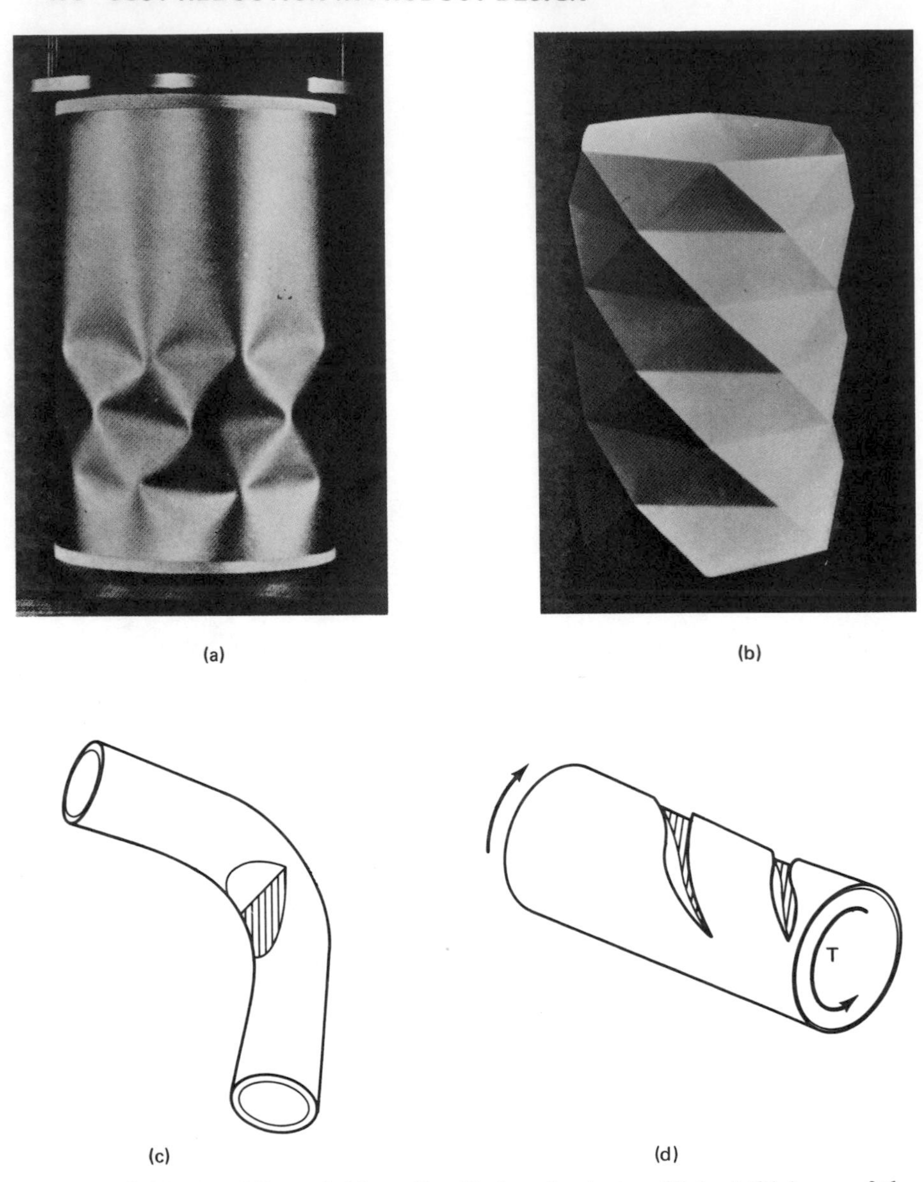

Figure 4.16. Buckling of thin-wall cylinders due to insufficient thickness of the walls. Photographs from Miura [43]. (a) shows local buckling caused by axial compression. (b) shows pseudocylindrical concave polyhedral buckling (PCCP), the ideal pattern of (a). In (c) we see local buckling caused by bending, and in (d), torsional buckling.

4.2. DESIGN CONSIDERING BUCKLING

The bending strength of a beam can be increased by increasing its depth and decreasing its thickness. Wouldn't a beam that is very deep and very thin be very strong? No! Though such a beam would not easily fail by yielding, it would easily buckle. Buckling is an instability failure due to insufficient stiffness. Thin, slender, and flimsy structures are prone to buckling. However, we tend to use thin sections to get more efficient beams and columns. Buckling is usually the dominant criteria for designs with a high strength-to-weight ratio. The same is true for a torsional member. By replacing a solid rod with tubing, the efficiency in carrying the torsional load is increased. Further, by using a thin-wall tubing of large diameter, more efficiency is gained. Such a design strategy is also limited by buckling failure.

An example of compression buckling is shown in Figure 1.24. Figure 4.16 shows three kinds of local buckling for thin tubings: com-

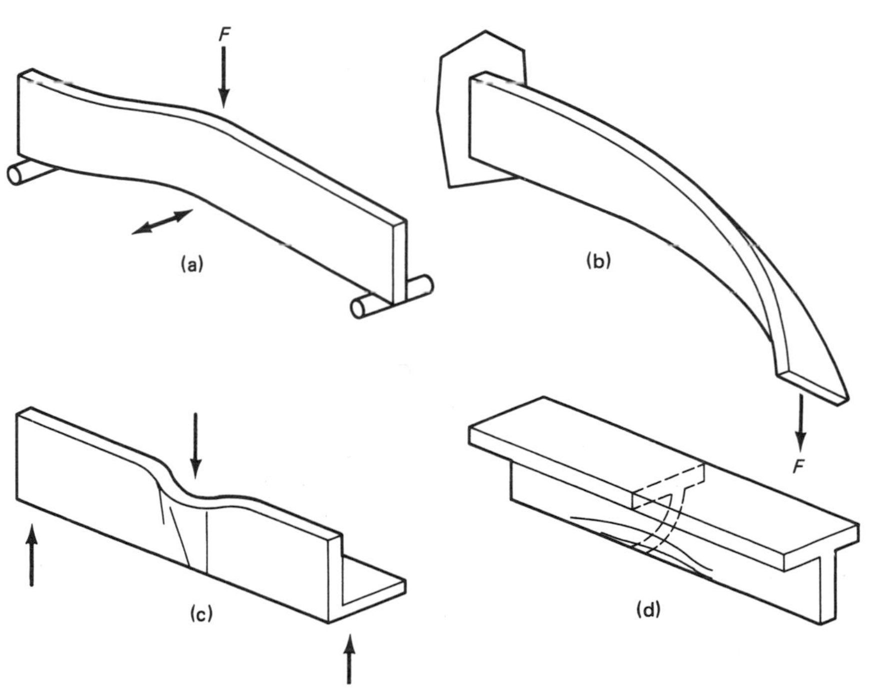

Figure 4.17. Buckling of beams due to insufficient stiffness. (a) lateral buckling of a deep beam; (b), torsional buckling, (c), buckling of a compression web, and (d), lateral buckling of a tension web.

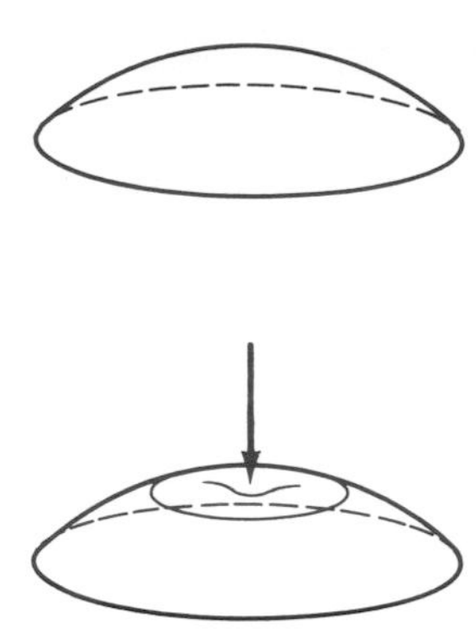

Figure 4.18. Elastic buckling of thin shells is used in calculator keyboards to provide a positive contact and with a clicking sound.

pression, bending, and torsion. Several modes of buckling for beams are shown in Figure 4.17. Thin shells also buckle if the stiffness is not sufficient (Figure 1.27 and Figure 4.18).

The cause of buckling is insufficient stiffness, which is also called rigidity or spring rate. Preventive measures are summarized as follows:

(1) Use a more rigid material with a higher Young's Modulus E and a higher shear modulus G.
(2) Use a low-strength material (lower S_y) such that yielding, rather than buckling, becomes dominant. High-strength materials will eventually fail in buckling because they will not yield.
(3) Increase the thickness of the wall. A small-diameter tubing with

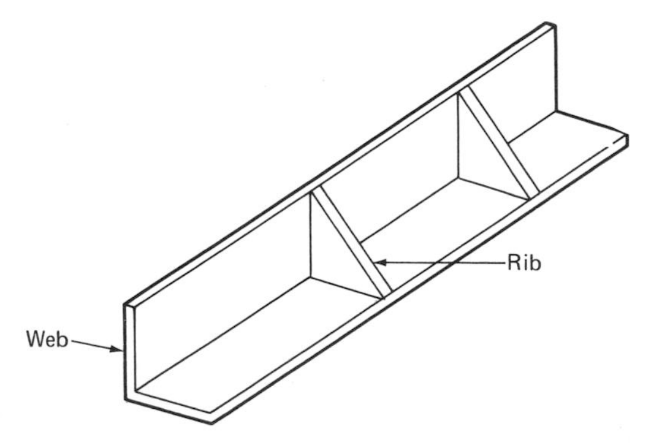

Figure 4.19. Using ribs to prevent lateral deflection and eliminate buckling of webs.

Figure 4.20. The isogrid is machined from a flat sheet of aluminum and bent into a curved plate. Resin is used to stiffen the ribs during bending and to prevent buckling. After the bending, the resin is reclaimed by melting. See Figure 4.62 for more details about the isogrid material. (Courtesy General Dynamics, San Diego, Calif.; photo by Slysh.)

a thick wall of ductile material is less likely to buckle under bending.

(4) Apply laterial support to prohibit lateral deflection. This method is very effective against the buckling modes shown in Figure 4.17(a) and (b).

(5) Use stiffeners such as ribs, corrugations, honeycombs, and surface textures to prevent local buckling. Figure 4.19 shows how ribs can be used to prevent the buckling of compression and tension webs of beams (Figure 4.17(c) and (d)).

(6) Use a filler such as sand, wax, or resin when a thin-wall tubing has to be bent. The filler will stiffen the walls and create hydrostatic pressure during bending. After the bending, the filler can be melted, reclaimed, and reused. This method is also used in bending isogrid plates [55] (Figure 4.20). For bending deep, thin plates, fiber glass stiffeners can be used to sandwich the plate. These stiffeners, which are temporarily clamped on both sides of the plate, spring back to a straight shape when released.

4.2.1. Overall Buckling of Columns

For a beam subjected to end moments, M, its stiffness, K, is given by the following equation

$$K = \frac{M}{\delta} = \frac{8EI}{l^2} \left(\frac{\text{in.-lb}}{\text{in.}}\right) \tag{4.4}$$

See Figure 4.21 for notations.

Since buckling is due to insufficient stiffness, it is not surprising that Euler's Equation for the buckling load is directly related to the stiffness defined above

$$\text{maximum load without buckling} = P_{cr} = \frac{\pi^2 EI}{l^2} = \frac{\pi^2}{8} K \tag{4.5}$$

The allowable stress (load/cross-sectional area) is

$$\sigma_{\text{allowable}} = \frac{P_{cr}}{\text{area}} = \frac{\pi^2 E}{\left(\frac{l}{r}\right)^2} \tag{4.6}$$

where

r is the radius of gyration of the section area, $Ar^2 = I$,
l is the length of the member, and
l/r is the slenderness ratio.

A column that is flexible in bending has a low buckling strength regardless of the material yield strength (Figure 4.22). On the other hand, Euler's Equation predicts infinite strength against buckling for extremely short columns. This prediction is overridden by the material's yield strength. In fact, Euler's Equation becomes inaccurate when the column gets short. The column-strength plot in Figure 4.22 can be broken down into four sections: for very short columns, allowable stress = S_y; for intermediate length columns (with slenderness ratio between 10 and about 120*),

*The transition point from elastic to plastic buckling is: slenderness ratio $l/r < 1.414\pi\sqrt{E/S_y}$. The slenderness ratio of 120 is based on steel with a yield strength S_y of 40 ksi. Note that a high strength steel has a lower transition point and is more likely to buckle.

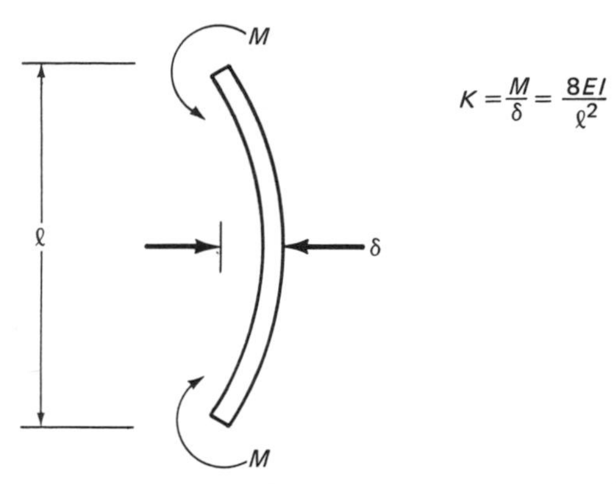

Figure 4.21. The stiffness that indicates the likelihood of column overall buckling in a column is the ratio of end moment to lateral deflection at midlength, where

$$K = \frac{M}{\delta} = \frac{8EI}{l^2}$$

E = Young's Modulus, I = the section moment of inertia, M = the bending moment, and δ = deflection at midpoint.

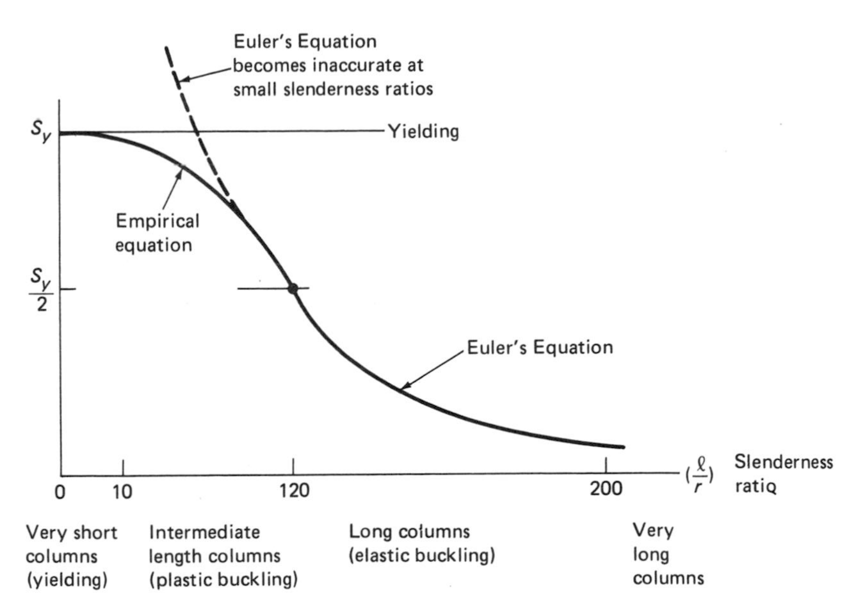

Figure 4.22. The allowable stress versus slenderness ratio.

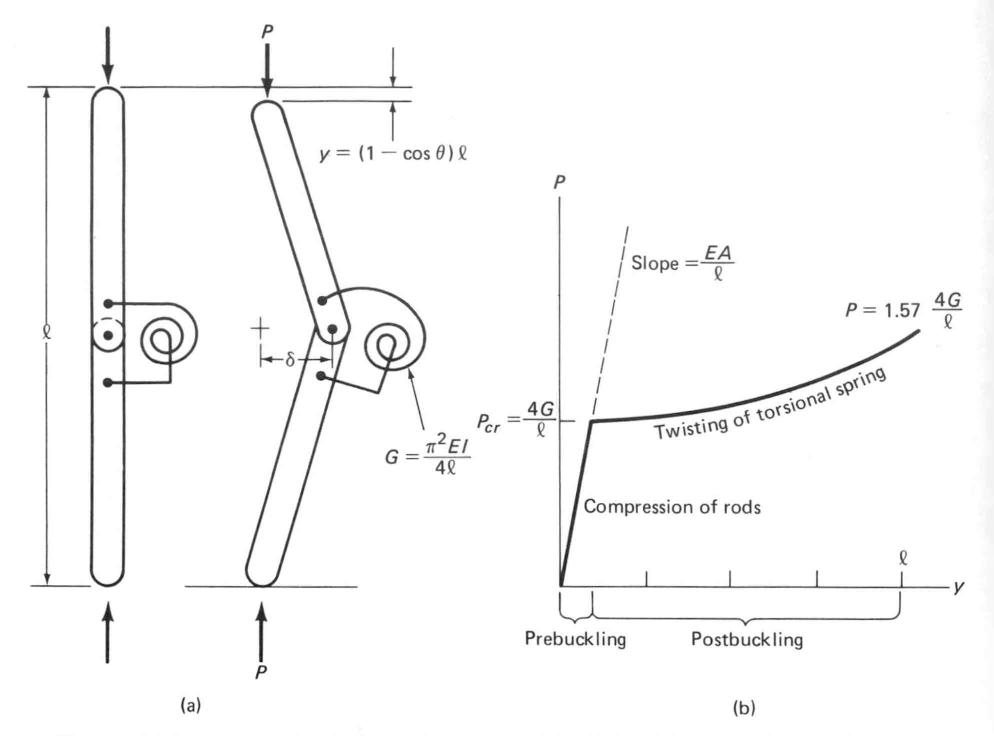

Figure 4.23. A model of elastic buckling. Elastic buckling can be modeled by two rods and a torsional spring as shown in (a). The rods are pinned together and the torsional spring aligns the two rods.
The bending moment at the pin is $M = P\delta = P(l/2) \sin \theta$.
The counter moment by the spring $M = G\theta 2$.
When buckling occurs, $Pl/2 \sin \theta = G\theta 2$

$$P = 4G(\theta/\sin \theta)/l$$

for small θ, $\theta/\sin \theta = 1$.
Initial buckling load $P_{cr} = 4G/l$.
This model column is stable after buckling. P increases with θ up to a maximum at $\theta = 90^\circ$, where $P = 1.57(4G/l)$.
P is plotted against y in (b). The prebuckling characteristic is the compression of the rods, and the postbuckling characteristic is the twisting of the torsional spring.

$$\text{allowable stress} = S_y - \frac{S_y{}^2}{4\pi^2 E}\left(\frac{l}{r}\right)^2 \qquad (4.7)$$

(this equation is a parabolic formula, and is also known as the J. B.

Johnson formula); for long columns (slenderness ratio between 120* and 200),

elastic buckling: Euler's Equation

$$\text{allowable stress} = \frac{\pi^2 E}{\left(\frac{l}{r}\right)^2} \tag{4.6}$$

for very long columns (slenderness ratio larger than 200), the column is practically useless, yet its strength can be predicted accurately by Euler's Equation. If the slenderness ratio is over 120, the buckling is elastic and the column has postbuckling strength, as depicted in Figure 4.23.

A column can be made stiffer by replacing pin-joints with weld-joints, or by adding lateral supports. The buckling strength depends on the effective curvature of the beam. Figure 4.24 shows how additional end constraints and lateral supports change the effective bending length, curvature, and stiffness, and, thus, the allowable load. A column that is driven into the ground can support a tremendous axial load because it is completely supported on the sides by the earth. Buckling cannot occur when lateral deflection is prohibited.

4.2.2. Lightweight Columns

Buckling strength can be increased by increasing the radius of gyration r of the cross-section (for example, by using a hollow tube instead of a solid rod). Using the same amount of material, a thin-wall tubing with a larger diameter will give a very high buckling strength. As the wall is thinned down to gain strength, another problem arises: the local buckling of thin walls (Figure 4.16).

For the most efficient design, two or three modes of failure should be equally likely. If a column fails in overall buckling, the tube can be made larger in diameter and smaller in wall thickness. If a column fails in local buckling, the wall thickness should be increased and the diameter decreased.

*see footnote on page 174.

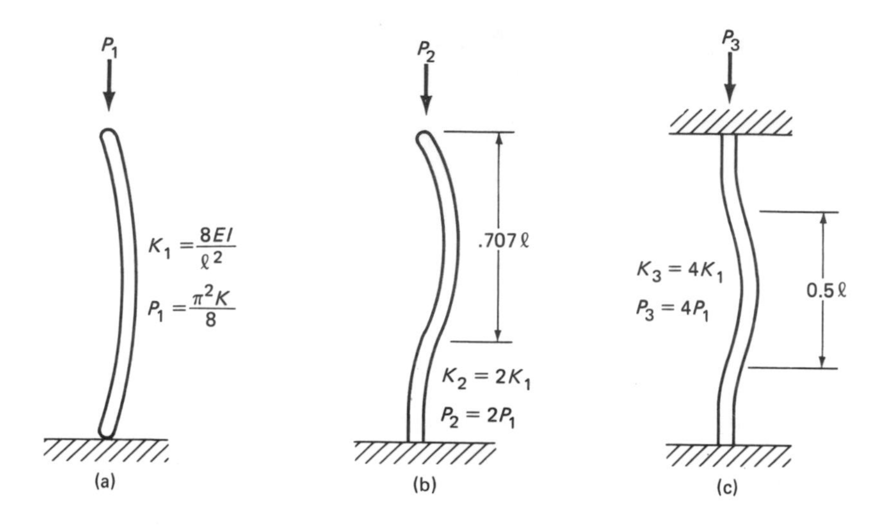

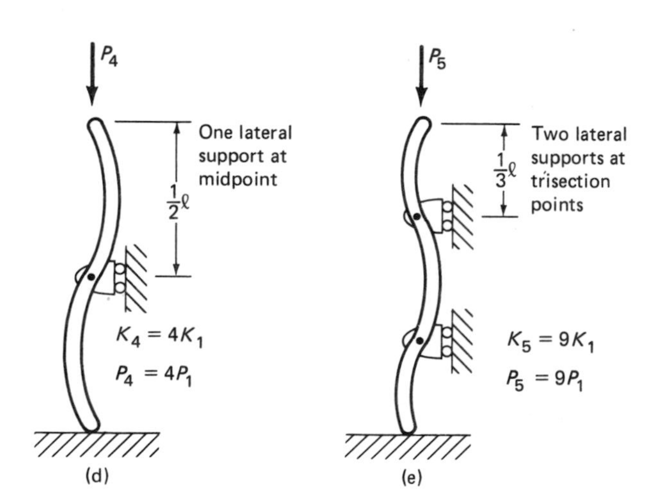

Figure 4.24. Adding end constraints and lateral supports increases stiffness because effective bending length is decreased. In actual practice, the increase in strength in (b) and (c) is not fully realized because the idealized end condition is not dependable.

For any given set of material properties (Poisson's ratio ν, Young's Modulus E, and yield strength S_y), there is a critical wall thickness t_{cr}, where all three modes of failure are equally unlikely. If $t > t_{cr}$, local buckling will not occur. If the $t < t_{cr}$, the failure modes are between local and overall buckling. Reference [44] gives the formula for t_{cr} as:

$$t_{cr} = \frac{r_o}{0.5 + \dfrac{E}{S_y} \dfrac{0.577}{\sqrt{1 - \nu^2}}} \tag{4.8}$$

Since the diameter and wall thickness of most practical tubings are fixed from standardization, the design objective is to find the maximum length where two modes of failure can coexist. Figure 4.25 shows the domains of failure as functions of: (1) the ratio of the length to the outside radius l/r_o, and (2) the ratio of the inside radius to the outside radius r_i/r_o. With a given (r_i/r_o), the maximum length is determined from one of the following two equations:

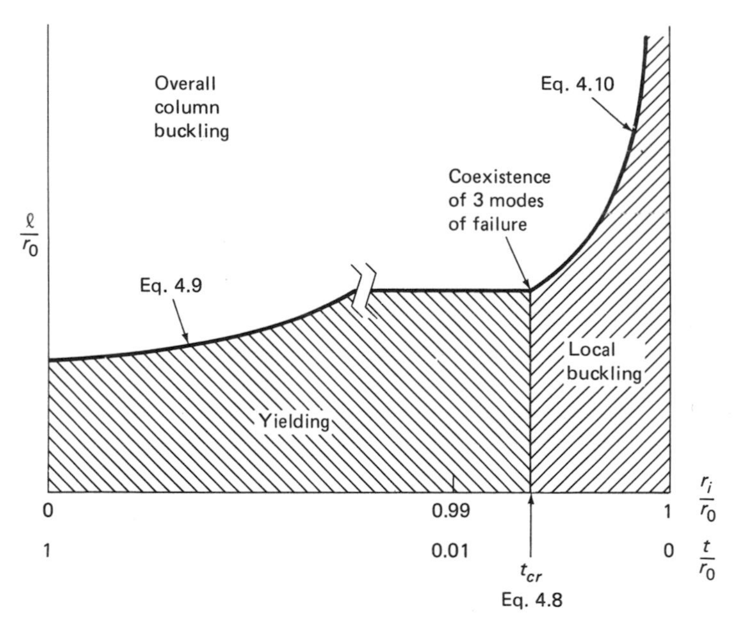

Figure 4.25. The domains of failure modes. The three types of failure for tubings are yielding, local, and overall buckling. (Marshek and Rosenberg [44].)

If $t > t_{cr}$, the tubing is "thick," and the following equation gives the coexistence of yielding and overall buckling

$$l_{max} = \frac{r_o \pi}{2} \sqrt{\frac{E\left[1 + \left(\frac{r_i}{r_o}\right)^2\right]}{S_y}} \tag{4.9}$$

If $t < t_{cr}$, the tubing is "thin," and the following equation gives the coexistence of local and overall buckling

$$l_{max} = \frac{r_o \pi}{2} \sqrt{\frac{\sqrt{3(1 - \nu^2)}}{2}} \sqrt{\frac{(1 + r_i/r_o)\,[1 + (r_i/r_o)^2]}{[1 - (r_i/r_o)]}} \tag{4.10}$$

4.3. DESIGN BEYOND THE ELASTIC LIMIT (YIELD POINT)

4.3.1. Plastic Deformation

Ductile materials can undergo plastic deformation without failure. Figure 4.26 shows that the stress-strain curve reaches a plateau where the stress remains constant with plastic strains. This plastic behavior may result in stress redistribution and insensitivity to stress concentra-

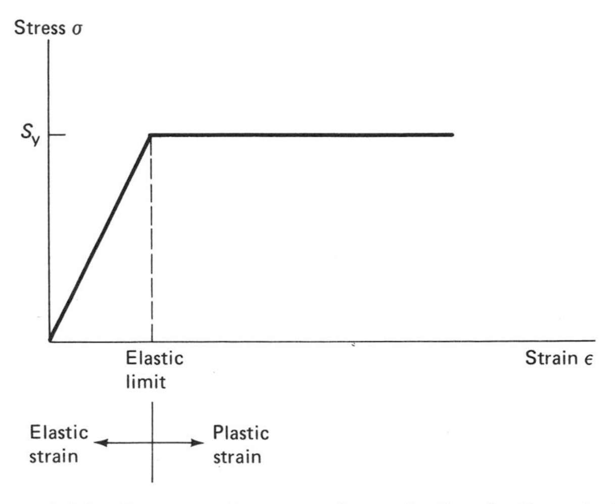

Figure 4.26. Stress-strain curve of a perfectly plastic material.

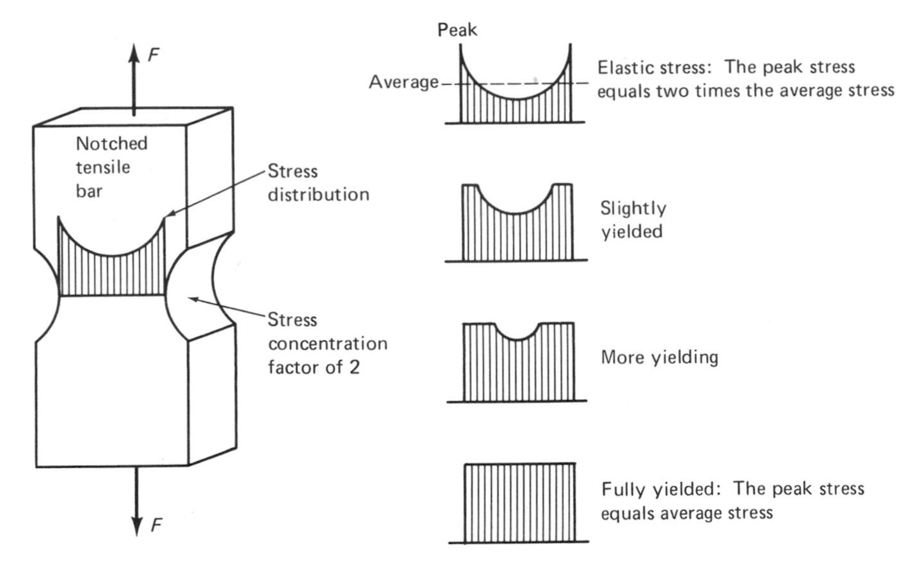

Figure 4.27. Plastic deformation can eliminate the effects of stress raisers.

tion. Figure 4.27 shows a notched bar with a stress concentration factor of 2. If the material is brittle, failure may occur when the average stress is half the material strength. If the material is ductile, stress redistribution will occur through yielding. The bar will fail when the average stress is equal to the material yield strength (Figure 4.27). Stress concentration points are less damaging in a ductile material. In the design of bolt-joints (Figure 1.12), once the allowable stress in one of the bolts is exceeded, there will be plastic deformation (or slipping) and stress redistribution. Experimental results show that the real strength is more than 10% higher than that indicated by theoretical elastic stress calculations.

Structures with a nonuniform stress distribution can have a higher strength rating if plastic deformation is allowed. Figure 4.28 shows two parallel tensile bars each capable of taking 100 lbs of load. Since the shorter bar is stiffer (its spring rate is twice that of the longer bar), the load and the stress will be two times those in the longer bar. When the shorter bar is fully loaded to 100 lbs, the longer bar has only 50 lbs of load. Because of this incompatibility in stiffness, the total load for elastic design is 150 lbs. If the shorter bar is allowed to deform plasticly, more and more load will be shifted to the longer bar. By plastic design, the total load can be increased to 200 lbs.

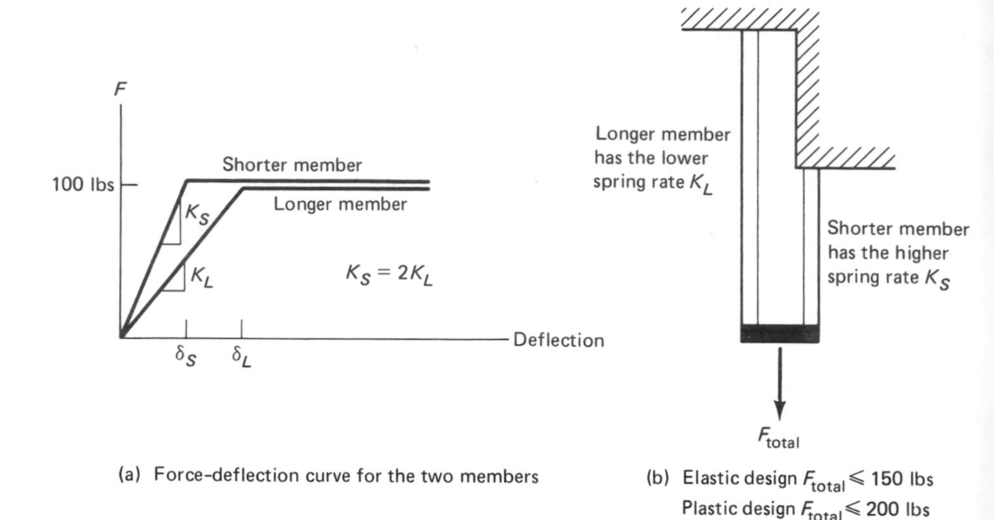

(a) Force-deflection curve for the two members

(b) Elastic design $F_{total} \leqslant 150$ lbs
Plastic design $F_{total} \leqslant 200$ lbs

Figure 4.28. The strength of a structure with nonuniform stress distribution depends on the evaluation criterion: whether elastic design or plastic design.

This shows that statically indeterminate structures become determinate in plastic design. If the stress is uniform in a structure, there is no additional strength beyond the yield point (note: perfectly plastic material will not work-harden). The increased strength from plastic design calculations is a measure of the nonuniformity of stress and the redistribution of stress through plastic deformation.

When designing for impact strength, bolts and eye-bars are made to undergo plastic deformation as much as possible. Such a member is stronger because the energy capacity of the material (toughness) is fully utilized. Figure 1.22 showed how a slender member could be stronger than a bulkier member if plastic deformation were considered.

In elastic design, a structure is considered to have failed if part of it yields. In plastic design, failure is defined as the physical crumbling or tearing of the structure. The actual strength of the structure is independent of the definition of failure or the evaluation criteria. Plastic design offers a more rational and realistic evaluation of the situation. Appropriate safety factors can then be applied to maintain a consistent margin of strength for different members and different designs.

Yielding is really very common. Many manufacturing processes such as bending, rolling, drawing, forging, and extruding subject materials to large plastic deformations. There may be very high residual stresses in these members: as much as 80% of the yield stress. A 20% additional stress would produce plastic deformation. If elastic design is carried out with the assumption of zero residual stress, yielding easily occurs even though the calculated stresses are below the elastic limit.

4.3.2. The Plastic Moment of a Beam

A beam has higher stresses at the top and bottom surfaces and very low stresses near the neutral axis. Since the stress is nonuniform, local yielding does not represent the collapse of the entire beam. The ratio of the moment, M_p, which causes full yielding, to the moment, M_y, which causes initial yielding, is the *shape factor*. An ideal beam, having all the material concentrated at the flange and zero thickness at the web, has a uniform stress distribution and a shape factor of 1.* The I-beam, being the next best in bending strength, has a shape factor of about 1.17. A rectangular section has a shape factor of 1.5. The higher the shape factor is, the less efficient is the beam in elastic bending and, thus, there is more reserve strength beyond the yield point. Unequal elastic safety factors for rectangular and I-beams produce consistent collapse safety factors. In view of the "reserve strength" after the initial yielding, a smaller safety factor may be used for inefficient beam sections.

Figure 4.29 shows a plot of the bending moment versus the curvature for a beam with a rectangular section. This is similar to a force-deflection plot for a spring. Since the stress is not uniform, the plot gradually tapers from the initial yielding point M_y to the asymptote of $1.5\,M_y$. The asymptotic value is obtained by assuming all the material above the neutral axis is under a uniform compressive stress S_y from full yielding. All the material below the neutral axis is under a uniform tensile stress S_y. Referring to Figure 4.29:

$$\text{the plastic moment } M_p = F \cdot r = S_y\left(\frac{A}{2}\right)r$$

$$= S_y\left(\frac{bh}{2}\right)\frac{h}{2} = S_y\frac{bh^2}{4}$$

*We ignore shear stresses for the purpose of simplifying the discussion.

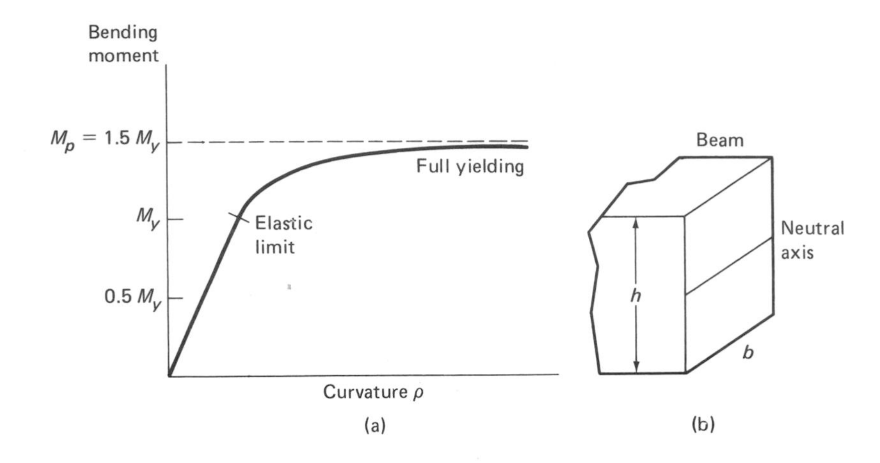

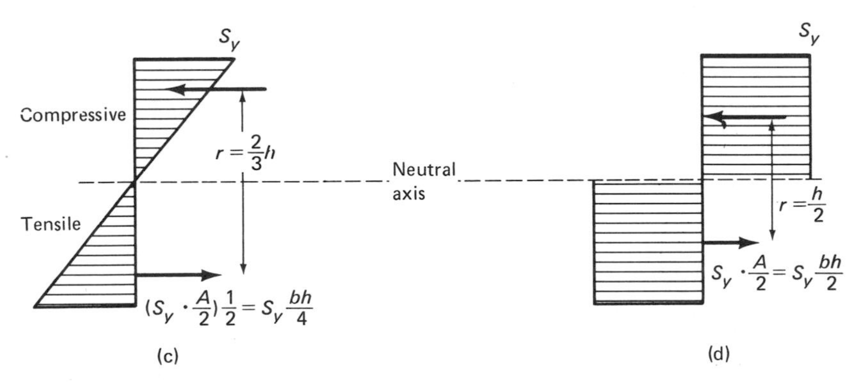

Figure 4.29. Elastic and plastic moments of a beam with a rectangular cross-section.

r = the distance between the center of tensile stress and the center of compressive stress,

b = width of the beam, and

h = depth (height) of the beam.

(a) Bending moment versus curvature.

(b) The beam section.

(c) Stress distribution in elastic design

$$M_y = S_y \frac{bh^2}{6}$$

(d) Stress distribution in plastic design

$$M_p = S_y \frac{bh^2}{4}$$

$$\text{for elastic bending, } S_y = \frac{M_y c}{I} = \frac{M_y\left(\frac{h}{2}\right)}{I}$$

$$\text{maximum elastic moment } M_y = S_y \frac{I}{(h/2)} = S_y \frac{bh^2}{6}$$

$$\text{shape factor} = \frac{M_p}{M_y} = 1.5 \qquad (4.11)$$

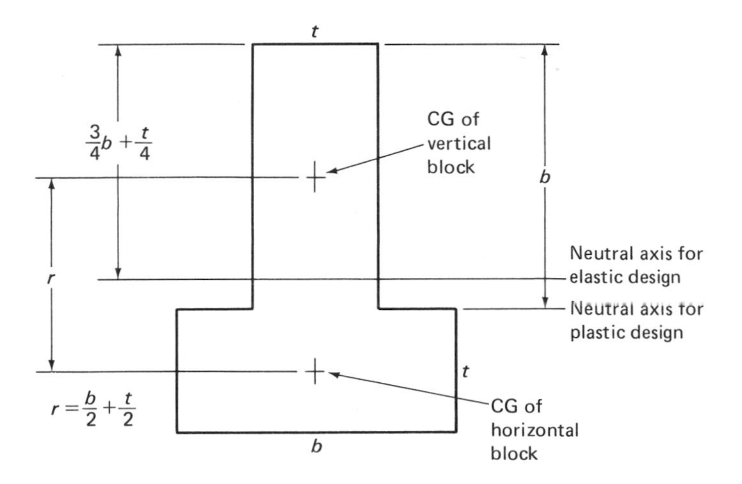

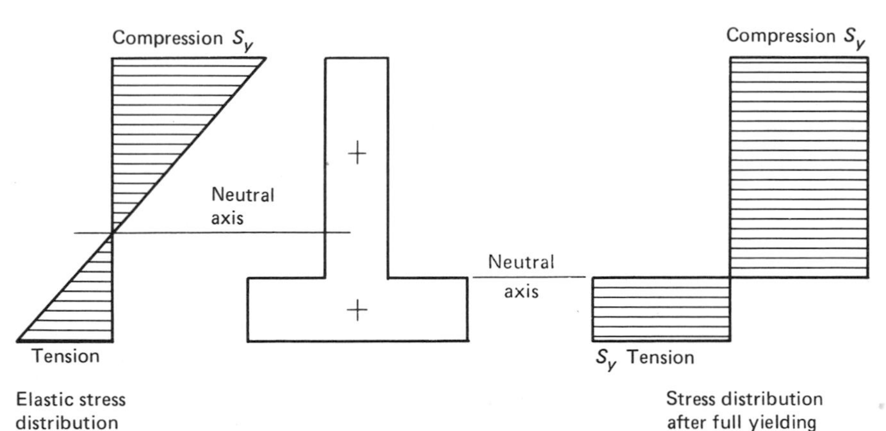

Figure 4.30. The neutral axis of a T-Section. The neutral axis shifts when an asymmetrical section undergoes plastic deformation.

TABLE 4.1. COMPARISON OF ELASTIC AND PLASTIC DESIGNS [46].*

		Failure criterion based on		Strength ratio $\frac{\text{plastic}}{\text{elastic}}$
		Elastic design	Plastic design	
Tensile bar with stress concentration		$F = \frac{S_y \cdot A}{K}$	$F = S_y \cdot A$	K
		M_y	M_p	Shape factor
Beams	Rectangular (b, h)	$M_y = S_y(\frac{bh^2}{6})$	$M_p = S_y(\frac{bh^2}{4})$	1.5
	Circular (R)	$M_y = S_y(0{\cdot}785R^3)$	$M_p = S_y(1.33R^3)$	1.7
	I-section	$M_y = S_y\frac{I}{C} = S_y(S)$ S = Section modulus	$M_p = S_y(Z)$ Z = Plastic modulus	$\frac{Z}{S} \approx 1.17$
	T-section (b, b) t is small	$M_y \approx S_y(\frac{5b^2t}{18})$	$M_p \approx S_y(\frac{b^2t}{2})$	≈ 1.8
	Hollow rect. (b, h) t is small	$M_y \approx t\left[bh + \frac{h^2}{3}\right]S_y$	$M_p \approx t\left[bh + \frac{h^2}{2}\right]S_y$	$\frac{bh + h^2/2}{bh + h^2/3}$
	Hollow circ. t is small	$M_y \approx S_y(\pi R^2 t)$	$M_p \approx S_y(4R^2 t)$	1.27

*Collapse safety factor = elastic safety factor × (shape factor).

Table 4.1 shows the elastic moments, the plastic moments, and the shape factors of several sections. Note that the extra allowance from plastic design depends on the uniformity of stress in the structure. The difference between initial yielding and complete collapse is large for sections that are less efficient. The neutral axis shifts after yielding, if the section is not symmetric about the horizontal axis (Figure 4.30). For elastic bending, the neutral axis passes through the center of gravity (CG) of the area. For complete plastic bending, the neutral axis bisects the sectional area. For the T-section, the line dividing the area in two halves is at the junction between the vertical and horizontal blocks. The plastic moment is one half the area times the yield stress times the moment arm r, which is the distance between the centers of gravity of the two half areas (Figure 4.30).

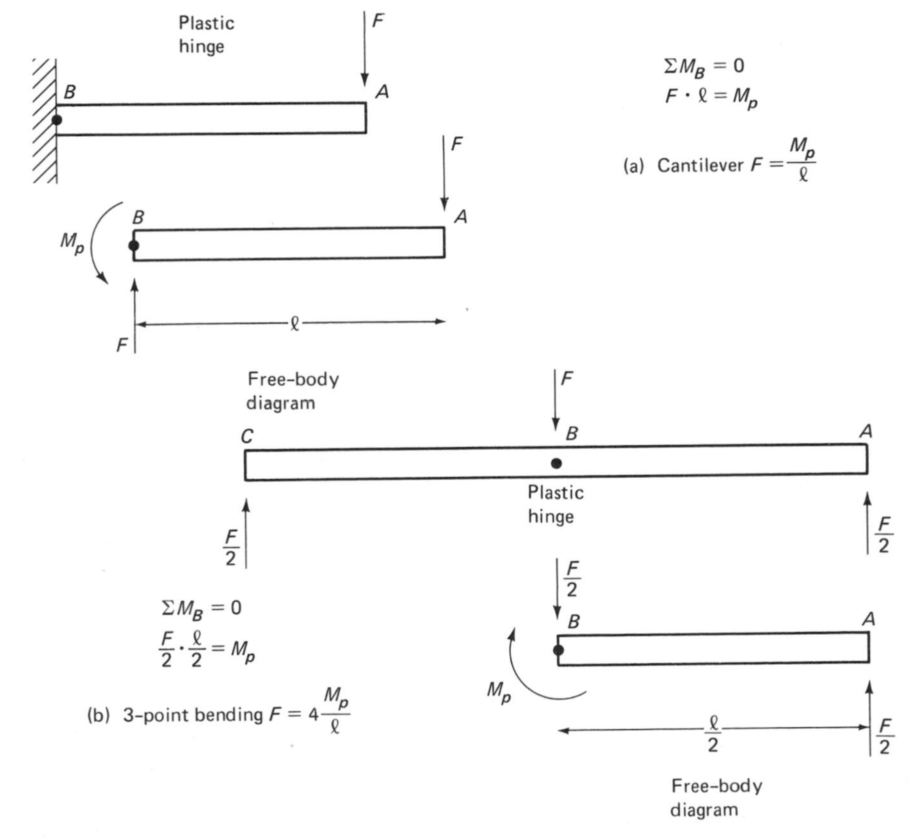

Figure 4.31. The maximum load F is expressed as a function of the plastic moment.

4.3.3. Plastic Hinge in a Beam

After knowing the plastic bending moment, we can further investigate the behavior of a beam. Once the plastic moment is reached, the beam will have an infinite curvature, and will basically rotate at a fixed point. Hence, this phenomenon is called a *hinge*. For the cantilever beam of Figure 4.31(a) the allowable force is the plastic moment divided by the beam length. For a beam loaded at the center and supported on both ends, the maximum moment is $Fl/4$ (Figure 4.31(b)). Thus, the maximum force is four times the plastic moment divided by the length.

The kinematic mechanism associated with plastic collapse calculations defines the locations of the hinges, the number of hinges, and the length of the moment arms. If there is more than one mode of collapse, the one associated with the lowest force or energy is realized. Figure 4.32 shows a centrally loaded beam with welded joint on the

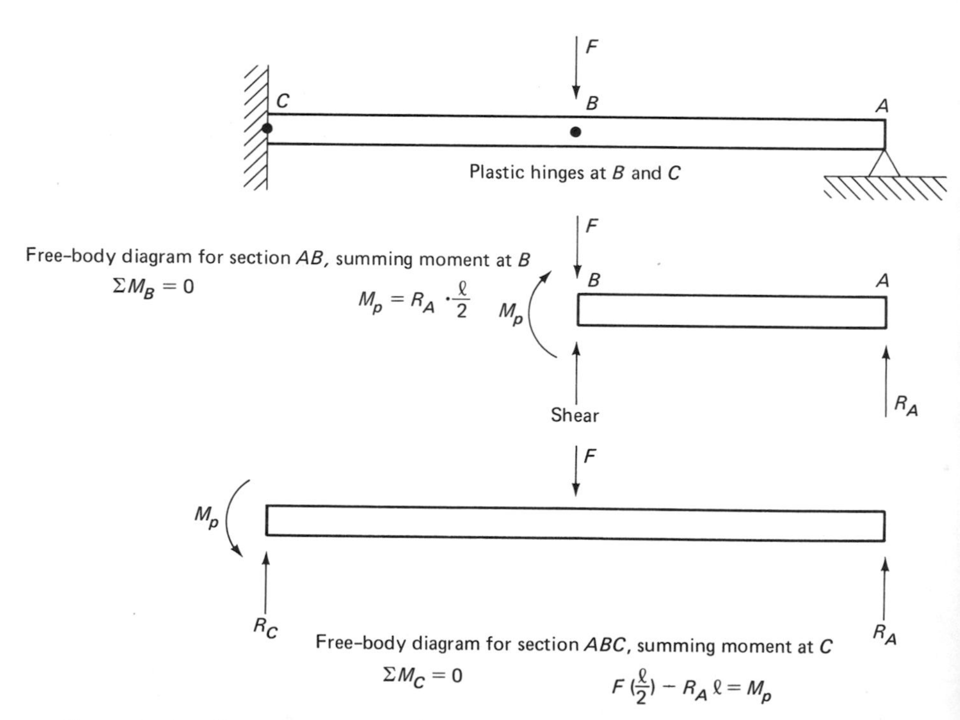

Figure 4.32. For a center-loaded beam with a welded support on one end and a pin-joint at the other end, collapse strength $F = 6\ M_p/l$.

left and a pin-joint on the right. Figure 4.33 shows a similar beam with welded joints on both ends. The collapse load increases as more and more plastic hinges are incorporated in the beam. To calculate the collapse load F, we can first draw a free-body diagram for section AB of Figure 4.32. Sum up all moments about point B:

$$R_A\left(\frac{l}{2}\right) - M_p = 0$$

Then we can draw the free-body diagram of the complete beam ABC and sum up all moments about C:

$$F\left(\frac{l}{2}\right) - R_A l - M_p = 0$$

$$F = \frac{2}{l}(M_p + R_A l) = 6\,\frac{M_p}{l} \tag{4.12}$$

Because of symmetry in Figure 4.33, only the free-body diagram of one half of the beam ABC is needed. The sum of all moments about point B is zero

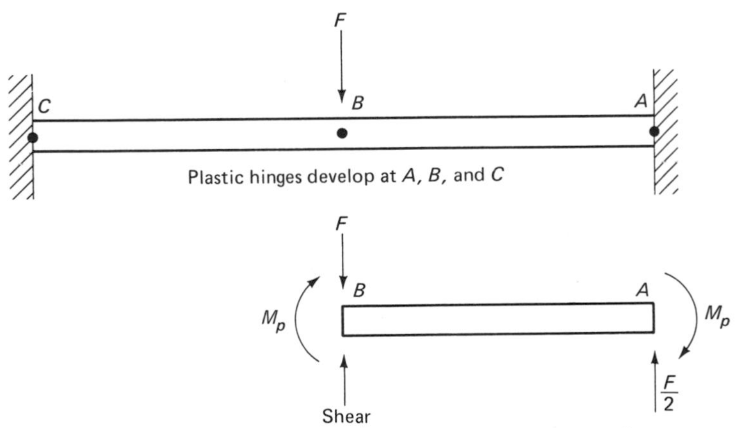

Because of symmetry, the reaction force at A is $\frac{1}{2}F$ and only section AB need be considered.

Free-body diagram for section AB, summing moment at B

$\Sigma M_B = 0$ $\qquad$ $\frac{F}{2}\left(\frac{\ell}{2}\right) = 2M_p$

Figure 4.33. For a center-loaded beam with welded supports at both ends, collapse strength $F = 8\,M_p/l$.

$$\frac{F}{2}\left(\frac{l}{2}\right) - 2M_p = 0$$

$$F = 8\,\frac{M_p}{l} \tag{4.13}$$

4.3.4. Discussions on Plastic Design

The basic assumption in these discussions is that the material is perfectly plastic. For any amount of strain larger than the elastic limit, the stress is equal to the yield stress (Figure 4.26). Real materials may work-harden, and the stress can get larger as the amount of plastic strain increases. This work-hardening will increase the collapse strength and make our calculations conservative. Real materials also will not satisfy the indefinite yielding of perfectly plastic material. Fracture may occur with real materials because of finite yielding.

The theory of plastic design predicts the collapse strength of a structure. This will not be applicable in situations where the structure collapses in buckling, or fails due to too large a deflection (not enough rigidity). In polymer materials, such as polystyrene, the material turns white when yielding occurs. This color change may be a reason for not allowing plastic deformations.

Plastic design is independent of stress concentrations and residual stresses. When the material is fully yielded, whether or not it started with any initial residual stress is immaterial. On the other hand, elastic design is dependent on stress concentrations and residual stresses. To prevent yielding, the sum of the residual stress and the stress due to the external load has to be less than the yield stress.

Figure 4.34 shows an example from Jones [45] that demonstrates the fallacy of elastic design. According to elastic design, the strength of the plain tube is twice that of the finned tube. The reason is that the depth of the section in the finned tube is larger (twice as large), while the moment of inertia remains unchanged (because the fins are too thin to contribute anything substantial)

$$\text{for elastic bending, } S_y = \frac{Mc}{I}, \quad M = S_y\,\frac{I}{c} \tag{4.14}$$

where c is one half the depth of the section.

$$\text{for the plain tube, } M = S_y \frac{I}{r}$$

$$\text{for the finned tube, } M = \frac{1}{2} S_y \frac{I}{r}$$

In reality, adding fins to a plain tube could not reduce the strength by a half. A plastic design theorem states: adding material to a structure cannot decrease its collapse strength.

Plastic design is especially useful in applications where occasional or minor bends and dents are allowed. Some temporary structures are blasted or destroyed after the intended use. Collapsible automobile and highway safety structures make use of the energy absorption capacity of plastic deformation. For shipping containers, dents from handling can be tolerated but designs must be strong enough to avoid tearing.

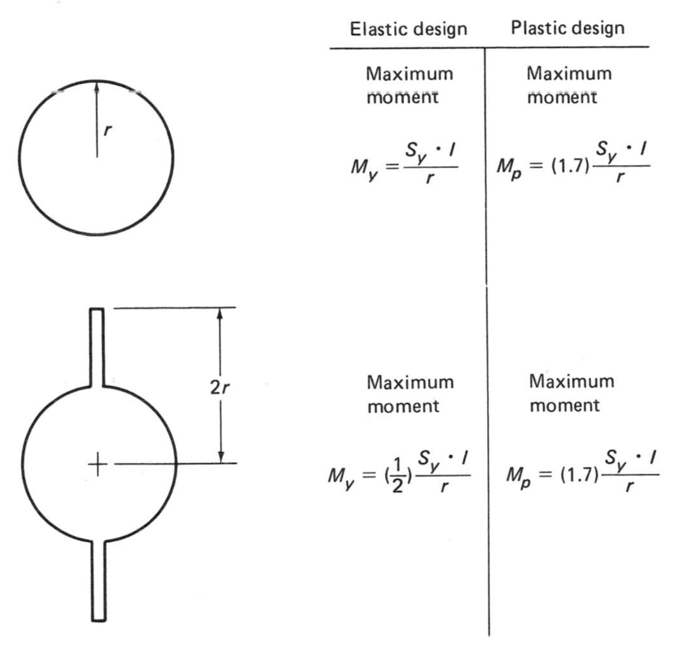

Figure 4.34. According to plastic design, the addition of a fin to a tube will not decrease its collapse strength, but elastic design calculations unrealistically indicate that the plain tube is twice as strong as the finned tube. The drawings show sectional views of a plain tube (a) and a finned tube (b). (Jones [45])

4.4. PRACTICAL CONCEPTS AND MODELS

Very rarely can a product be developed completely on an analytical basis. Judgment and the designer's personal preference add a touch of "style." In this section we present several practical concepts concerning strength and weight. Each concept should be viewed as a special interpretation or application of the general theories.

4.4.1. Carve-out Design

Carve-out means "shave material from where it is not needed." A good designer can see through the part and visualize the flow of forces. He can cut holes or thin down sections where there is little flow of force. Experimentally, if many samples of an existing product are available, holes can be cut and material can be removed in suspected low-stress areas. The original and modified samples will be tested to failure, and their strength-to-weight ratios will be compared. The one with the highest strength-to-weight ratio is the most efficient. This structure does not have to be the strongest. If the strength of the most efficient design is lower than the required strength, its shape should be retained and its sections can be scaled up.

In locating areas of low stress, the designer's intuition can be assisted by the rules set down from the Michell strain field theory. After all carving and destructive testing, the best design should contain very little waste material. Every section in the best design is loaded to its full capacity. Section A is as strong as section B, and section B is as strong as section C. If a large number of these "best designs" are tested to failure, there should be the same number of failures at sections A, B, and C. A balanced design is one in which at any point in the design, the strength is proportional to the stress. Conversely, the equal likelihood of all modes of failure means no excess strength in any section.

4.4.2. Reinforcement Design

Reinforcement means "patching-up" or "beefing-up" the weak spots. Through such an operation, a balanced design in which there are no weak spots, or every section has the same safety factor, can be obtained. Even though reinforcement can be achieved by adding webs of the same material, more often composite materials are used

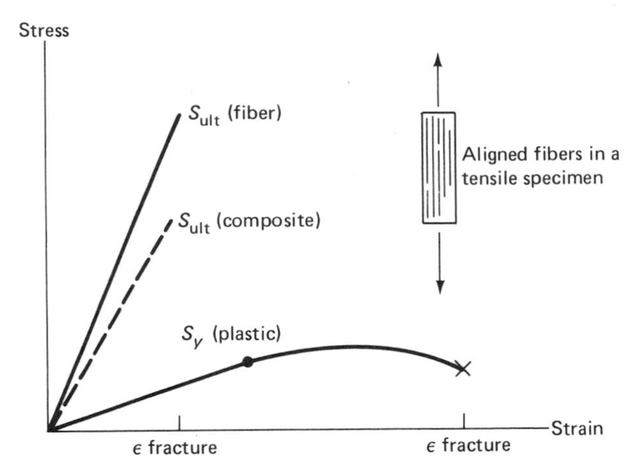

Figure 4.35. The behavior of a composite material in which the filaments are oriented in the direction of the tensile load.

to take advantage of the strengths of different materials. A material that is weak in tensile strength (e.g., concrete or plastics) may be strengthened by steel, glass, or carbon fibers. In order for the two materials to "cooperate," they must have a strong interface (bond together well) and a proper stiffness ratio. The stiffer material will carry most of the force. The reinforcing material should be stiffer than the reinforced material (the matrix). A typical relationship for fiber and plastic is shown in Figure 4.35. The uniaxial tensile strength

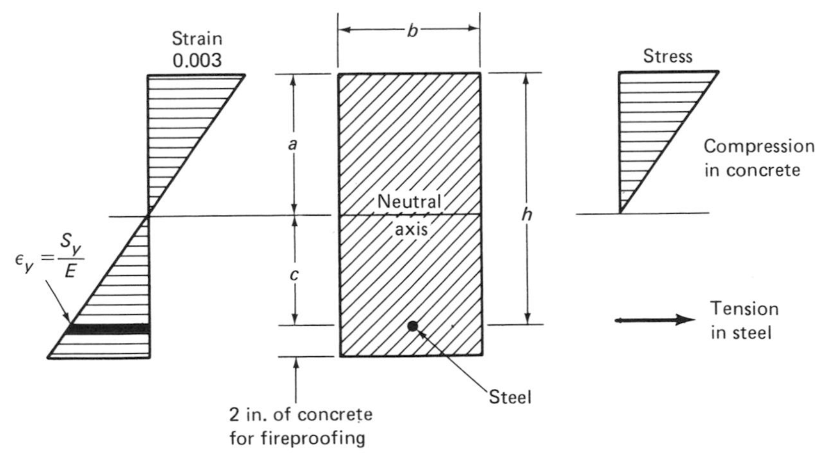

Figure 4.36. A concrete beam with steel reinforcement in areas of tensile stress.

and modulus are in between the values for the fiber and the plastic. The fracture strain is about the same as that for the fiber. The fiber-reinforced plastic has a smaller fracture strain than the plain plastic. The composite is stronger and more rigid than the matrix, but may not improve in impact strength. Fiber-reinforcement is well known for increasing resistance to fatigue and creep in plastics such as acetals. The two examples given below, reinforced concrete and the Captive Column, illustrate the concepts of reinforcement design.

Since concrete is weak in tension, steel bars are embedded at places of high tension to strengthen the concrete (Figure 4.36). Ridges are made on the surface of the steel bars to facilitate force transfer from concrete to steel by surface shear. Because of the weakness of concrete in tension, it is assumed that all the tensile force is carried by the steel bars. The sectional area of the concrete below the neutral axis is assumed to contribute nothing to the overall strength. The designs in Figure 4.37 are equally strong from the simplified calculations.

The location of the neutral axis can be determined by considering the strain. The top surface of the concrete would be strained to 0.003 at fracture. The steel bar would be strained to its yield point $\epsilon_y = S_y/E$; the actual strain in the bar depends on the strength of the steel. Since strain distribution is linear (Figure 4.36),

$$\frac{0.003}{a} = \frac{S_y}{E} \cdot \frac{1}{c} \qquad \text{(similar triangles)}$$

and $a + c = h$, the position of the neutral axis from the top surface of the beam is

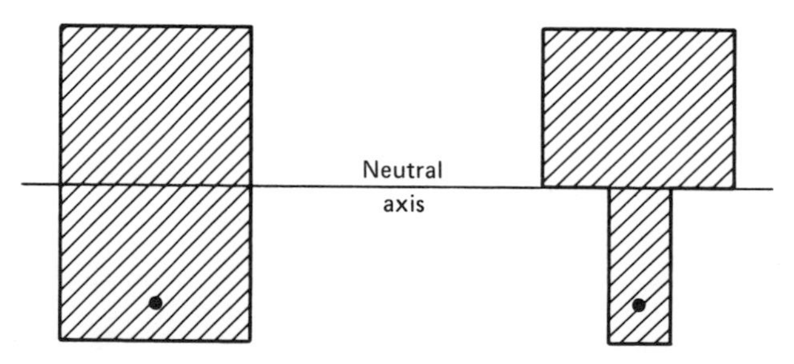

Figure 4.37. Sections of two reinforced concrete beams. These two designs are equally strong by the simplifying assumptions. Since concrete is very weak in tension, the area of concrete below the neutral axis is assumed to contribute nothing to the overall strength.

$$a = \frac{h}{1 + \frac{S_y}{E} \cdot \frac{1}{0.003}} \tag{4.15}$$

By the ultimate-strength theory, the compressive force in the concrete is replaced by a single force of magnitude $0.7225(S_{ult})ab$ acting at a distance $0.425a$ from the top surface of the beam (Figure 4.38). S_{ult} is the fracture strength of the concrete. To satisfy the force and moment equilibrium about the neutral axis: (1) the tensile force in the steel bar equals the compressive force in the concrete, and (2) the sum of the moments of the tensile and compressive forces about the neutral axis equals the external moment. A balanced failure is one in which the steel bars and the concrete fail simultaneously

$$\text{ultimate strength of concrete} = 0.7225(S_{ult})ab \tag{4.16}$$

$$\text{yield strength of steel} = S_y \cdot A \tag{4.17}$$

Equating 4.16 and 4.17:

$$\text{the sectional area of steel required} = A$$

$$= 0.7225\left(\frac{S_{ult}}{S_y}\right)ab \tag{4.18}$$

$$\text{the external moment at the time of failure} = S_y \cdot A(h - 0.425a) \tag{4.19}$$

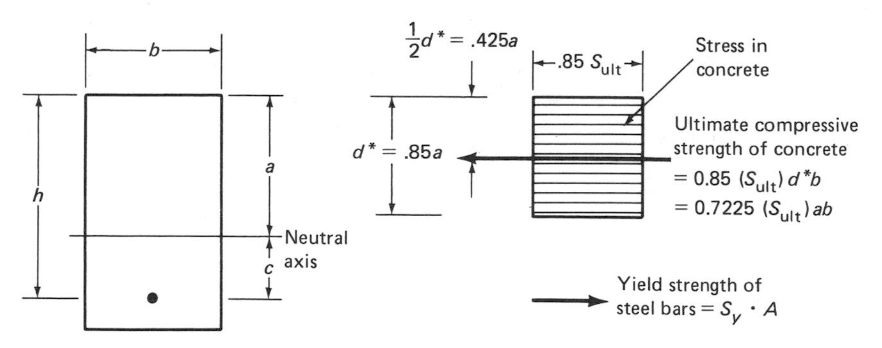

Figure 4.38. The strength of the reinforced beam and the sectional area of the steel bars can be calculated from the ultimate-strength theory.

*For high strength concrete ($S_{ult} > 4000$ psi), the value of d is reduced to $[0.85 - (S_{ult} - 4000)(0.00005)]a$.

In the actual design of concrete beams, there are two common and necessary practices that are used despite structural inefficiency. First, the steel bar is placed about 2 in. from the bottom of the beam to provide fire protection for the steel. Second, the building codes forbid a design with balanced failure because the failure of concrete is sudden and catastrophic, whereas the steel can stretch into the inelastic region and show signs of pending danger. The steel members are purposely made 25% weaker so that in case of an overload, signs of insufficient strength are observable before the total collapse. The concrete is "overstrengthened." Thus, a balanced design is sacrificed for reliability.

Figure 4.39 shows a "Captive Column," designed by Lawrence Bosch (U. S. Patent No. 3501,880) [48]. The three components of the column are each subjected to loads in their strongest directions: (1) the cores are subjected to radial compression; (2) the columns are subjected to axial tension or compression: and (3) the skin filaments are subjected to tension. The Captive Column is so named because it resembles a column driven into the ground. When a column is compressed from the sides (by the earth or by the skin filament),

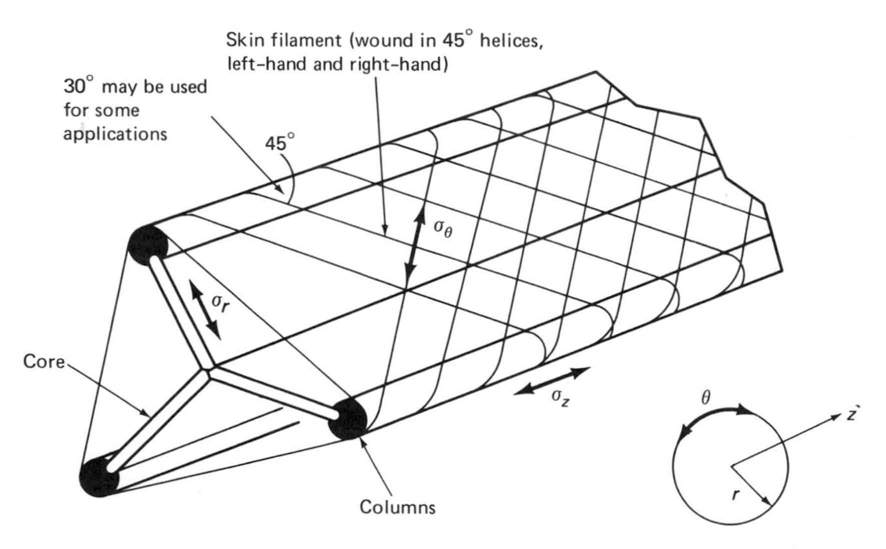

Figure 4.39. Lawrence Bosch's Captive Column. The captive column is strong because each component is subjected to stresses in its strongest direction: core, radial compression σ_r; column, axial tension or compression σ_z; and the skin filament, circumferential tension σ_θ.

buckling cannot occur, and its strength becomes extremely high. A captive column weighing 3.25 lbs/ft is equivalent in strength to a high-strength steel beam (A441 steel) that weighs approximately 14 lbs/ft.

4.4.3. Deflections and Forces

A determinate structure can be converted into a mechanism by reducing the number of members or constraints by one. Conversely, a linkage system, such as a folding chair or an umbrella, can be locked into a structure by adding one constraint (a latch). The question is where to place the locking device.

In a system where energy is not dissipated, the products of deflections and forces are constant (equal to the energy)

$$\text{energy in} = \text{energy out} \tag{4.20}$$

$$F_a \cdot \delta_a = F_b \cdot \delta_b = T_c \cdot \theta_c = T_d \cdot \theta_d$$

where a, b, c, and d are arbitrary points on the linkage system, F is the force in the direction of the deflection δ, and T is the torque in the direction of the angular deflection θ.

The product of velocity and force is also constant (equal to the power)

$$\text{power in} = \text{power out} \tag{4.21}$$

$$F_a \cdot V_a = F_b \cdot V_b = T_c \cdot \omega_c = T_d \cdot \omega_d$$

where V is the linear velocity, and ω is the angular velocity.

If an external force applied to a linkage causes deflections δ_a, δ_b,

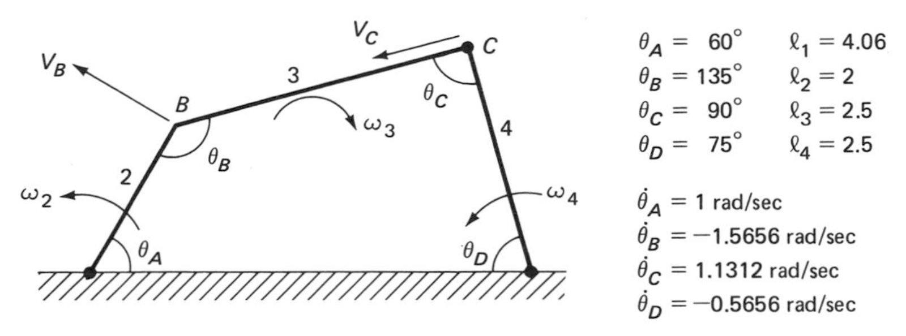

Figure 4.40. A locking device placed at joint B would freeze the linkage into a structure with the minimum amount of torque.

δ_c, and δ_d in the linkage at points a, b, c, and d, the reactions at these points needed to stop the movement are inversely proportional to the deflections. A locking device placed at the point of maximum deflection, δ, or angular deflection, θ, will have to bear the least force or torque. In the 4-bar linkage shown in Figure 4.40, the locking torque is minimum at joint B because $\dot{\theta}_B$ is largest.

In a drive train, it is advantageous to put a clutch at the high-speed shaft because the torque and the rating of the clutch will be minimum.

4.4.4. Paper Model of Plate Structures

Thin plates have high strength against tension (F_x), shear (F_y), and bending (M_z) in the plane of the material, but the strengths against torsion (M_x), lateral bending (M_y), and lateral force (F_z) are very poor (Figure 4.41). In the design of structures with plates, every effort should be made to subject the members to their inherent strengths. The following discussions and photographs are based on an article from *Machine Design* (March 6, 1975) by O. M. Blodgett [49].

Paper models can be used to show the relative deflections and stress distribution in steel plate structures. Though the Young's Modulus of paper is greatly different from that of steel, the stress dis-

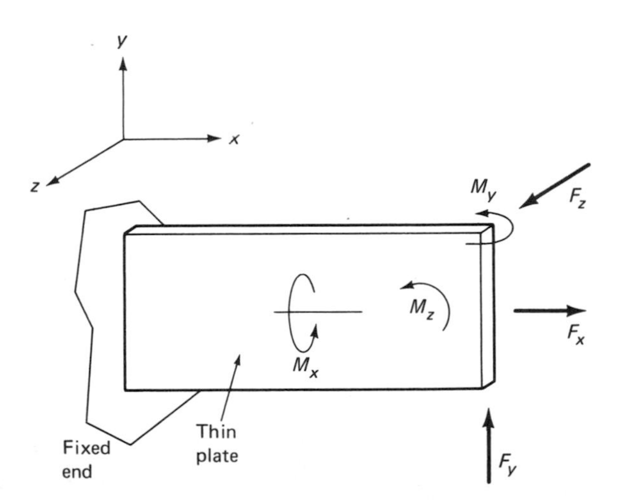

Figure 4.41. Strengths and weaknesses of thin plates. The thin plate is strong against tension F_x, in plane shear F_y, and in plane bending M_z. The thin plate is weak in torsion M_x, lateral bending M_y and lateral force F_z.

tribution is independent of Young's Modulus as long as the material behaves elastically. When two steel structures are modeled by paper, the ratio of deflections of paper models is the same as that of the pair of steel structures. Failure modes, weak spots, and stress concentrations can also be determined. Since paper models are inexpensive, such an experimental method is very useful. Deflections are especially easy to observe because of the flexibility of paper material.

It has long been observed that a box frame with diagonal braces (Figure 4.42(a)) has exceptional torsional rigidity. Figure 4.42(b) shows that torsion creates shear stresses along the horizontal and vertical lines on the surface of the cylinder. Maximum compres-

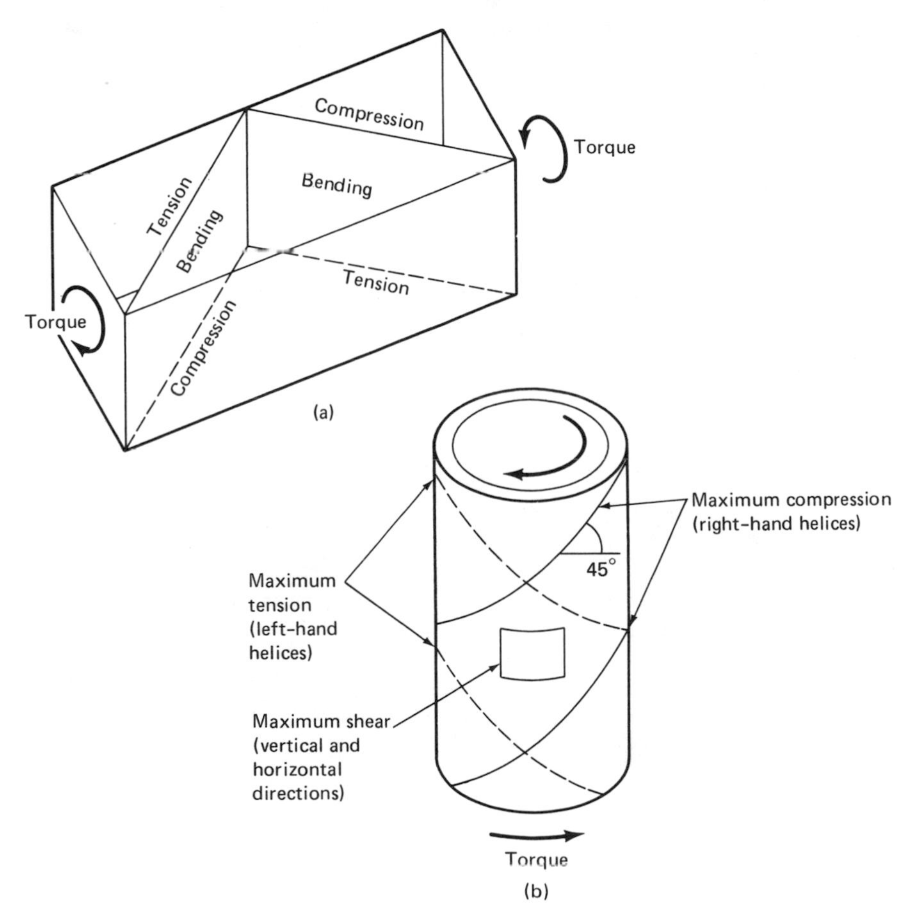

Figure 4.42. Structures under torsion. A box frame (a) and a tube (b).

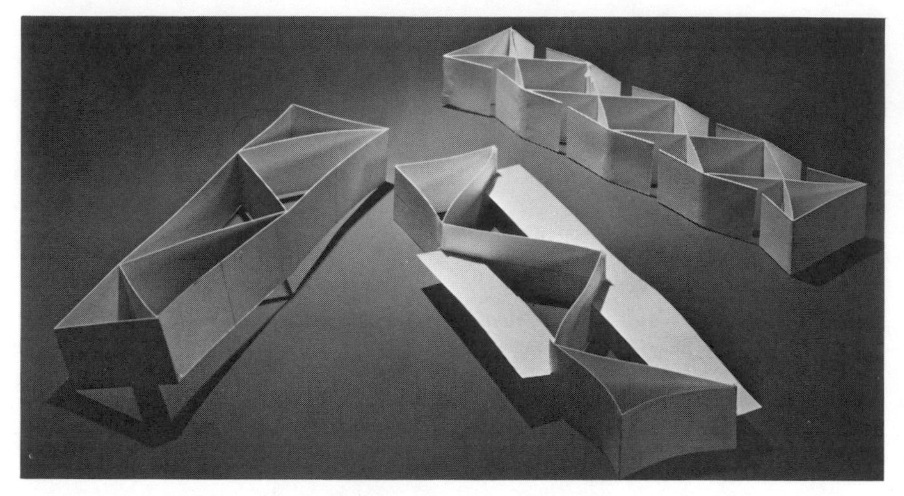

Figure 4.43. Several box frames modeled in cardboard. The loss of stiffness upon replacing or cutting a member indicates the member's contribution and function in the original structure. (Blodgett [49])

sive stresses occur along right-hand helices, and maximum tensile stresses occur along left-hand helices on the cylinder. When the braces in Figure 4.42(a) make a 45° angle with the sides, the braces are subjected to in-plane bending. The behavior of various members of the frames can be evaluated simply by cutting them or replacing them with more flexible members (Figure 4.43). When the side members of the single-diagonally-braced frame are replaced the

Figure 4.44. Cast-iron sliding beds studied by cardboard models. The structure on the left has diagonal bracing, and is very rigid under a torsional load. The structure on the right has parallel bracing, and is flexible when subjected to bending at 45° to the direction of the bracing. (Blodgett [49])

torsional and bending stiffness vanishes. For the double-diagonally-braced structure, cutting the side members does not change their torsional stiffness, but greatly decreases the bending stiffness and the axial tension stiffness. The loss of stiffness upon replacing a member indicates the member's contribution and function in the original structure. This method of testing is based on the superposition of parallel systems.

In Figure 4.44, a cast-iron sliding bed for a printing press is studied with the aid of cardboard models. The models verify that diagonal bracing (left) is superior to parallel bracing (parallel to sides) in torsional stiffness. Similar models can be constructed for the chassis of an automobile or a home trailer.

4.4.5. Soap Film Analogy of Torsional Shear Stresses

The stresses in a cylindrical member under a torsional load are easy to visualize (Figure 4.42(b)). It is a fortunate coincidence that most torsional members, such as shafts and pulleys, are, indeed, round in cross-section. If the section is noncircular, the stress distribution is so complex that most engineers lose all intuitive feel. Figure 4.45 shows the behavior of a split tube (roll pin) under torsion. A transverse plane that is initially flat becomes warped. The relative axial movement across the split can be explained by the fact that shear stresses in the axial direction are now relieved, leading to a shear movement. The membrane analogy by Prandtl (1903) can be used to

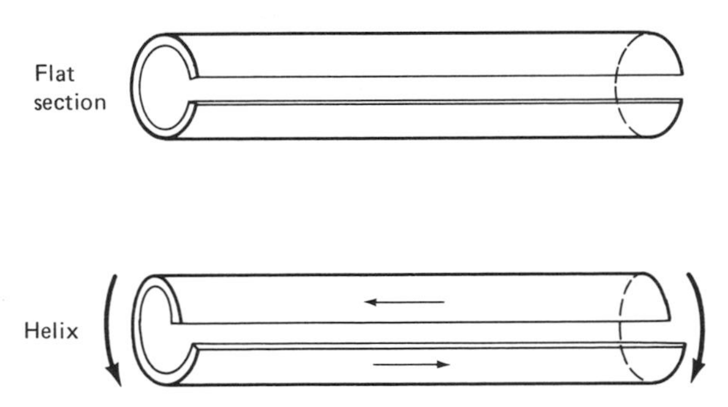

Figure 4.45. A split tube under torsion. There is axial movement across the split.

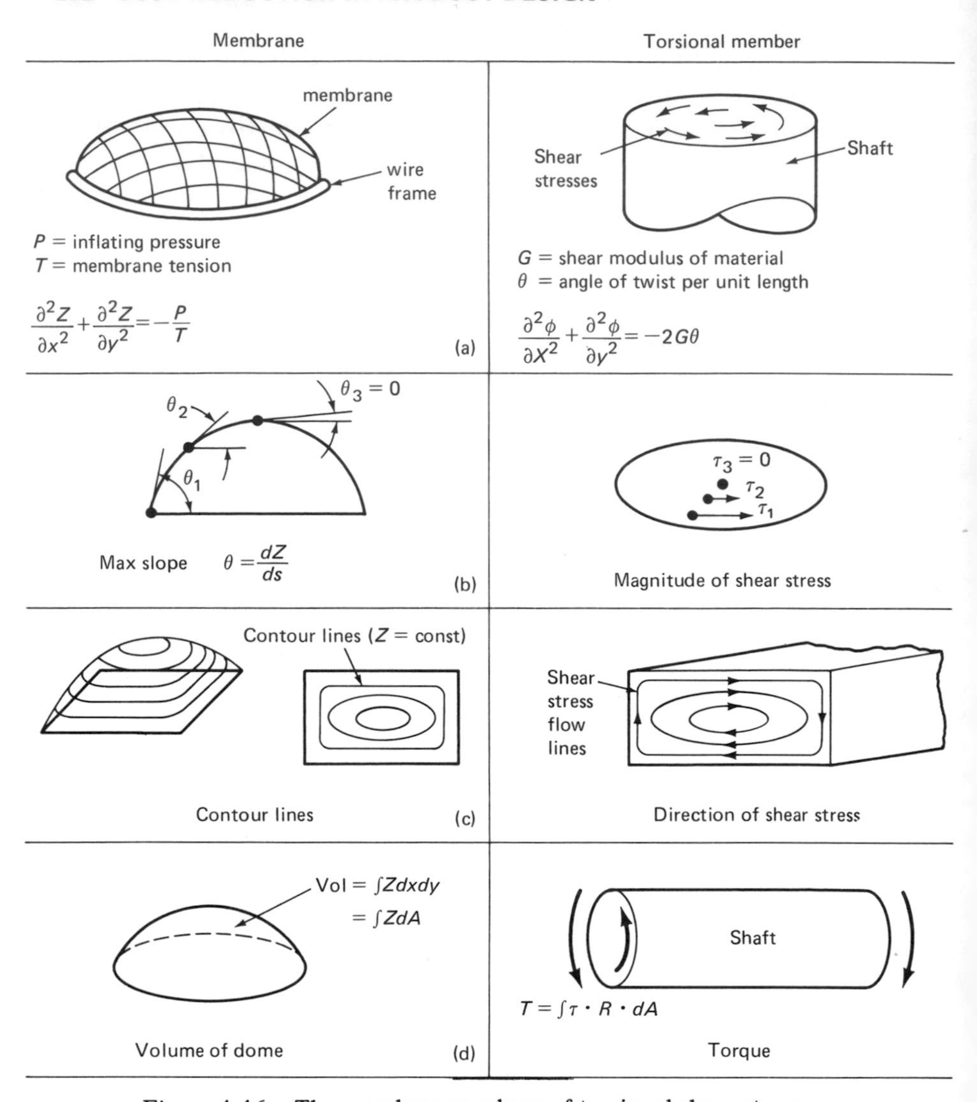

Figure 4.46. The membrane analogy of torsional shear stresses.

visualize the direction and distribution of the shear stresses, and the torsional strength of the member. Such understanding is essential for the design of noncylindrical torsional members. The discussion below is based on an article by H. W. Agrawal [50].

The analogy is based on the similarity of the stress function, ϕ, and the height function, Z, as shown in Figure 4.46(a). The cross-section of the member is replaced by a wire frame, which is an exact duplicate of the circumferential contour of the section. The wire frame is covered by a membrane. Pressure is applied to one side of the membrane so that a dome is formed. The maximum slope of the membrane at a point represents the shear stress at that point. Figure 4.47 shows that sharp protruding corners carry very little shear stresses whereas sharp concaving corners (recesses) carry high shear stresses. Sharp internal and external corners should be avoided. The center of a circle or the middle of a narrow web are locations of zero slope and zero stress. The direction of the shear stresses in the cross-section are contour lines (constant z lines) on the membrane. The

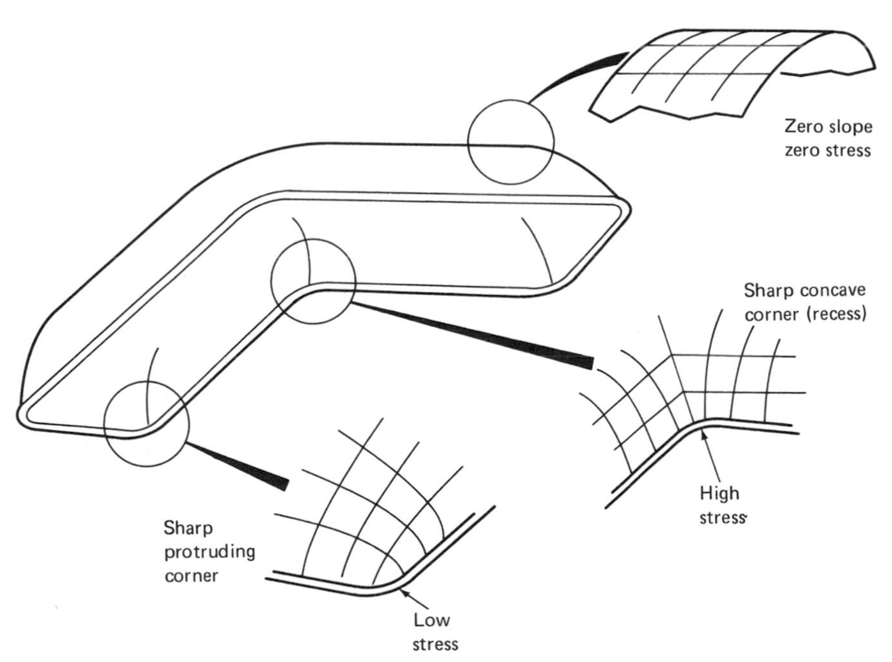

Figure 4.47. The shear stress at a point is proportional to the slope of the membrane [50].

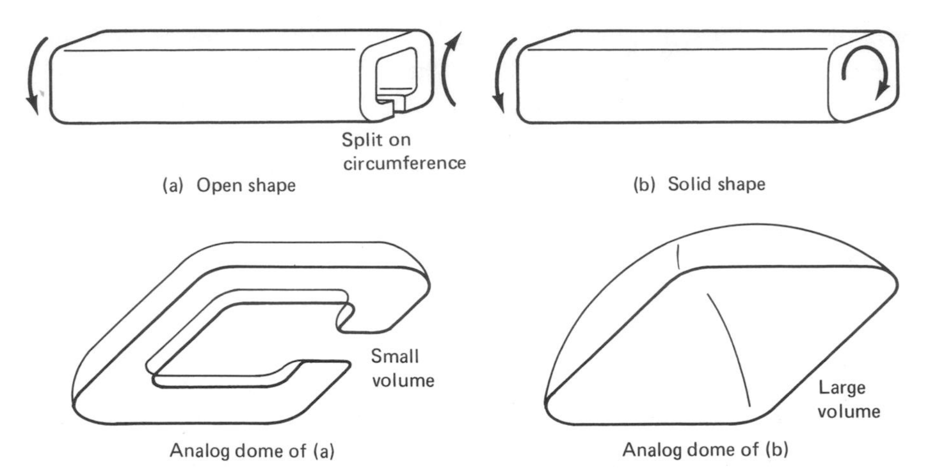

Figure 4.48. A section is greatly weakened by a split in the circumference (see text for details).

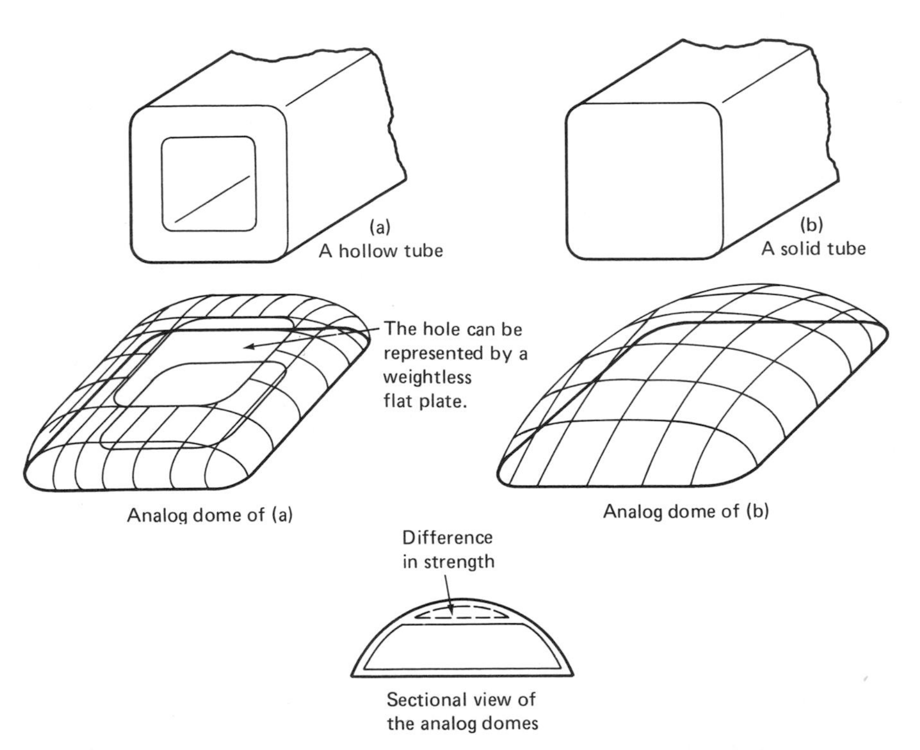

Figure 4.49. A hollow tube is slightly weaker than a solid tube. The strength-to-weight ratio of a hollow tube is higher than that of a solid tube.

volume under the membrane is proportional to the torsional strength of the member. Figure 4.48 shows that sections that are open (with a split on the circumference) are much weaker than sections that are closed: the volume under the dome increases greatly if the section is generally round or closed. Agrawal stated: "to increase strength, increase the volume of the analog bubble." If the section is an open shape, such as a thin blade, an L-section, an I-section, a T-section, a U-section, or a Z-section, the torsional strength is proportional to the product of the sectional area, A, and the web thickness, t. This implies that the shape of the section is rather unimportant.

If the section is hollow instead of solid, the loss in strength is very small (Figure 4.49). The internal hole can be represented by a thin weightless plate that is cemented to the membrane. Because of this plate, the hollowed area has a zero slope and no stress. The web in Figure 4.50(b) adds nothing to the torsional strength because it bisects an internal hollow space. The membrane model of the web is flat, and the volume under the dome remains constant regardless of the presence of the web.

The stiffness of a torsional member per unit length is the product of the shear modulus, G, and the polar moment of inertia, J, of the

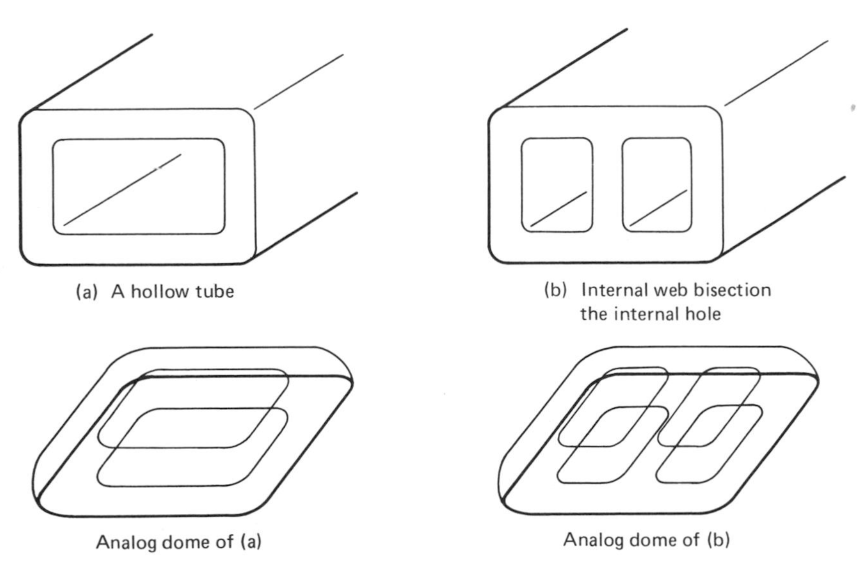

Figure 4.50. An internal web partitioning a symmetric shape adds nothing to the torsional strength. There is no shear stress developed in this web because the analog membrane is flat [50].

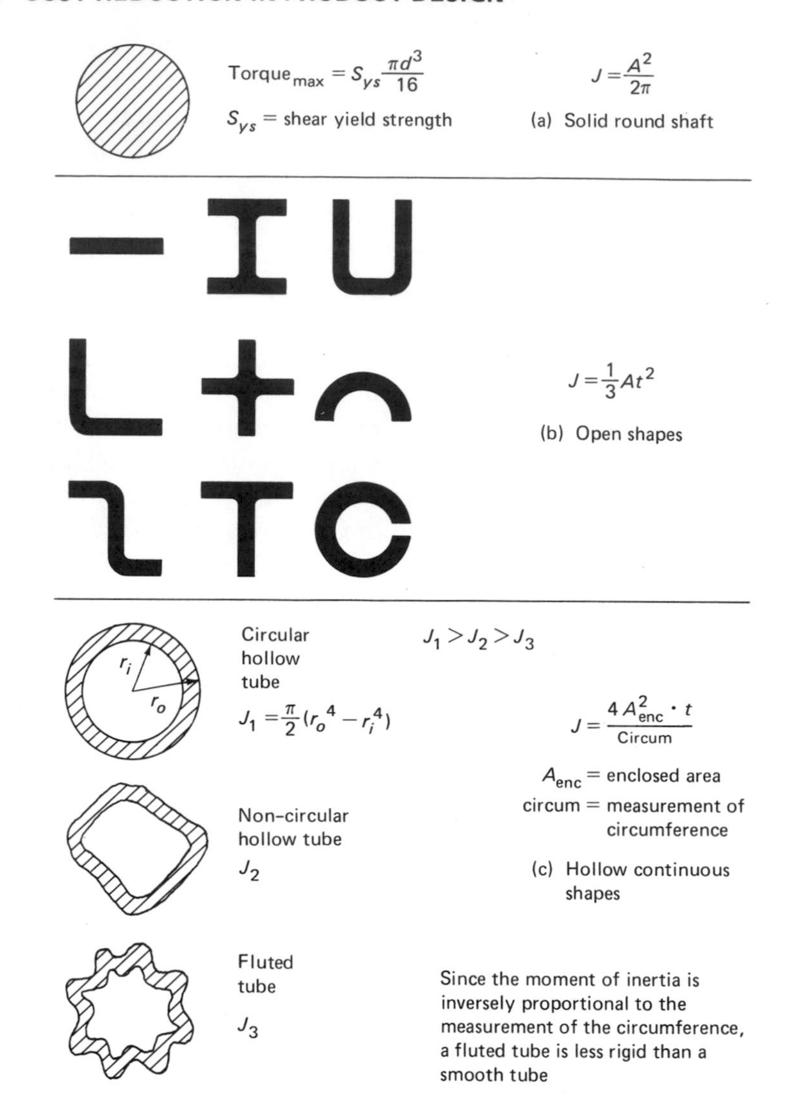

Figure 4.51. The polar moment of inertia, J, of several sections. Stiffness, $K = GJ/l$. Strength = $S_{ys} \cdot J/c$.

section. Several equivalent Js are shown in Figure 4.51 for different kinds of sections. For solid sections, rigidity is proportional to the square of the area. For open sections, rigidity is proportional to the area times the square of the web thickness. Again, the shape is unimportant. For hollow sections, the rigidity is proportional to the square of the enclosed area, but is inversely proportional to the measurement of the circumference.

4.4.6. Stress Concentration

Stress concentration points are major causes of inefficient designs. Minimizing such inefficiency can lead to material savings and strength improvements.

As discussed in Section 4.3.1, ductile materials tend to be insensitive to stress concentrations because local yielding can redistribute stresses. Brittle materials, on the other hand, are sensitive to stress concentration. Experimental determination of stress concentrations can be made by comparing the material's ultimate strength, S_{ult} to the nominal stress, σ_{nom} at fracture of the part (Figure 4.52). Plaster models of the part can be made, and the failure load can be measured. The effect of stress concentration on a particular geometry can be measured by comparing the failure load to a similar part without the stress raiser (Figure 4.53).

Qualitative evaluation of stress concentration can be made on a complex part by brittle-coating. After the part is coated it is sub-

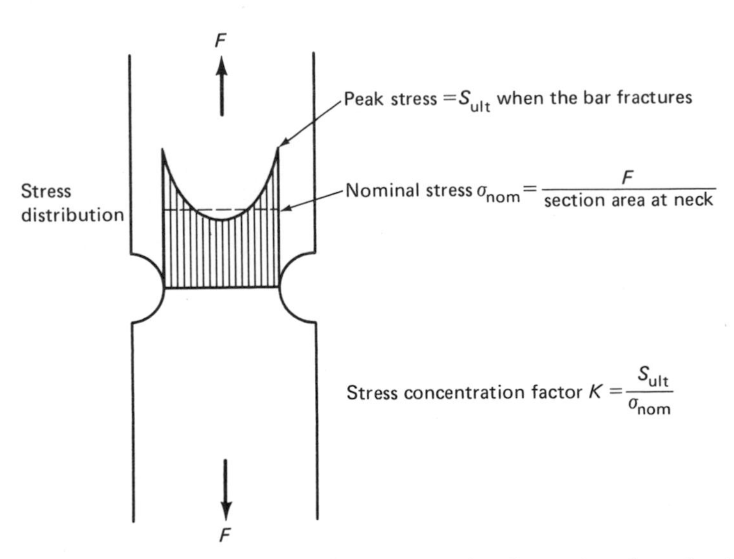

Figure 4.52. The stress concentration factor is the ratio of peak stress to nominal stress.

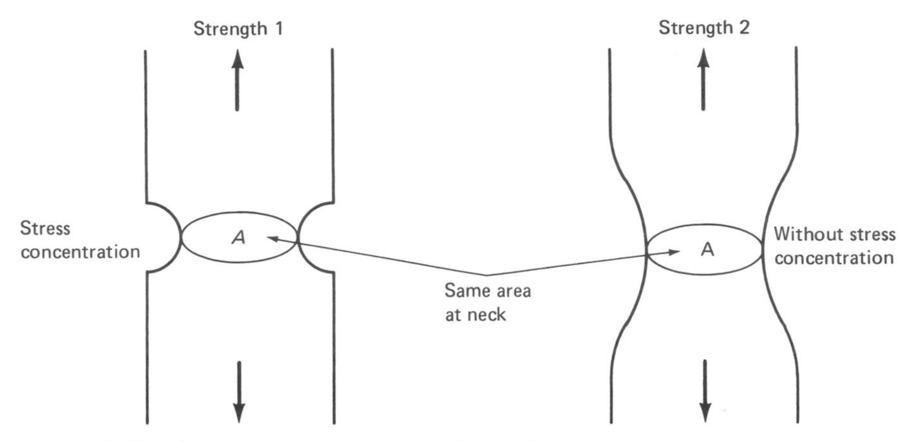

Figure 4.53. The stress concentration factor K is the ratio of strengths without and with the geometric discontinuity.

$$K = \text{strength } 2/\text{strength } 1$$

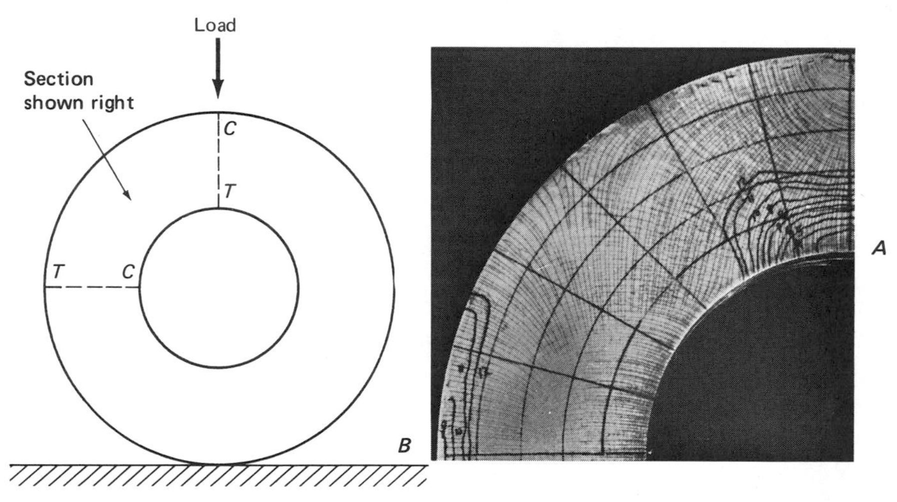

Figure 4.54. An aluminum ring under compression being studied by brittle-coating method. The photograph shows the upper-left quadrant of the test specimen. The ink-marked contours are boundaries of the crack areas as the load increases. A larger number is marked at a higher load level. The region with the highest tensile strain is on the inner radius directly under the load (Region A). Another region with tensile strain is on the outer boundary (Region B).
(From Introduction to the Theoretical and Experimental Anlaysis of Stress and Strain, by Durelli, Phillips, and Tsao. Copyright 1958. Used with permission of McGraw-Hill Book Co.)

jected to loads. When the tensile strain at a point exceeds a threshold value, the coating will show cracks. By gradually loading the member and marking the propagation of cracks (the boundary of the crack pattern), the stress gradient can be determined (Figure 4.54). The concentration of crack propagation contours for equal increments of load gives a quantitative indication of the stress concentration. The stress is obtained by multiplying the strain by the modulus of elasticity of the base material (e.g., steel).

Brittle-coating is also a good method to determine the direction and location of the principal strains, so that electrical strain gages can be placed for dynamic measurements.

Figure 4.47 illustrates the concentration of stress due to a sharp radius. The usual solution to stress concentration is to apply a generous radius (Figure 4.55). Round corners are cost-reduction features in themselves because less material is needed (volume-to-surface ratio is the largest in a sphere). The force-flow concept is often used to visualize stress concentration. The lines of force tend to concentrate at places with geometric discontinuities. Sometimes additional grooves and holes can divert the force-flow to a more favorable pattern (Figure 4.56). A design feature that involves abrupt geometric change should be located at areas of low stress. An identification number punched at critical locations might become a cause of failure. Inserts made of a low-modulus material can be used to cushion the stress transition between two members. Cement-joints and weld-joints are considered superior to bolt-joints and rivet-joints because they have larger areas of contact and a gradual transfer of stress across the joints. It is essential that a ductile and elastic weld material be used to allow for the relief of shrinkage stresses.

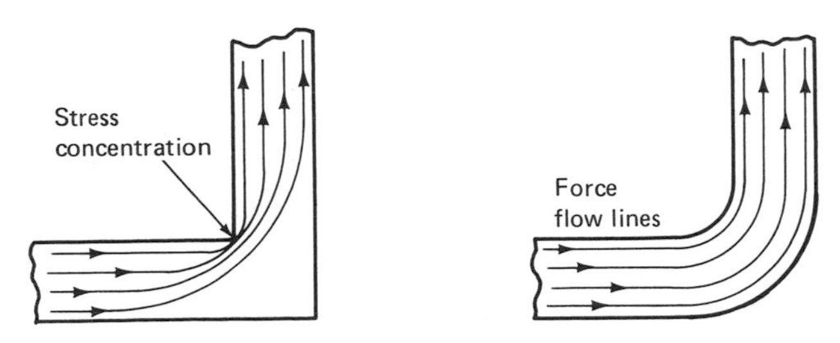

Figure 4.55. Stress concentration can be avoided by using a large radius.

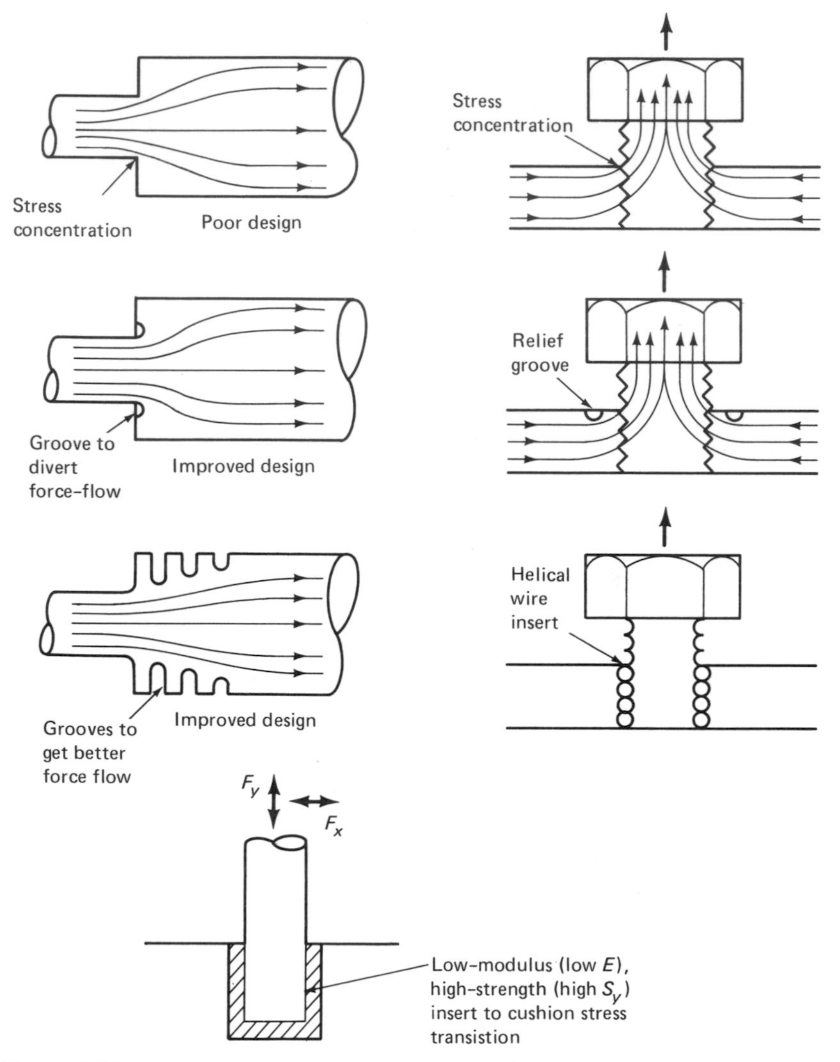

Figure 4.56. Stress concentration can be reduced by removing material to obtain a more favorable force-flow pattern.

4.5. SKINS, SANDWICHES, AND TEXTURED MATERIALS

Trusses, beams, and frames are components of skeleton structures. They resemble bones, with which vertebrates support loads. Insects do not have bones; they support loads with stiff skins. The fact that an ant can carry objects many times its weight shows that stiff-skin

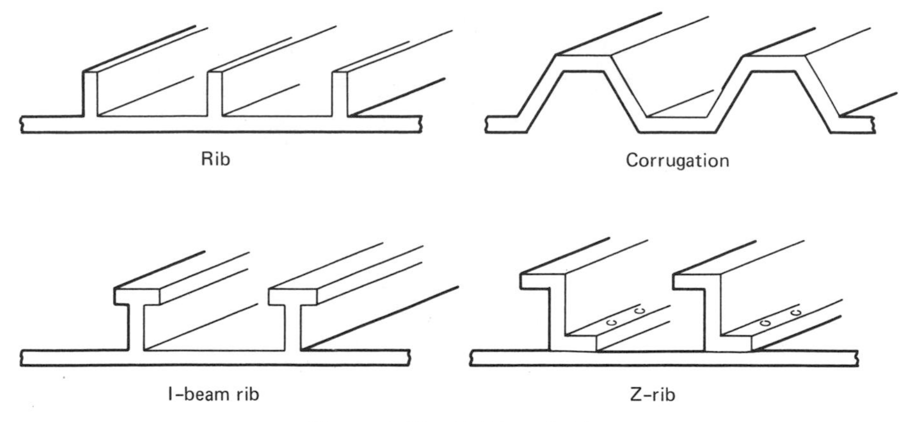

Figure 4.57. Stiffened skins.

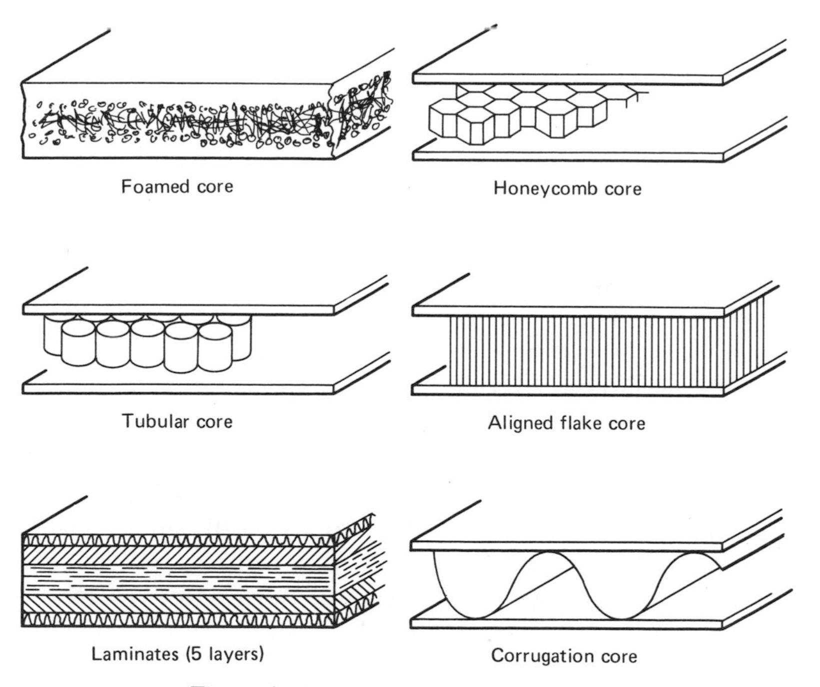

Figure 4.58. Sandwiches and laminates.

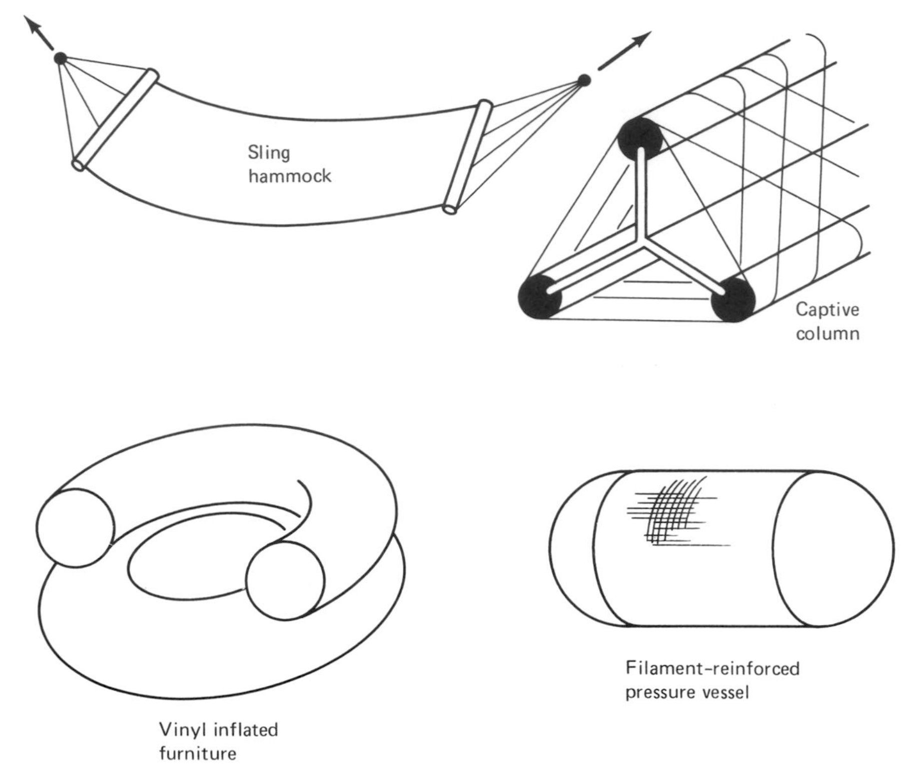

Figure 4.59. Tension skins.

structures can be very efficient. Bridges, buildings, cranes, and bicycles are examples of skeleton structures. Shells, pressure vessels, space vehicles, and ocean vessels are examples of stiff-skin structures. Manufacturing methods such as wire-forming are well adapted to producing skeleton structures. Blow-molding and rotation-molding produce stiff-skin structures. Four classes of materials are discussed in this section: (1) stiff skins (Figure 4.57), (2) sandwiches and laminates (Figure 4.58), (3) tension skins (Figure 4.59), and (4) textured materials (Figure 4.60). The efficiency of a structure depends on the type of load. Therefore, it is difficult to say what is best if the load is unspecified or unknown. As an example, for a space vehicle, there may be several "best" structural candidates, depending on the proposed set of external loads.

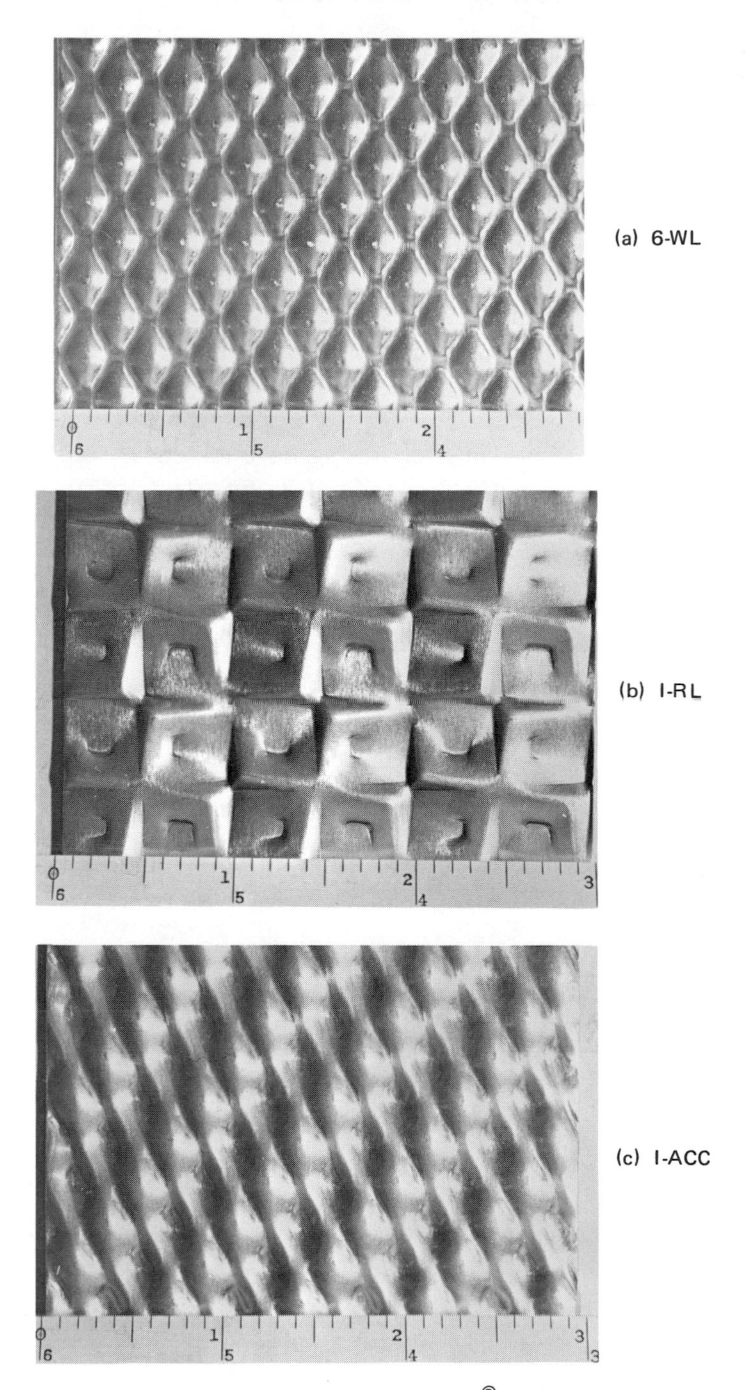

Figure 4.60. Several textured materials of Rigid-Tex,® made by Rigidized Metals Corp.

4.5.1. Stiff Skins

A hollow tube is more efficient than a solid tube in bending and torsion because it has larger moments of inertia about the x, y, and z axes (I_x, I_y, and J) for its weight. A stiff-skin structure has the additional advantage of surface protection against external impact and abuse. Stiffening is the cure for buckling, which is the common mode of failure of thin shells. Stiffening is achieved by adding ribs or corrugations. The ribs bear the lateral bendings and lateral forces (see Figure 4.41). The forces and moments in the plane of the skin are still carried mostly by the skin. The following is a detailed description of rib design and corrugation design. Such a design analysis is based on elastic bending, assuming no buckling and no plastic deformation. The strengthening and stiffening effects of ribs and corrugations are attributed to the increased distance of material from the neutral axis of bending.

Rib Design

The following analytical method was introduced by Lifshey [54]. Usually, the rib thickness, T, is less than the thickness of the wall, W, to avoid shrinkage marks (Figure 4.61). For low-shrinkage material,

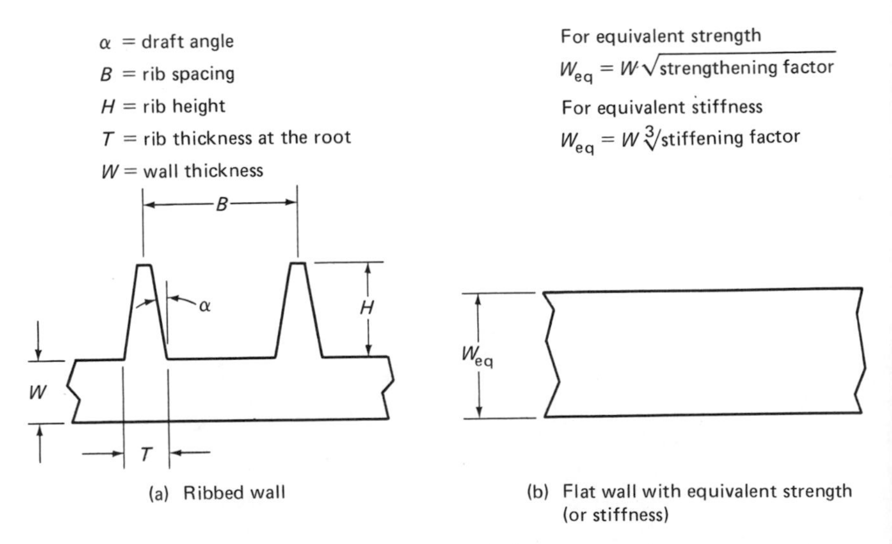

Figure 4.61. A ribbed wall and a flat wall of the same mechanical properties.

¾ W thickness and 1° draft is used for the ribs. For high-shrinkage material, a combination of ½ W thickness and ½° draft is more appropriate. Tables 4.2 and 4.3 show the strengthening and stiffening effects when rib spacing B and rib height H are known. The derivations and equations are given in Appendix 3.

The bending strength of a beam is proportional to its section modulus (I/c), where c is the distance from the neutral axis to the tip of the rib. The tip of the rib is the first point to yield because it is farthest from the neutral axis. If the part is made of a material that is weak in tension (e.g. plastic, cast iron, or glass), the skin should be subjected to tension and the ribs to compression. As an example, if the wall is a part of a pressurized housing, the ribs should be on the side with the higher pressure. The bending stiffness of a beam is proportional to its moment of inertia I.

The strengthening factor is the ratio of the strength before adding the ribs to the strength after adding the ribs:

$$\text{strengthening factor} = \frac{(I/c) \text{ with ribs}}{(I/c) \text{ without ribs}}. \tag{4.22}$$

The stiffening factor is the ratio of stiffnesses before and after adding the ribs:

$$\text{stiffening factor} = \frac{I \text{ with ribs}}{I \text{ without ribs}}. \tag{4.23}$$

Take an example from Table 4.2: when ¼-in.-high ribs are added to a 1/16-in. wall at a spacing of 5/16 in. per rib, the strength is increased 3.5 times and the stiffness is increased 26 times. A ribbed wall can be used to substitute for a thicker flat wall with savings in material. Since the section modulus (I/c) increases by the square of the wall thickness, a flat wall equivalent in strength to a ribbed wall (thickness W) would have a thickness W_{eq}:

$$W_{\text{eq. strength}} = W\sqrt{\text{strengthening factor}} \tag{4.24}$$

$$\% \text{ saving in material} = 1 - \frac{1 + \text{rib material}}{\sqrt{\text{strengthening factor}}}. \tag{4.25}$$

On an equal strength basis, replacing a 0.117-in.-thick unribbed wall with a ribbed wall would save 27% of material.

TABLE 4.2. PROPERTIES OF RIBBED WALLS–½° DRAFT ON BOTH SIDES OF RIB; THICKNESS OF RIB AT THE ROOT EQUALS ½ WALL THICKNESS.

$\frac{\text{rib height}}{\text{plate thickness}} = \frac{H}{W}$

$\frac{\text{rib spacing}}{\text{plate thickness}} = \frac{B}{W}$	1			2			4			6			8			10		
	str fctr[a]	stif fctr[b]	rib mat[c]	str fctr	stif fctr	rib mat	str fctr	stif fctr	rib mat	str fctr	stif fctr	rib mat	str fctr	stif fctr	rib mat	str fctr	stif fctr	rib mat
2	1.4	3.6	25%	2.9	12	48%	7.6	51	93%	14	128	134%	22	249	172%	30	418	206%
3	1	2.8	16%	2	8.8	32%	5.4	38	62%	10	98	90%	15	191	115%	21	321	138%
5	.77	2.2	10%	1.3	6	19%	3.5	26	37%	6.4	68	54%	10	134	69%	14	226	83%
7	.65	1.9	7%	1	4.8	14%	2.6	20	27%	4.7	53	38%	7.4	105	49%	10	177	59%
10	.55	1.6	5%	.79	3.7	10%	1.9	15	19%	3.4	40	27%	5.3	79	34%	7.5	135	41%
14	.49	1.4	4%	.62	3	7%	1.4	12	13%	2.5	30	19%	3.9	60	25%	5.5	102	29%
18	.46	1.3	3%	.53	2.6	5%	1.1	9.4	10%	2	24	15%	3.1	48	19%	4.3	83	23%
25	.42	1.2	2%	.44	2.1	4%	.83	7.2	7%	1.5	18	11%	2.3	36	14%	3.2	62	17%

[a]Strengthening factor = ratio of strengths with and without ribs.
[b]Stiffening factor = ratio of stiffnesses with and without ribs.
[c]Rib material = increase in weight after adding ribs.
See Appendix 3 for equations to calculate values in the table.

TABLE 4.3. PROPERTIES OF RIBBED WALLS–1° DRAFT ON BOTH SIDES OF RIB; THICKNESS OF RIB AT THE ROOT EQUALS ¾ WALL THICKNESS.*

$$\frac{\text{rib height}}{\text{plate thickness}} = \frac{H}{W}$$

$\frac{\text{rib spacing}}{\text{plate thickness}} = \frac{B}{W}$	1			2			4			6			8			10		
	str fctr	stif fctr	rib mat	str fctr	stif fctr	rib mat	str fctr	stif fctr	rib mat	str fctr	stif fctr	rib mat	str fctr	stif fctr	rib mat	str fctr	stif fctr	rib mat
2	1.8	4.6	37%	4	15	72%	10	64	136%	18	158	194%	28	304	244%	37	502	288%
3	1.4	3.6	24%	2.8	11	48%	7.4	49	91%	13	122	129%	20	234	163%	27	384	192%
5	.97	2.7	15%	1.8	8	29%	4.8	35	54%	8.6	87	77%	13	167	98%	17	273	115%
7	.79	2.2	10%	1.4	6.3	20%	3.5	27	39%	6.4	68	55%	9.7	132	70%	13	216	82%
10	.66	1.9	7%	1	4.9	14%	2.6	21	27%	4.7	52	39%	7	101	49%	9.6	166	58%
14	.57	1.6	5%	.82	3.9	10%	1.9	16	19%	3.4	40	28%	5.2	78	35%	7	128	41%
18	.52	1.5	4%	.68	3.3	8%	1.5	13	15%	2.7	32	22%	4.1	63	27%	5.6	105	32%
25	.46	1.4	3%	.55	2.7	6%	1.1	9.7	11%	2	24	15%	3.1	48	20%	4.1	79	23%

*The notations and equations are similar to those used for Table 4.2.

Similarly, savings in material can be realized by using ribbed plates instead of flat plates if stiffness is the main design criterion. The stiffness of a plate depends on I, which increases by the cube of the thickness. The flat wall equivalent in stiffness to the ribbed wall would have a thickness W_{eq}:

$$W_{\text{eq. stiffness}} = W \sqrt[3]{\text{stiffening factor}} \tag{4.26}$$

$$\% \text{ saving in material} = 1 - \frac{1 + \text{rib material}}{\sqrt[3]{\text{stiffening factor}}} \tag{4.27}$$

The strengthening and stiffening effects of the ribs are only for bending in the plane of the ribs. For two-directional bending, such as that in the wall of a fluid container, cross ribs are needed. The percentage saving in material then becomes:

equivalent strength % saving in material

$$= 1 - \frac{1 + 2(\text{rib material})}{\sqrt{\text{strengthening factor}}} \tag{4.28}$$

equivalent stiffness % saving in material

$$= 1 - \frac{1 + 2(\text{rib material})}{\sqrt[3]{\text{stiffening factor}}} \tag{4.29}$$

From Figure 4.44, we can see that even cross ribs are weak when bending is applied at a 45° angle to the ribs. The ribbing needed to produce a plate strong in all directions is the triangular isogrid (Figure 4.62). With this type of ribbing, the load is distributed among all ribs regardless of the direction of bending [55].

An interesting effect, shown in Tables 4.2 and 4.3, is that adding ribs sometimes decreases strength (strengthening factors often fall below 1 when $H/W < 2$ and $B/W > 5$). This happens because the increase in c outweighs the increase in I. Such a phenomenon was previously discussed and demonstrated in Figure 4.34. If inelastic deformation is allowed, strengthening factors will not be less than 1. The designer should also avoid using H and B values larger than 10 times W because buckling of thin ribs may occur. A design that is prone to buckling, but actually has no buckling failure is very efficient.

If a change in material is involved, such as replacing a zinc plate

Figure 4.62. The isogrid–king of light weight design. The triangular isogrid has a very high strength-to-weight ratio because every member contributes to the overall strength. The design is well balanced against all modes of failure. The diagram shows a fabricated cylindrical section suitable for aircraft and aerospace applications. (Courtesy General Dynamics, San Diego, Calif.; photo by Slysh.)

with a ribbed plastic plate, these factors should be multiplied by the appropriate ratios of material properties.

$$\text{Total strengthening factor} = \text{strengthening factor} \times \left(\frac{W}{W_{\text{original}}}\right)^2 \times \frac{S_y(\text{ribbed})}{S_y(\text{original})} \qquad (4.30)$$

For equivalent strength, the total strengthening factor should be equal to 1.

Total stiffening factor = stiffening factor

$$\times \left(\frac{W}{W_{\text{original}}}\right)^3 \times \frac{E_{\text{ribbed}}}{E_{\text{original}}} \tag{4.31}$$

For equivalent stiffness, the total stiffening factor should be equal to 1.

Besides the straight-sided ribs, I-beam ribs can further increase the strength-to-weight ratio. I-beam ribs can be produced by machining or investment casting. One-directional I-beam ribs can be produced by extrusion. Integral ribs can be formed with a bead at the tip of the ribs.

Corrugation Design

Corrugating is a method for strengthening and stiffening, in which a flat sheet is curved into waves to increase the moment of inertia, I. Figure 4.63 shows corrugations obtained by molding, drawing (pressing), and rolling (bending). For the drawn corrugation, the web is thinner than the crest because the material stretches during drawing. The sheet maintains the same overall dimensions before and after the corrugating. The rolled corrugation causes reduction in the overall length of the sheet, but maintains a uniform thickness at the web. A drawn-corrugated sheet has a higher strength-to-weight ratio than the rolled type because of the thinner webs. The corrugated material is stronger and stiffer only in the direction of the corrugation.

An efficient method of analyzing corrugation was given by E. A. Phillips [56]. His analysis was based on a drawn corrugation. A corrugation that is symmetrical about the neutral axis is stronger than one that is not, because equal amounts of material are placed on both sides of the neutral axis. Asymmetrical corrugations have one side weaker than the other.

For the shapes shown in Figure 4.64(a), the following equations give the strengthening and stiffening factors:

$$\text{strengthening factor} = \frac{H^2(1 + 4F/B) + T^2}{T(H + T)} \tag{4.32}$$

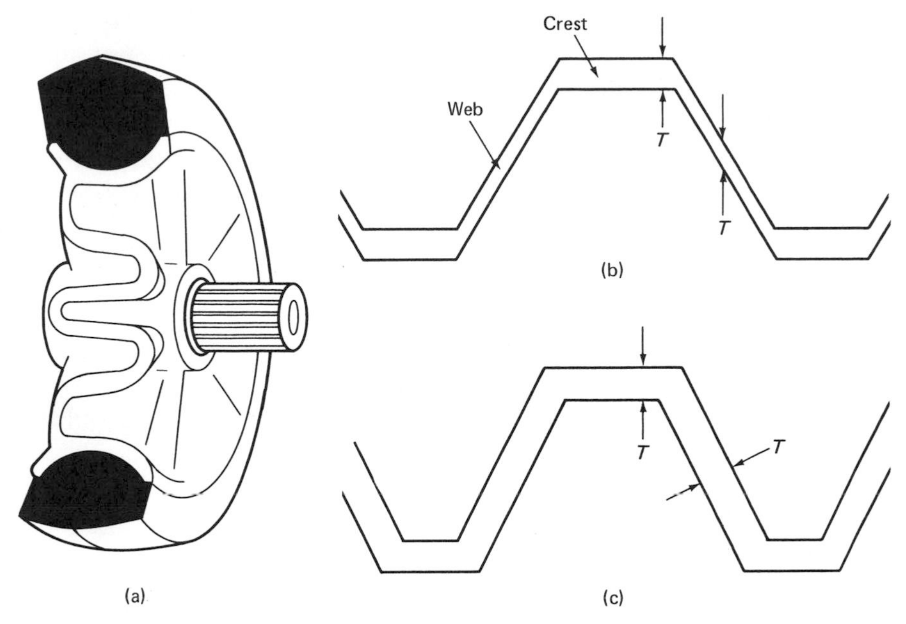

Figure 4.63. Corrugation can be manufactured by molding, drawing, and rolling. (a) shows a wheel hub molded from high-density polyethylene by Wagner Products Corp., Hustisford, Wis. This wheel has replaced spot-welded steel wheels for motorized mowers. (b) shows corrugation by drawing (pressing), and (c) shows corrugation by rolling or molding.

$$\text{stiffening factor} = \left(\frac{H}{T}\right)^2 \left(1 + 4\frac{F}{B}\right) + 1 \qquad (4.33)$$

where

H = wave amplitude,
B = length of wave,
T = material thickness at flange, and vertical measurement of material thickness at web, and
F = length of crest.

The derivation of these equations is given in Appendix 4. The method is a good approximation for corrugation shapes that have uniform thicknesses and round crests (peaks of waves). The square wave is the strongest shape, and a trapezoidal wave approximating a square is the second strongest. The efficiency of these shapes results from the increased material at the crest: the largest amount of material is farthest away from the neutral axis. The above formulas are not applicable to a square wave. A square wave cannot be made by drawing because the web thickness would approach zero when the web is close to vertical. The following equations apply to a square wave of uniform wall thickness (molded or rolled) as shown in Figure 4.64(b):

$$\text{strengthening factor} = \frac{2T}{(H+T)}\left[\frac{F}{B} + 3\frac{F}{B}\left(\frac{H}{T}\right)^2 + \frac{(H-T)^3}{BT^2}\right] \tag{4.34}$$

$$\text{stiffening factor} = \frac{2F}{B} + 6\frac{F}{B}\left(\frac{H}{T}\right)^2 + \frac{2(H-T)^3}{BT^2} \tag{4.35}$$

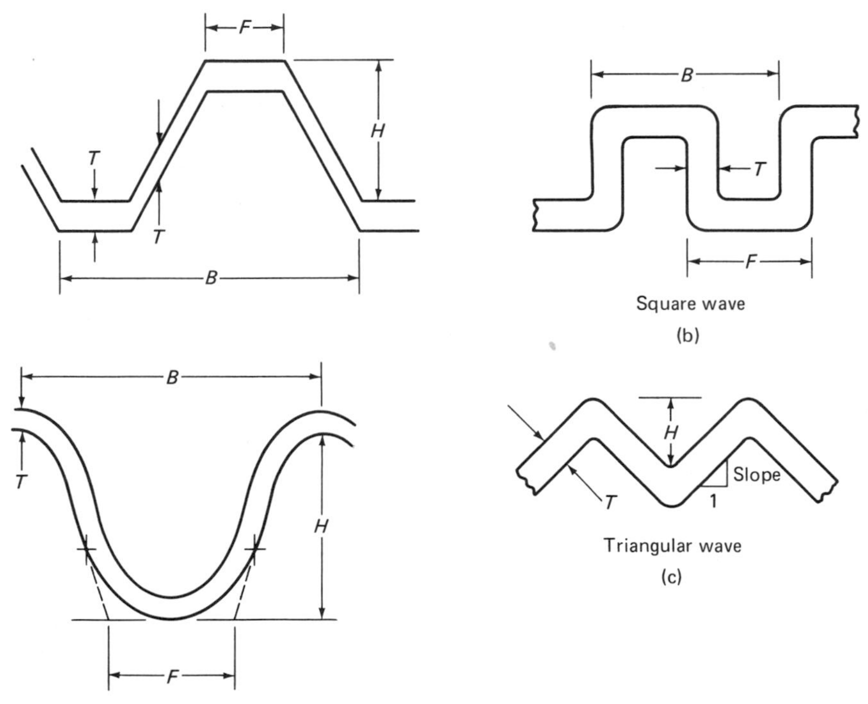

Figure 4.64. The characteristic dimensions for several common corrugations.

A triangular wave is less efficient than both the square wave and the trapezoidal wave. The strength of a triangular wave can be increased by increasing the wave amplitude H. For a triangular wave (Figure 4.64(c)) with T measured in the vertical direction, the factors given by Equations 4.32 and 4.33 are accurate. However, if T is the thickness of the material, a correction factor is needed

$$\text{strengthening factor} = \frac{H^2 + T^2}{T(H+T)}(\text{slope}^2 + 1) \tag{4.36}$$

$$\text{stiffening factor} = \left[\left(\frac{H}{T}\right)^2 + 1\right] \cdot (\text{slope}^2 + 1)^{1.5} \tag{4.37}$$

4.5.2. Sandwiches and Laminates

Laminates are made to take advantage of a combination of material properties. The following may be obtained from lamination: (1) a higher strength-to-weight ratio, (2) corrosion resistance, (3) fireproofing, (4) reflective and decorative surfaces, (5) sound-damping, (6) friction properties, (7) thermal conductance or resistance, and (8) electrical properties. Laminates of plastics and glass produce shatterproof glass. Laminates of plastics on photographs give scratch protection. The directional properties of wood are eliminated in plywood. Stainless steel pots and pans are laminated with copper to increase conductivity. Aluminum pans are coated with Teflon to reduce food sticking. Cast-iron bathtubs are coated to improve appearance, prevent rusting, and decrease heat conduction (so they will not feel chilly).

A general discussion of laminates is impossible because of the diversity of laminated structures. The thermal and mechanical properties of different layers are very important. Unequal thermal expansion can produce stresses that deflect a bimetallic laminate, although this property is useful for thermometers and temperature controls. Due to the different mechanical properties, high shear stresses may develop between the layers and cause delamination. The bonding between two layers should be strong enough to prevent delamination (shear failure of glue joints). Plywoods are made with an odd number of layers, so that shrinkage on side layers will balance at the center layer.

A sandwich is a three-layer laminate consisting of two strong face-plates and a relatively weaker, lighter core. Structurally, a sandwich is like an I-beam: the face-plates carry most of the bending stresses and the core takes the bending shear and supports the face-plates against crushing caused by loads normal to the faces. Several types of sandwiches are shown in Figure 4.58. Paper and cardboard are used as face-plates, with gypsum or foam plastic as the core. Aluminum faces are used with wood or plastic cores.

Honeycomb is a core material that is very successful in aircraft and aerospace applications. It is now applied to trucks, marine tanks, housing, sports equipment, and chassis for electronic assemblies. The core material is usually aluminum or plastic-treated paper. Honeycomb is first made as a stack of bonded sheets, then expanded into a cellular form, and finally the face plates are bonded to both sides of the core.

An interesting expanded-core process was developed by W. Smarook of Union Carbide about 1970 [57]. A common form of the material (NorCore*) is shown in Figure 4.65. A sheet of plastic about ⅛-in. thick is heated until it is sticky. The plastic is compressed slightly

*Trademark of Norfield Corp., Danbury, Connecticut.

Figure 4.65. Two samples of expanded thermoplastic core material, NorCore. (Courtesy Norfield Corp., Danbury, Conn.)

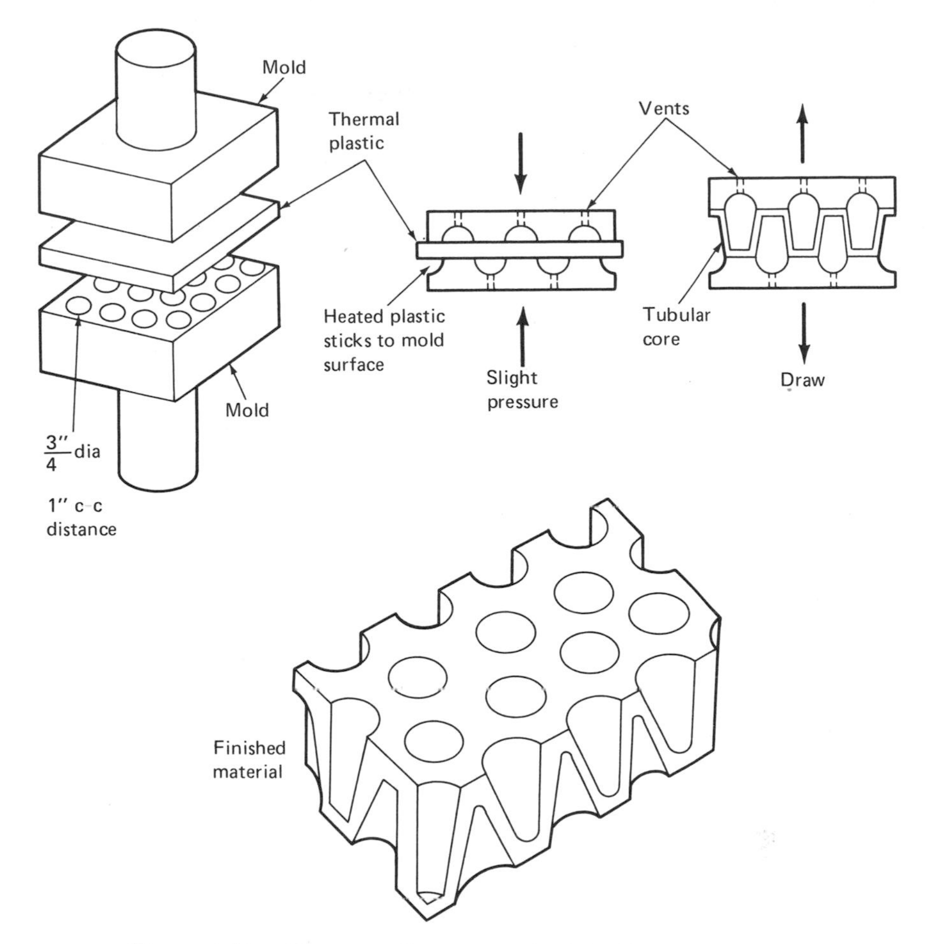

Figure 4.66. The expanding process for NorCore material.

between the hot mold plates, and the press is drawn apart (Figure 4.66). The plastic is drawn into two faces connected by tubular cores. Usually, the holes (¾-in. dia.) are staggered so that the core tubes are conical. Another version uses a patterned plate to form hexagonal vertical core. Drawing a typical sheet of ⅛-in. thick plastic to over 1 in. thick stiffens it by 100 times. There are numerous applications of the NorCore material including highway safety impact structures, cabinets, watercraft, etc. NorCore can be made transparent, which opens up applications as light diffusers, signs, and stage decorations.

4.5.3. Inflatable Structures

An inflated structure is a tension skin enclosing a volume of slightly compressed fluid (usually air). Provided that the skin is strong enough, the stiffness and the collapse strength of the structure increase with higher inflating pressures. The configuration of the structure is in a state such that the internal volume is maximum for the amount of skin area. Because of this, the structures generally have spherical surfaces and rounded edges and corners. Examples of inflated structures are blimps, life rafts, and automobile safety air bags. To understand the characteristics of an inflated structure, we have to investigate the properties of membranes.

A membrane is a flexible sheet of material that has only tension strength in the plane of the material. It has neither compression strength nor bending strength. Because a membrane is very thin, stresses are measured as force per unit length (instead of the standard practice of force per unit sectional area). Figure 4.67 shows stresses revealed by taking three sections of some standard geometries: (1) stresses in spherical surfaces, (2) circumferential stresses in a cylinder, and (3) axial stresses in a cylinder. The membrane stresses revealed by these sections are $rP/2$, rP, and $rP/2$, where P is the inflation pressure and r is the radius of curvature of the surface. Because the surface tension of a soap film is the same regardless of the area or thickness, the pressure in a soap bubble is inversely proportional to the

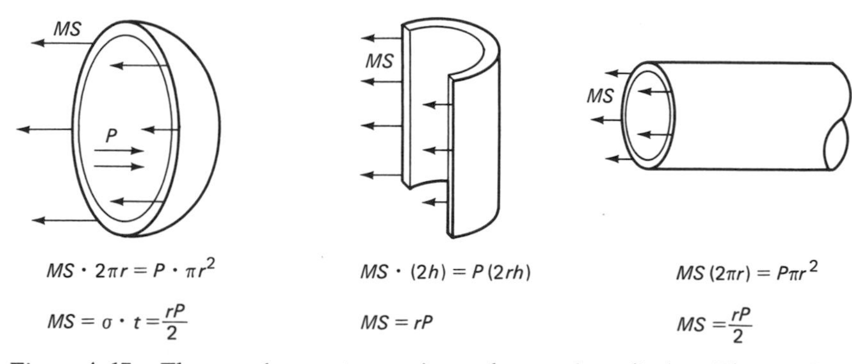

Figure 4.67. The membrane stresses in a sphere and a cylinder. The membrane stress (MS) can be revealed by drawing a free-body diagram for portions of the sphere or cylinder; membrane stress = stress × thickness of membrane; P = inflating pressure; and r = radius.

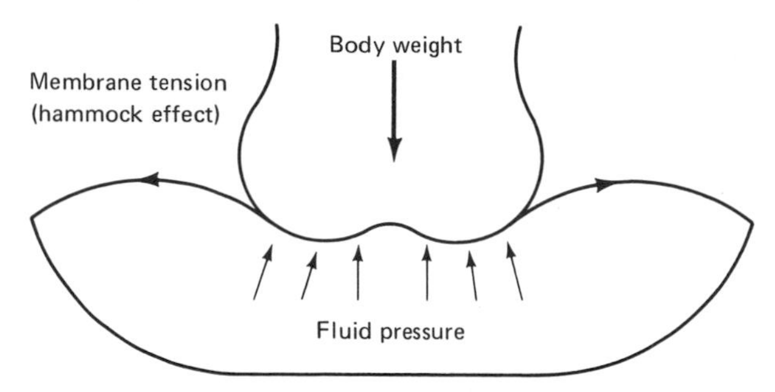

Figure 4.68. A person sitting on a fluid cushion. The weight of the person is supported by fluid pressure and the hammock effect of membrane tension. Because the membrane stress can be very large compared to the fluid pressure, the membrane tension has a significant effect on the pressure distribution.

radius. For a rubber balloon, the inflation pressure is very high initially and decreases after the balloon is moderately inflated. (That is why it is so difficult to get the first mouthful of air into a balloon.) For most structural devices, the membranes do not stretch significantly. The pressure in a football or a vehicle tire increases as more and more air is injected. The only way to get a flat surface with a membrane is to have no pressure difference on the two surfaces of the membrane.

When an external load is applied to an inflated structure, the volume of the structure decreases. The external work is stored in the form of increased pressure in the fluid and tension in the membrane. Figure 4.68 shows the local equilibrium at the interface between a person and an inflated cushion. The body weight is supported by fluid pressures in the cushion and by membrane tension. The membrane supports weight similar to a hammock. Even though the internal pressure of the fluid is rather uniform, the pressure exerted on the person may be quite nonuniform due to the hammock effect of the membrane.

4.5.4. Surface Textures

Surface textures are two-dimensional corrugations that strengthen and stiffen a sheet metal. As an example, after a thin-wall cylinder has experienced local buckling (Figure 4.16(a)), the axial stiffness decreases but the radial stiffness increases. Such a structure (pseudo-cylindrical concave polyhedral) can resist 5 to 10 times higher hydrostatic pressure (Figure 4.69). Table 4.4 shows the test results of two texture-patterns of Rigid-Tex.* The improvements from texturing may be different for different materials and thicknesses. Strength and rigidity increase more prominently in thinner sheets. The stiffness and strength of textured metal sheets are often directional, even if the pattern looks the same in the longitudinal and transverse directions. Weight reduction can be achieved by using a thinner gage textured metal instead of a thicker flat sheet. The reduction in weight depends on the design criterion being either strength or stiffness. The bending strength is proportional to the square of the thickness, whereas the bending stiffness is proportional to the cube of the thickness. The following two formulas show the weight reduction using strengthening and stiffening factors from Table 4.4.

*Registered trademark of the Rigidized Metals Corp.

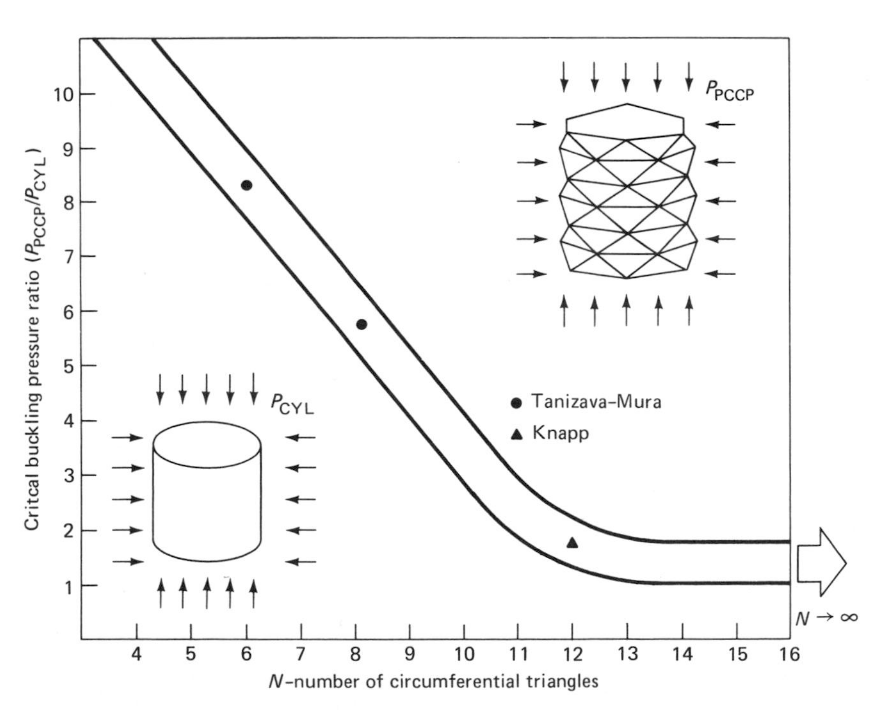

Figure 4.69. A pseudocylindrical concave polyhedral structure has a higher radial compression stiffness and, thus, a higher buckling strength. It is suggested for undersea habitats, observatories, and aquarium applications. (Knapp [58])

TABLE 4.4. STRENGTHENING AND STIFFENING OF TEXTURED MATERIALS. (COURTESY RIGIDIZED METALS CORP.)

			Pattern 1-RL[a]				Pattern 6WL[d]			
			Strengthening Ftr[a]		Stiffening Ftr		Strengthening Ftr		Stiffening Ftr	
Material	Gage	Thickness, in.	Long[b]	Tran[c]	Long	Tran	Long	Tran	Long	Tran
Stainless steel 302 (annealed) U. S. standard gage	26	.0188	2.33	2.00	4.69	4.69	3.00	1.00	2.00	1.56
	22	.0312	1.87	1.47	2.25	1.56	1.67	0.87	1.06	0.81
	20	.0375	3.18	1.73	1.64	1.53	1.41	0.82	1.05	0.78
SAE 1010 Carbon steel (cold-rolled) Manufacturer's standard gage	24	.024	2.67	1.89	4.16	3.33	1.89	1.11	2.50	1.00
	18	.048	1.40	1.05	1.07	1.82	1.51	1.13	1.20	1.03
	16	.060	1.11	0.87	0.74	1.53	1.35	1.10	1.00	0.93
3003-0 Aluminum		.020	2.34	2.00	3.75	3.75	2.84	2.16	1.75	1.25
		.032	2.13	2.00	2.48	2.48	2.25	2.00	1.66	1.21
		.051	2.75	2.60	1.59	1.59	2.40	1.86	1.19	0.915
3003 H-14 Aluminum		.025	1.20	0.80	2.60	2.27	1.80	0.90	1.77	1.30
		.032	1.30	0.76	2.16	1.94	1.35	0.76	1.26	1.00
		.051	1.23	0.79	1.41	1.77	0.98	0.77	1.02	0.79

[a] Ftr = Factor.
[b] Long = Longitudinal.
[c] Tran = Transverse.
[d] See Figure 4.60 for photographs of the patterns.

TABLE 4.5. APPLICATIONS OF TEXTURED MATERIALS. (COURTESY RIGIDIZED METALS CORP.)

Applications	Increase rigidity	Reduce weight	Improve impact resistance	Reduce surface friction	Reduce cost of maintenance	Increase heat transfer and dispersion	Provide a distinctive beauty	Hide scratches and dents	Reduce glare, provide light diffusion	Improve acoustics	Deter vandalism	Nonslip, nonskid	Typical patterns specified
Aircraft flooring and walls	X[a]	(X)[b]	X									X	6WL 5WL
Gasoline pumps	X	X	X		X		X	(X)			X		5WL 2WL
Seatbacks/wainscoting	X	X	X		X		X	X			(X)		1CS 5WL
Conveyor systems	X	X	X	(X)	X								6WL
Pipe and tank insulation	(X)	X	X		X			X					5WL
Telephone booth panels	X	X	X		X		X	X	X	(X)	X		Various
Curtain wall panels	X	X	X		X		X	X	(X)				Various
Vending machines	X	X	X		X		X	(X)					Various
Jet engine shrouds	X	(X)	X			X				X			6WL
Metal cabinetry	X	X	(X)				X	X					2FL 5WL
Elevator cab panels	X	X	X		X		X	X	X		(X)		Various
Appliances	X	X	X				(X)	X					Various

Fire engine panels	X	X	ⓧ		X		X	X			X		6WL 5WL
Paper processing machines	X	X	X	ⓧ	X								6SL 5WL 2WL
Ceiling panels	X	X	X				X	X	X	ⓧ			1NA 1UN
Light reflectors	X	X	X						ⓧ				1HM 5WL
Scuff plates	X	X	X		X			ⓧ		X			5WL
Furniture	X	X	X		X		ⓧ						Various
Space heaters	X	X	X			ⓧ		X					1CS
Doors	X	X	X		X		X	X	ⓧ				1NA 5WL
Commercial ovens	X	X	X			ⓧ							6WL
Food service equipment	X	X	X	ⓧ	X			X					1NA 5WL 2WL 1HM[c]
Cargo containers	X	X	ⓧ		X			X					6WL ACC-4-1
Textile machinery	X	X	X		X							ⓧ	6WL

[a] X indicates secondary benefit.
[b] ⓧ indicates primary benefit.
[c] Approved for application in food-zone areas by the National Sanitation Foundation.

Figure 4.70. A textured fiberglass material for temporary landing of light aircraft and movement of trucks. (Courtesy Air Logistics Corp.)

Weight saving when substituting flat sheets of equivalent strength

$$= 1 - \frac{1}{\sqrt{\text{strengthening factor}}}$$

Weight saving when substituting flat sheets of equivalent stiffness

$$= 1 - \frac{1}{\sqrt[3]{\text{stiffening factor}}}$$

Textured metals have many other uses. They can increase heat dissipation, reduce glare, and reduce friction. Contact between a textured surface and a flat surface occurs only at the tip of the "bumps." The friction, in this case, is less than that between two flat surfaces. Such a property makes textured metals useful for conveyor systems and chutes. When soft materials are pressed against a textured surface, the two surfaces would interlock over a large area. Texturing can increase friction and provide a firm grip. The outstanding patterns hide small surface imperfections, scratches, and fingerprints. In the appliance industry, textured metals are increasingly used because they are easier to maintain and clean. Many other advantages and applications of textured metals are shown in Table 4.5. Blow-molded plastic parts can be cost reduced by surface textures. In an aluminum mold, a textured surface is less expansive than a polished surface. Texturing also conceals the inherent roughness of blow-molded surfaces.

A textured fiber glass matting called Mo-Matt (mobility matting) is marketed by Air Logistics Corp. (Figure 4.70). The matt allows helicopters and light planes to land without stirring up excessive dust, and permits trucks to drive over muddy land surfaces. Wheel loads are spread out over a larger area of land surface, which provides a firm support even on very soft soil. The material can also be used as roofing and flooring for shelters and temporary warehouses.

5. INTEGRAL DESIGN WITH MULTIFUNCTIONAL PLASTICS

5.1. EXAMPLES OF MULTIFUNCTIONAL DESIGN

In Section 3.3.3, we saw that modular design offers flexibility, ease of maintenance, and higher production volumes. Other reasons for using modules may be savings in the cost of material or savings in processes of fabrication and assembly. At the other extreme, *integral design* aims at reducing the number of parts and the amount of interface between components, and eliminating the assembly of individual parts. The resulting parts have to perform the same overall functions. Thus, some of the parts perform multifunctions. In an electronic assembly, the chassis may act as a supporting structure, an electrical ground, and a heat sink. The Bethlehem Steel Co. has succeeded in making insulated steel members to act as both structural members and electrical conductors. Figure 5.1 shows a one-piece piston-and-rod replacing an assembly of a metal rod onto a plastic piston (made of high-density polyethylene). The new piston-and-rod is molded with propylene. Because of the better wear characteristics of plastics, the eyelet is also eliminated.

An integral design not only reduces assembly effort, but also decreases weight and complexity. A one-piece structure has no bolts, no joints, and fewer points of stress concentration. A high-strength and lightweight design is achieved through structural continuity. Though integral design has been used with many materials, its real capacity emerged only after the advent of plastics. Nowadays, there are plastics available for making springs, bearings, cams, gears, fas-

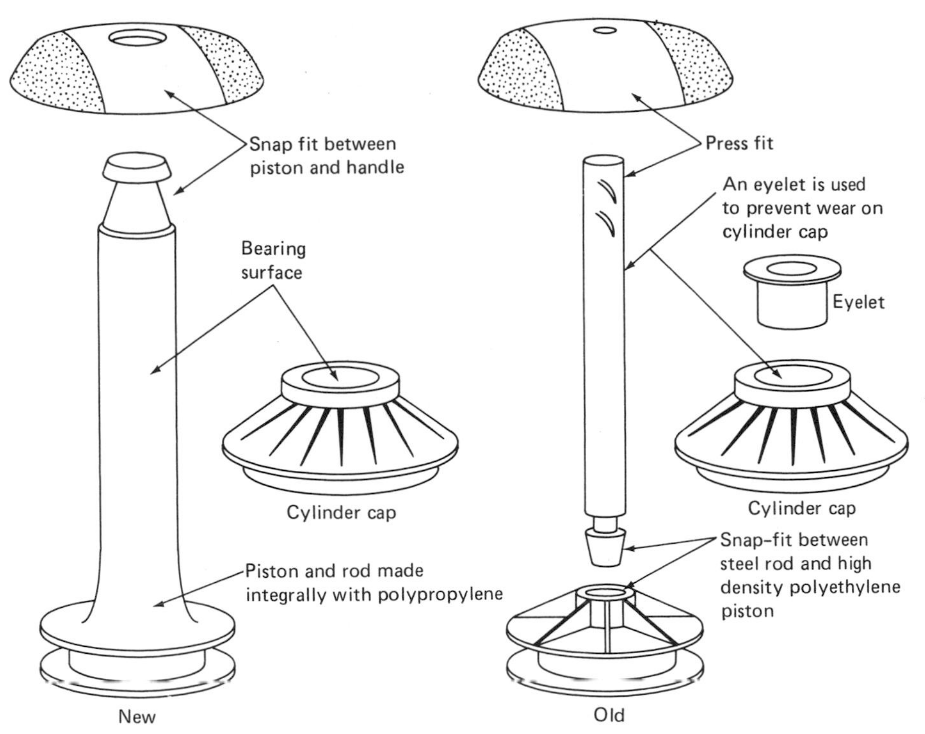

Figure 5.1. A pump assembly. The integral design eliminated two extra parts and two assembly operations: (1) the one-piece piston-and-rod eliminated the metal rod as well as the costly end-configuration of the rod and piston; (2) improved wear characteristics between the piston rod and the cylinder cap eliminated the eyelet. (Courtesy Kenner Products Co.; designed by W. Chow and H. March.)

teners, hinges, and optical elements. Plastic housings are used for protection from mechanical damage, shock, corrosion, and other factors. With a wide variety of molding, assembling, and decorating techniques, multifunctional design in plastics naturally advances rapidly.

A striking example of multifunctional design is shown in Figure 5.2. A one-piece, medical tubing-clamp, molded with acetal resin, has succeeded in replacing a similar metal unit assembled from five parts. The cost is $.055 for the acetal clamp and $1.95 for the metal clamp. Acetals, which are the best plastics for springs and wear surfaces, have found many applications in integral designs. The door-catch in Figure 5.3 is spring-loaded, such that after the door is closed, the

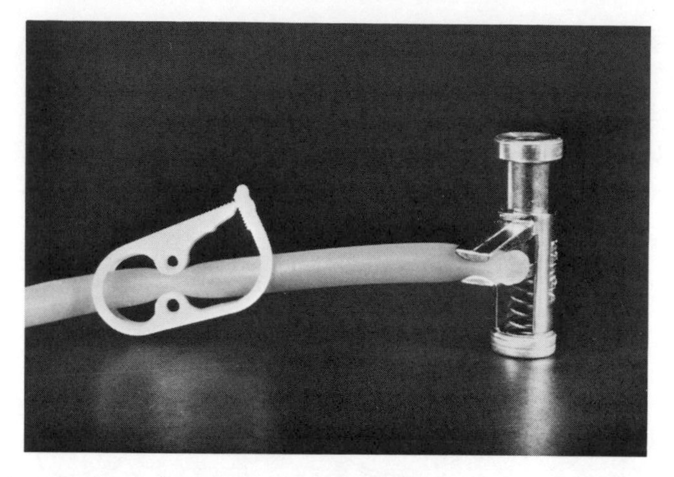

Figure 5.2 The acetal tubing-clamp (left) is molded in one piece and, thus, requires no assembly. The metal clamp is assembled from 5 pieces. (Courtesy DuPont.)

catch protrudes again to keep the door shut. Originally, the plastic slide was actuated by a metal coil spring. Later, the metal spring was replaced by a plastic spring made integral with the plastic slide.

Every key on a pocket calculator is spring-loaded to pop up as soon as the pressure is released. Since the housing is made of plastic and the keys are plastic, it is natural to make the springs of plastic too.

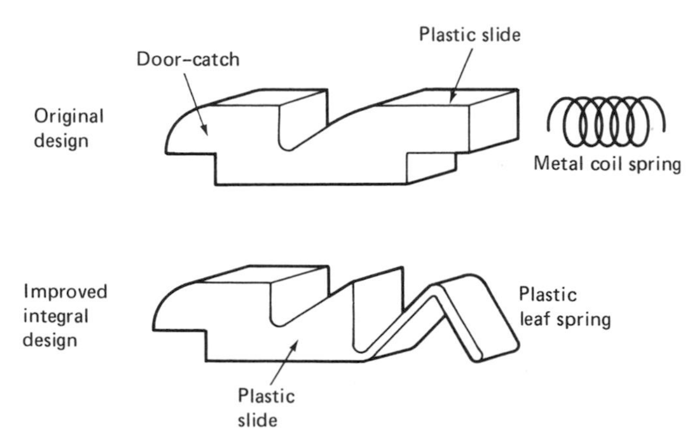

Figure 5.3. A door-catch improved by integral design. (Courtesy Ajax Hardware Corp.; designed in cooperation with DuPont [60].)

In one calculator (Texas Instruments TI-1200), all 19 keys, 19 springs, and the housing are molded in one piece.

In the following sections, we will explore the capacity of plastics to be used in different machine elements.

5.2. PLASTIC SPRINGS

Every material deflects under external forces. This property is described by the spring-rate of the material (the ratio of force to deflection of the material). Ceramics and glass are not suitable for making springs because they are brittle. When made with sufficient strength, their spring-rates become very high. Metals, especially steels, are good spring materials because they have very high strengths and resiliences ($R = 1/2\ S_y \epsilon_y = 1/2\ S_y^2/E$). Plastics are in general very poor spring materials because they deform inelastically at rather low levels of stress and because they behave viscoelastically. If one wants to use plastics in springs, one must carefully consider the stress level, the stress duration, and the operating temperature of the spring. A low stress can prevent inelastic deformation. A short duration of the load (intermittent operation) is commonly designed for plastic springs to prevent relaxation.

Viscoelastic behavior is a combination of solid behavior (elasticity) and liquid behavior (viscosity). Four important aspects of viscoelastic behavior follow.

(1) The rate of application of the load affects the measurement of the elastic modulus.
(2) The material behavior is influenced by all its past loadings, though the most recent loadings have the strongest effects. This is called the fading memory of a viscoelastic material.
(3) Given a constant stress, the strain will increase with time. This is called "creep."
(4) Given a fixed deformation (i.e., a constant strain) the stress will decrease with time. This is called "relaxation."

All the above properties are closely related and the measurement of one can give useful clues about the others. Creep and relaxation can be studied by mathematical models consisting of springs and "dashpots" (viscous units). Figure 5.4 shows such a model and the conse-

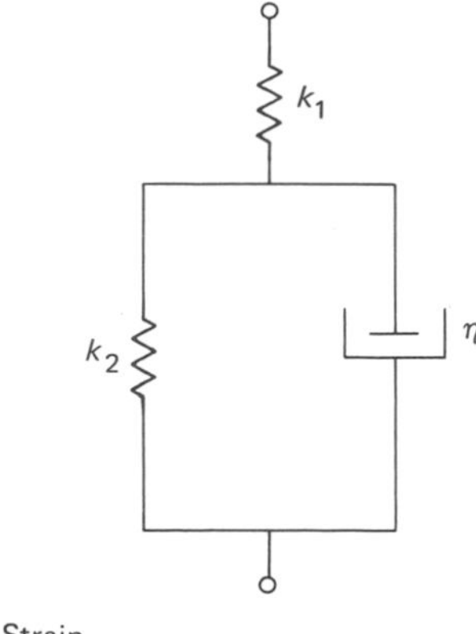

(a) The model

The spring with spring-rate k_1 represents immediate elastic behavior, and the one with k_2 represents the long-term elastic behavior. The dashpot with damping constant η represents the viscous behavior. The viscous behavior of polymers can be interpreted as the uncoiling of long molecules and the reorientation of molecular structures.

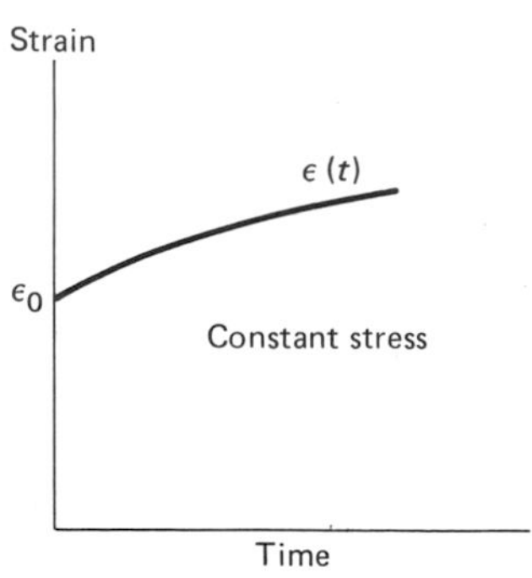

(b) Creep

$$\epsilon_0 = \frac{1}{k}\,\sigma$$

Steady-state strain

$$\epsilon\,(t = \infty) = \left(\frac{1}{k_1} + \frac{1}{k_2}\right)\sigma$$

Compliance modulus

$$J\,(t) = \frac{\epsilon\,(t)}{\sigma}$$

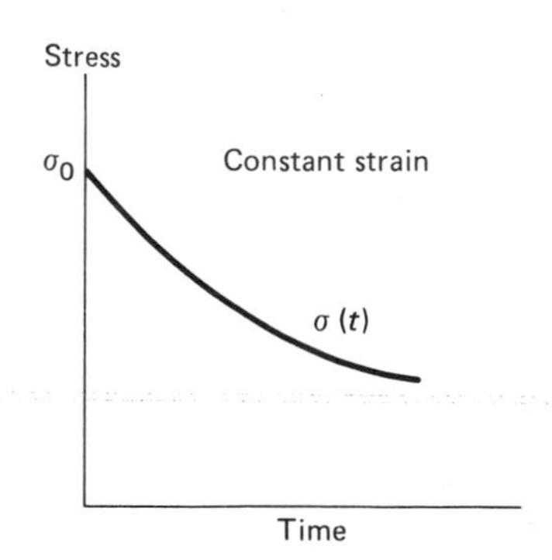

(c) Relaxation

$$\sigma_0 = k_1\,\epsilon$$

Steady-state stress

$$\sigma\,(t = \infty) = \left(\frac{k_1\,k_2}{k_1 + k_2}\right)\epsilon$$

Relaxation modulus

$$E\,(t) = \frac{\sigma\,(t)}{\epsilon}$$

Figure 5.4. Modeling of viscoelastic behavior. Elastic behavior is a solid behavior where stress is proportional to strain. Viscous behavior is a fluid property where stress is proportional to the strain rate. The immediate response is dominated by k_1 and the steady-state response is equal to the response with the viscous unit removed. In a creep experiment, the steady-state strain may go to infinity. In a relaxation test, the steady-state stress may approach zero. The steady-state behavior depends on molecular sliding, which is possible when the material is close to its melting point.

quent predictions. Note that the spring is the elastic component and the energy storage unit. The model predicts the same behavior regardless of the stress level. For many real plastics on a real time scale (e.g., 10 yrs), viscoelastic behavior becomes insignificant below a certain stress level. In order to predict a termination of viscoelasticity at a low stress, dry friction components must be added to the mathematical model. The viscoelastic effects are formidable obstacles to many engineering designs. If a nylon gear is transmitting load without rotating, the load may cause the gear teeth to creep. This permanent setting may then cause irregular spacing and vibration in the gear train.

In all plastics, viscoelastic behavior is more prominent at high temperatures. As the temperature approaches the melting point of the material, more molecular movement occurs, and the material becomes more "liquidy." The viscoelastic behavior is suppressed at low temperatures and disappears when the temperature drops below the glass-transition point. Many plastics behave similarly if the temperature scale is normalized as a percentage between the melting point and the glass-transition point. Plastics with cross-linking chemical bonding will not melt, and they behave significantly different from thermoplastics. Properly cured rubber is not viscoelastic at room temperature.

Strictly speaking, viscoelastic materials should not be used for springs. When the need for plastic springs arises, the design engineer has to validify the design by proving that creep or relaxation is insignificant in the particular application, or that the spring can operate at the steady-state (postrelaxation) stress level. High-strength engineering plastics are more favorable candidates for plastic springs because the stress level can be kept relatively low compared to the yield stress. If styrene, propylene, or other low-strength thermoplastics are used, the duration of the load must be very short. In Figure 5.1, the spring in the snap-fit is used only during assembly and the load lasts less than 1/2 sec. In Figure 5.3, the load is also applied only momentarily. If the load is applied over a long period of time, the part will deform and "take a set" (creep). If the plastic spring has to bear a prescribed deflection and the force is not important, making the spring thinner will make it more durable. For the same deflection, the stress is often lower in a softer spring. (See Section 1.3.2.)

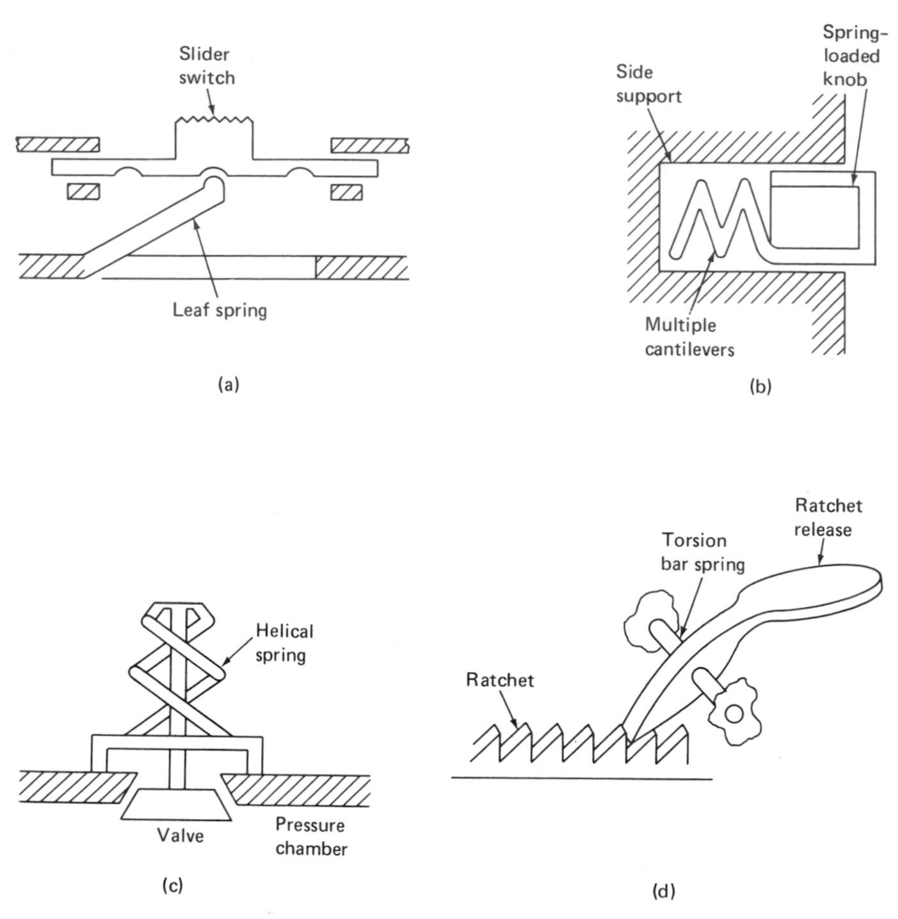

Figure 5.5. Plastic springs. A leaf spring (a) is simple in shape. Multiple cantilevers (b) increase the deflection and decrease the spring rate. A check valve with a helical plastic spring (c), made by Dab-O-Matic Corp., Yonkers, N.Y. A reversible ratchet with a torsion bar spring (d).

Figures 5.5 and 5.6 show several types of plastic springs and washers. The leaf spring is the most popular because of its simple geometry. The width of a leaf spring should be tapered so that the stress is more uniform and the material is better utilized, as explained in Figures 1.5 and 1.20. Multiple cantilevers and helical springs may buckle under compression if the axial dimension is excessive, and lateral support is absent. Disk springs are easily obtained by stamping sheet materials (such as cellulose acetate).

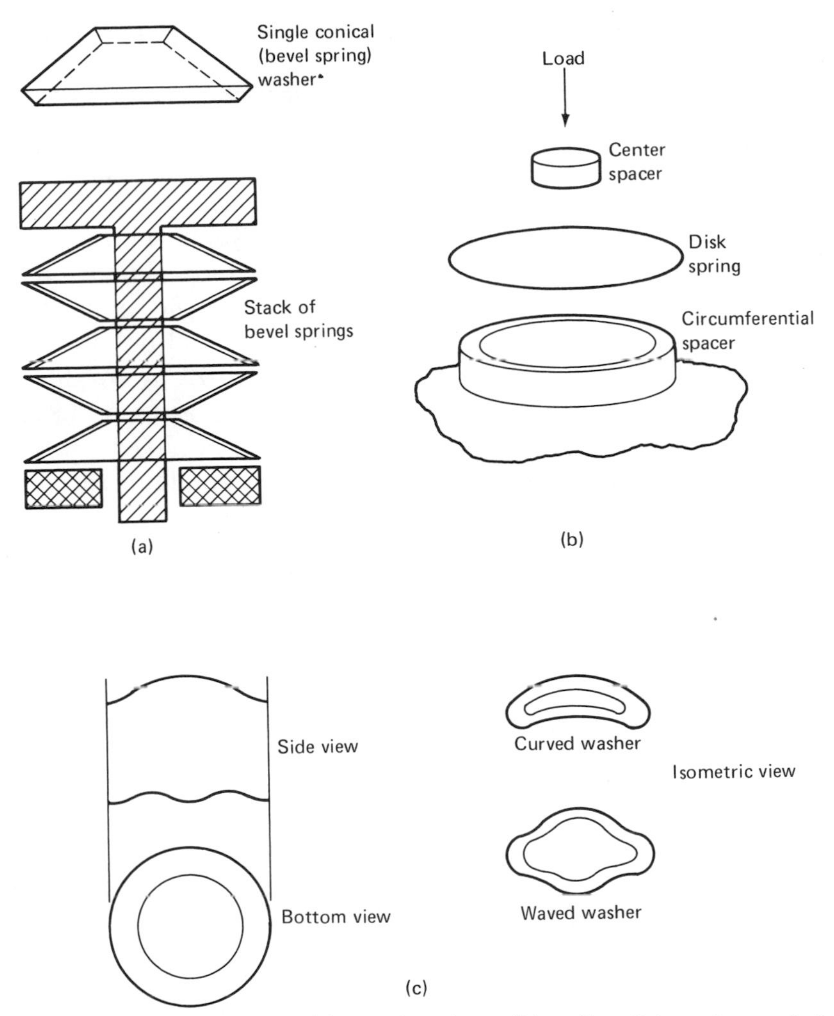

Figure 5.6. Spring washers. (a) bevel springs, (b) a flat disk spring made by stamping, (c) spring washers.

5.3. PLASTIC BEARINGS

Plastics make inexpensive bearings that can easily outperform wood, steel, and zinc bearings. There are situations in which plastic bearings are used out of necessity. For example, plastic bearings are strong candidates in corrosive environments, in nonlubricated bearings, and in lightweight bearings. In this section, we are more interested in cost reduction with integral plastic bearings that are convenient.

Plastics suitable for bearing applications include phenolics, nylon, fluorocarbon, polycarbonate, acetal, and ultrahigh molecular weight polyethylene. The above materials are used with steel shafts of fine finish (4 to 12 μin. rms). When plastic rubs against plastic, the wearing is higher. Plastics that can dissolve in solvents should not be mated with another component of the same material, but should be mated with solvent-inert plastics. For some specific applications, plastics such as urethane, polyester, polysulfone, and polyimide are mixed with fillers. Fluorocarbon resins (15% to 20% by weight) and silicone lubricants (2% by weight) are mixed with plastics to decrease friction. Glass fibers are also used (30% by weight) to increase strength and dimensional stability and to decrease molding shrinkage.

The design objectives for bearings are usually a long life or a high efficiency, or both. In other words, wear and friction are the big concerns in the design of bearings. The factors that control the performance of a bearing are pressure, velocity, material hardness, lubrication, surface finish, and temperature. Table 5.1 shows the operational limits of several typical materials. Wear and friction are not directly related. Brake-packing materials produce very high friction, but have low wear. A material with a low friction coefficient may wear rapidly. Wear is defined as the volume rate of material loss

TABLE 5.1. OPERATING LIMITS FOR PLASTIC BEARINGS MATING WITH STEEL SHAFTS. (COURTESY *MACHINE DESIGN* [65].)

	Maximum pressure (psi)	Maximum temperature (°F)	Maximum speed (fpm)	Maximum power per unit area *PV* (pressure) (speed)
phenolics	6000	200	2500	15,000
nylon	1000	200	1000	3000
fluorocarbon	500	500	50	1000
reinforced fluorocarbon	2500	500	1000	10,000
fabric fluorocarbon	60,000	500	150	25,000
polycarbonate	1000	220	1000	3000
acetal	1000	180	1000	3000

TABLE 5.2. FRICTION COEFFICIENTS AND WEAR FACTORS OF PLASTICS. (THEBERGE, ARKLES, AND CLOUD [67].)

Bearing material	Shaft material	Wear factor $\times 10^{-10}$ $\frac{\text{in./hr}}{\text{psi} \times \text{fpm}}$	Coefficient of friction	
			Static	Dynamic
acetal	acetal	10,200	0.19	0.15
acetal	steel	65	0.14	0.21
acetal (20% TFE)*	steel	17	0.07	0.15
nylon 6/6	nylon 6/6	1150	0.12	0.21
nylon 6/6	steel	200	0.20	0.26
nylon (20% TFE)	steel	12	0.10	0.18
nylon (30% glass) (15% TFE)	steel	16	0.19	0.26
polysulfone (30% glass) (15% TFE)	steel	55	0.12	0.10
polycarbonate	steel	2500		
polycarbonate (30% glass) (15% TFE)	steel	30		

radial wear (in./hr) = wear factor $\times P \times V$
The wear factors in the above table were measured at $P = 40$ psi, $V = 50$ fpm.
Wear factors of 200 or less are considered very good for most design purposes.
*TFE = fluorocarbon
All fillers are percentage by weight

(in.^{3}/hr). Table 5.2 shows the wear and friction data for acetal, nylon, and polysulfone. Wear factors of 200 or less are considered very good for most design purposes. The pressure at the bearing is the load divided by the projected area of the bearing (length $\times$ diameter). Nylon and fluorocarbon bearings have a tendency to cold-flow under moderate pressures. The contact pressure can be decreased by increasing the length of the bearing, provided that the dimensional stability is enough to maintain contact. The product of pressure and velocity, PV, is the power rating of the bearing per unit area. The product of PV and the coefficient of friction gives the energy dissipation (or the rate of heat generation). The wear rate is the product of the wear factor and PV. Increasing the material hardness by glass

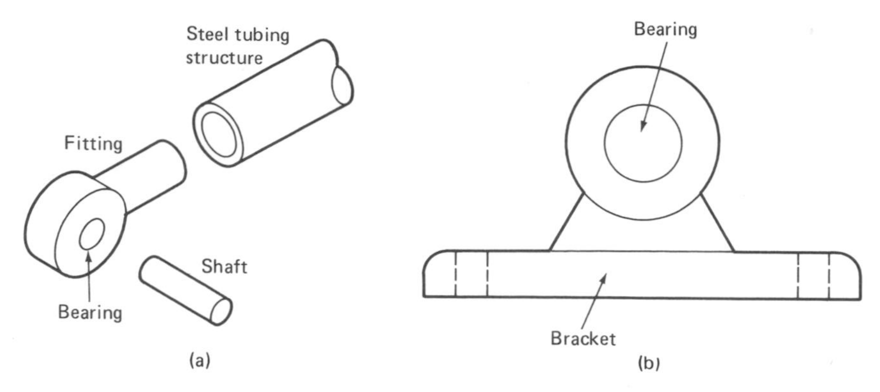

Figure 5.7. Integral plastic bearings. (a) is a bearing with a fitting; (b) is a bearing with a bracket.

filler or metal backing can reduce friction and wear. Plastic bearings are sensitive to temperature. When the temperature rises above a certain point, wear becomes extremely fast. The bearing surface temperature is usually unknown and impossible to measure. Because of the slow heat conduction through plastics, the best method of heat removal is by a fluid such as oil or water.

If a bearing operates at a very low speed or for short, intermittent durations, wear is usually not a problem. Nylon sliding pads and rollers are used for furniture drawers. Fluorocarbon sliding pads are used between overhead highways and their supporting pillars to allow for thermal expansion. Acetal bearings are used in door hinges, electrical appliances, and light-duty wheels. Fluorocarbon bearings are used in automobile steering linkages and food-processing equipment. Phenolics have the highest load and speed ratings. They are used in marine propeller shafts, hydroelectric turbine shafts, etc. Figure 5.7 shows integral designs of bearing-fittings. Plastic bearings are ideal for all light-duty applications because they are inexpensive, quiet, clean, and carefree. For heavy-duty applications, an extensive testing of material behavior in the appropriate environments is essential.

5.4. PLASTIC GEARS

Making an accurate plastic gear is more difficult than making a similar machined gear because the engineer has to consider errors in the machining of the mold, in the molding of the product, and in the postmolding shrinkage of the particular material. However, once a design is proven to be accurate enough, the cost of obtaining the accuracy becomes a fixed cost that can be amortized over the production volume. The cost of accuracy in a machined gear is a variable cost that is added to each production piece. Plastic gears usually have a diametral pitch between 16 and 48, a pressure angle of 20°, and a commercial accuracy (AGMA quality number of 5 to 10). Since plastic gears are inexpensive, light, corrosion-resistant, and "quiet," they are very competitive with die-cast gears, stamped gears, and sintered gears. Plastics are strong against impact. The softer teeth allow load sharing between adjacent teeth, which lowers the stress. In addition, plastic gears can be made integrally with bearings, pinions, cams, shafts, read-out dials, etc. Plastic gears are popular in light-duty applications such as timing devices, toys, accounting machines, instruments, tools, and small appliances.

A gear material should have the low friction and low wear of a bearing material. Moreover, the strength required for gears eliminates the use of soft plastics. The most widely used gear materials are nylon, acetal, and phenolics. Polycarbonates filled with fiber glass and fluorocarbon are also used. Dimensional stability can be a serious problem for plastic gears. Besides the large thermal expansion, some plastics swell in the presence of solvents; for example, nylon can absorb water moisture. Part of this expansion can be compensated for by using the same plastic material for the mounting plates. If an integral gear is made with styrene, the mating gear should be made with a plastic that would not weld with styrene.

One easy way to get a mold cavity with a gear profile is: (1) start with a brass blank and cut a gear form by hobbing; and (2) burn the cavity in the mold with an EDM (electric discharge machine), using the brass gear form as an electrode. The EDM will overburn on all surfaces: 0.002 in. on the radius and each side of the teeth. The final molded plastic part will shrink and become smaller than the mold cavity. The shrinkage is about 2% (0.020 in./in.) with acetals. However, the shrinkage will not be uniform: there will be no shrinkage at the top lands of the teeth because the plastic there is chilled from three sides. The diameter of the gear will shrink 2%. Unless precautions are taken, the teeth will be too fat (the thickness of the teeth at the pitch circle will be greater than the gap at the pitch circle) and the pressure angle will be larger than specified. Figure 5.8 shows five ways to put a gate in the mold. Methods shown in (a) and

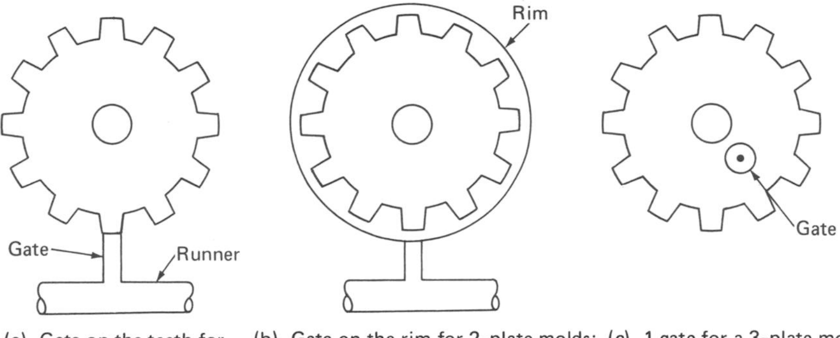

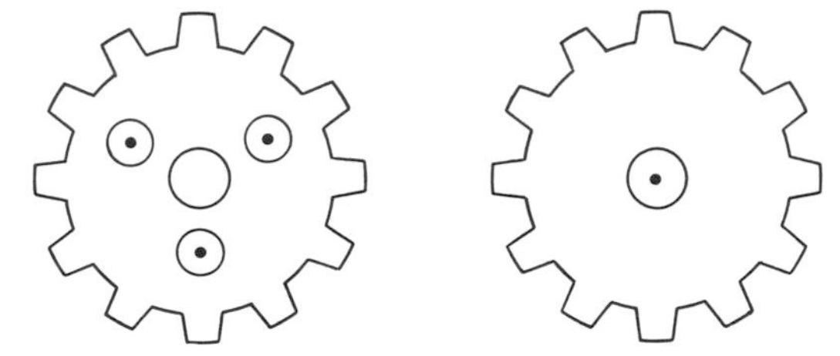

Figure 5.8. Gate position for gears. (a) is the worst and (e) is the best. The location of the gates influences the roundness, eccentricity, and wobble of gears because the material shrinks most in the direction of flow.

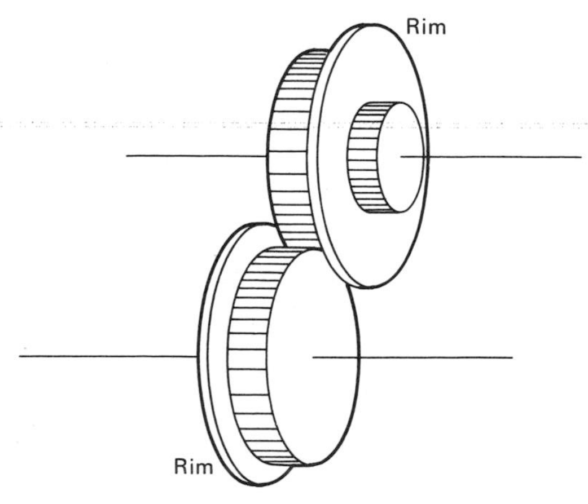

Figure 5.9. Gears with rims have to be arranged as shown.

(b) apply to a 2-plate mold; (b) is an improvement over (a). With method (a), the gate would inevitably be located on a tooth. The shrinkage near the gate would distort the tooth and the de-gate process would also change the tooth form. The gating shown in Figure 5.8(b) will move the gate away from the gear teeth. It also necessitates a special arrangement in the gear train (Figure 5.9). Two-plate molding has an inherent deficiency because the flow pattern is not symmetrical with respect to the center of the circle. Shrinkage is anisotropic and is usually the largest in the direction of flow, the smallest in the direction transverse to the flow. A 3-plate mold that is gated in the dead center is the best (Figure 5.8(e)). Usually, shrinkage can be decreased by packing the mold, which means holding the pressure while the material is cooling so that shrinkage loss can be replenished.

The diameter of the mold cavity should be the final diameter of the piece multiplied by $(1 + s)$, where s is the shrinkage in in./in. In order to have the same number of teeth on a larger diameter, the pitch on the mold has to be nonstandard [68]. The diametral pitch in the mold has to be decreased by the factor $1/(1 + s)$. The ratio of the cosines of the pressure angles in the mold and in the final piece is also $(1 + s)$. The pressure angle in the mold should be smaller than the standard angle in order to get a molded piece with the standard pressure angle. As an example, the shrinkage in acetal (Celcon produced by Celanese) is 0.023. In order to get a 2-in. dia gear of 24 diametral pitch and 20° pressure angle, the mold should be made with 2.046 in. dia, with a diametral pitch of 23.46 and a pressure angle of 15.99° The nonstandard gear form in the mold cavity can make it expensive. If the gear form in the mold cavity is made standard, the plastic molded piece will be nonstandard. If all the gears are made with the same material, they will be nonstandard in the same way and they will run compatibly with each other. To get more tolerance for the gears, the teeth can be made thinner (Figure 5.10). A decrease in tooth thickness will weaken the tooth. Therefore, this modification is used only on the gear and not on the pinion. The pinion teeth are inherently weaker than the gear teeth. The backlash can also be increased by moving the gears apart so that the center-to-center distance is greater than the theoretical value. The engineer should be careful not to overdo any modifications. If the gears are moved too far apart, the gears will disengage (skip) or

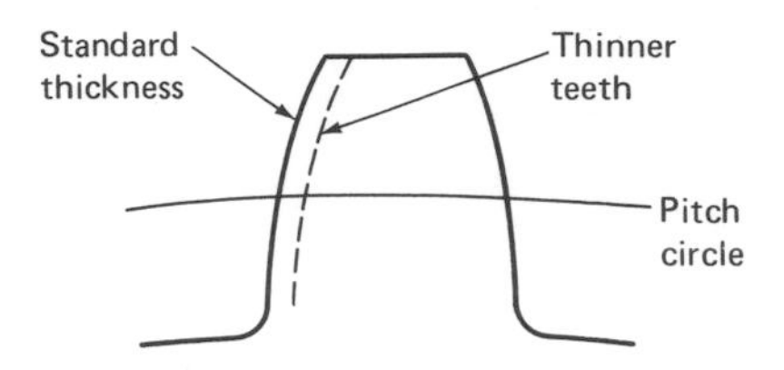

Figure 5.10. Thinner gear teeth can increase tolerance. The thinning should be used on the gear and not on the pinion.

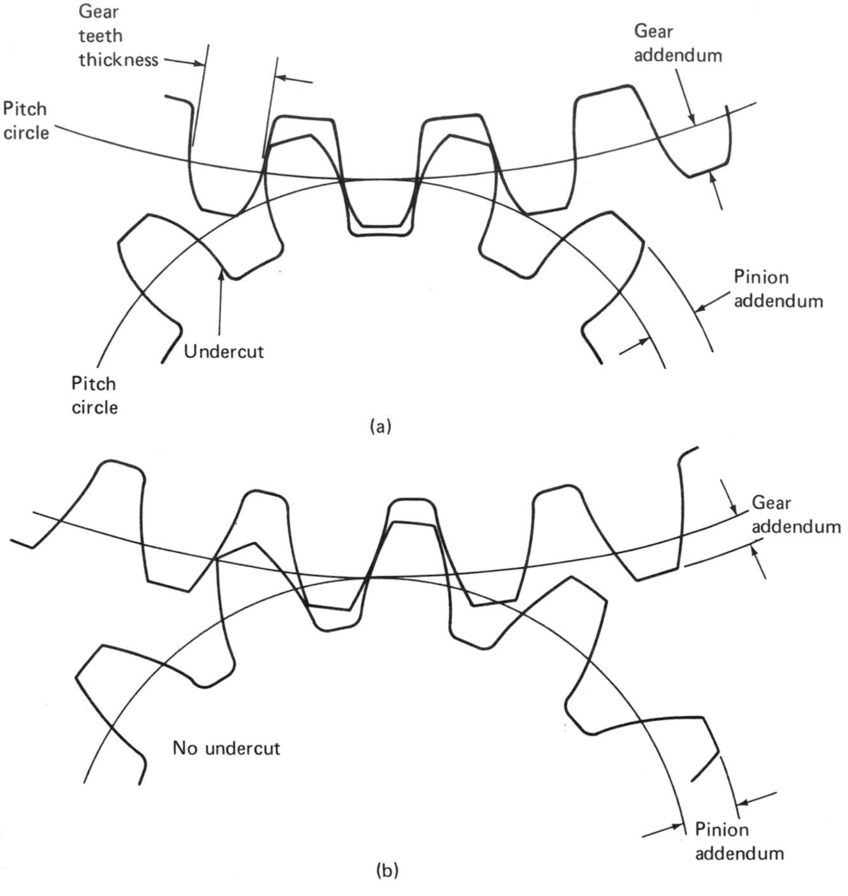

Figure 5.11. The long-and-short addendum modification. This modification does not change the center-to-center distance or the pitch circles. In the standard gear and pinion (a), the gear addendum is the same as the pinion addendum, and the gear teeth thickness equals the pinion teeth thickness. In the modification (b), the gear addendum is less than the pinion addendum, and the thickness of the gear teeth is less than the thickness of the pinion teeth, thereby strengthening the pinion.

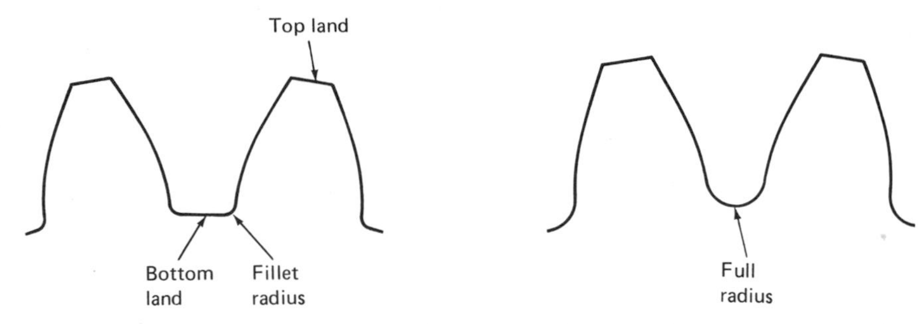

Figure 5.12. A full radius at the bottom land can reduce stress concentration [68].

jam with teeth tips hitting. Figures 5.11, 5.12, and 5.13 show three modifications of plastic gears: (1) using a long addendum on the pinion and a short addendum on the gear, so that the pinion strength is increased; (2) replacing the bottom land of a tooth a full radius curvature, so as to reduce stress concentration; and (3) putting a rim on the sides of gears to prevent adjacent gears from rubbing at the teeth.

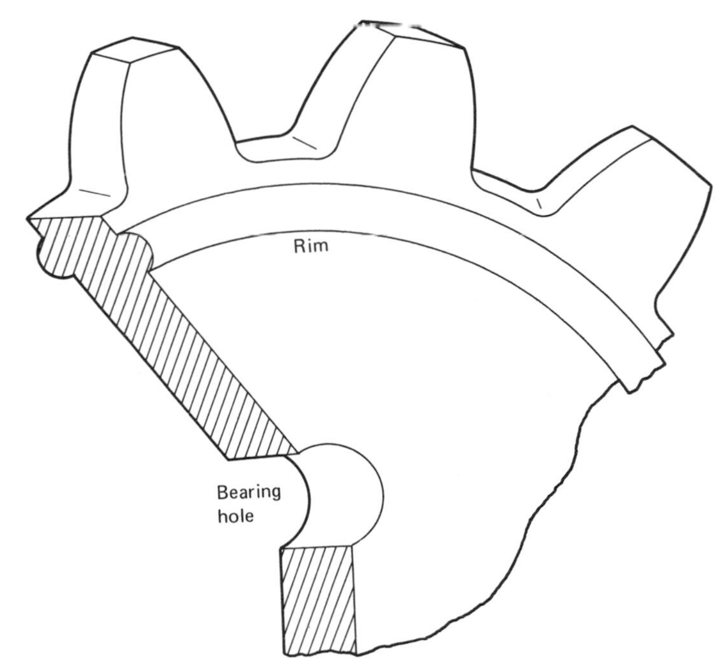

Figure 5.13. Rims on sides of gears can prevent adjacent gears from rubbing at the teeth.

5.5. PLASTIC FASTENERS

Research and development with plastics have produced several new fasteners. These plastic fasteners are specially made for non-plastic components. The techniques of joining plastics will be examined in the next chapter. There are five general kinds of plastic fasteners: (1) screw assemblies, (2) snap- and press-fits, (3) inserts, (4) ties and twists, and (5) shrink wraps.

Nylon is commonly used for fasteners because of its high strength and toughness (defined in Figure 1.15). However, special properties are needed for some applications such as the shrink wraps.

Screw fasteners (Figure 5.14). Nylon nuts and bolts are used for fastening automobile license plates because they are corrosion-resistant and vibration-proof. Plastic studs and self-threading sheet-metal nuts are also economical and easy to apply. Plastic nuts and screw bottle caps are difficult to mold because they require screw ejection or collapsible cores. One-thread plastic nuts can be easily molded and are often used despite their inferior holding strength.

Snap fasteners (Figures 5.15, 5.16, 5.17, and 5.18). Plastic snap buttons are used successfully on baby clothes. Several kinds of loop-and-hook fasteners are on the market. These fasteners resemble certain kinds of seeds that attach themselves to the furs of animals and to the clothes of people. Hedlok (by the 3M Company) is another alignment-tolerant snap-fastener with interlocking stems and balls. When using such fasteners, the design considerations are: (1) the number of cycles of operation, (2) the reliability of the holding force,

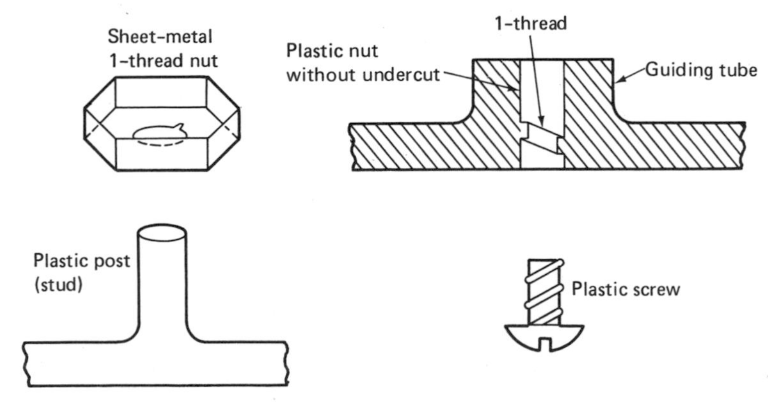

Figure 5.14. Plastic screw assemblies.

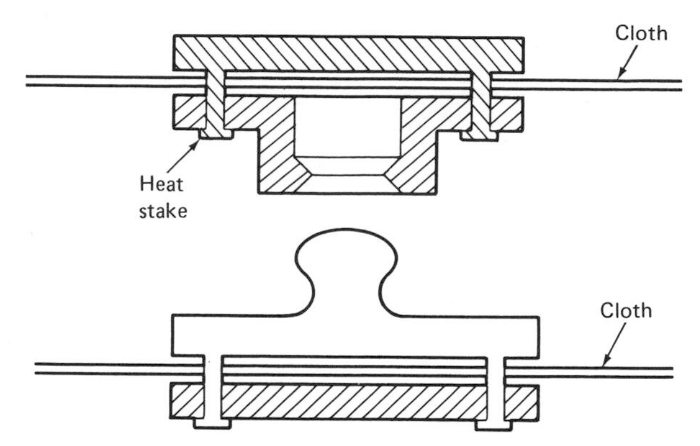

Figure 5.15. A plastic snap button.

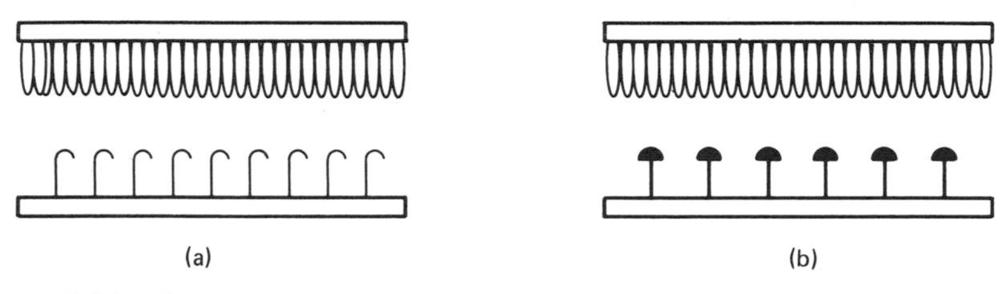

Figure 5.16. Two kinds of plastic hook fasteners. (a) loop and hook fastener; (b) loop and mushroom fastener.

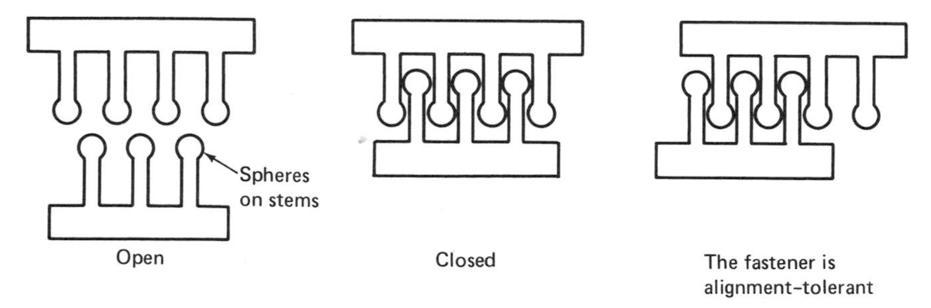

Figure 5.17. Hedlok II, by the 3M Company, has identical members mating together.

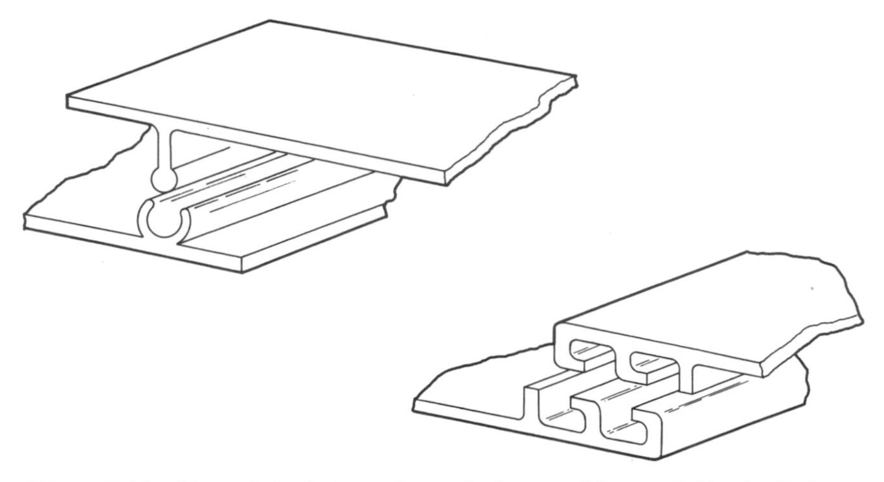

Figure 5.18. Extruded ethylene zipper designs used in resealable plastic bags.

(3) the temperature range, (4) the shear strength and the peel strength, and (5) the vibration-holding strength (or the shock strength).

Plastic zippers are successfully made from polyethylene. These zippers make resealable plastic bags that are air-tight. Plastic zippers are also used in office supplies and briefcases. Nylon zippers similar to conventional zinc and aluminum types are used for clothing.

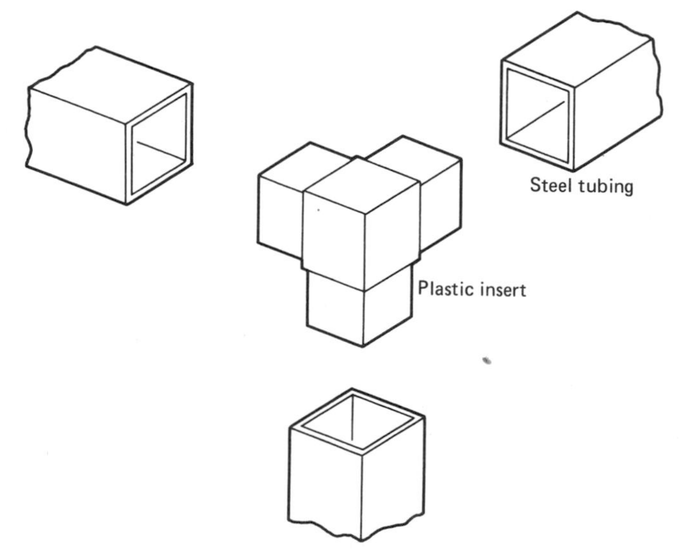

Figure 5.19. A plastic insert can be used to align and join other materials.

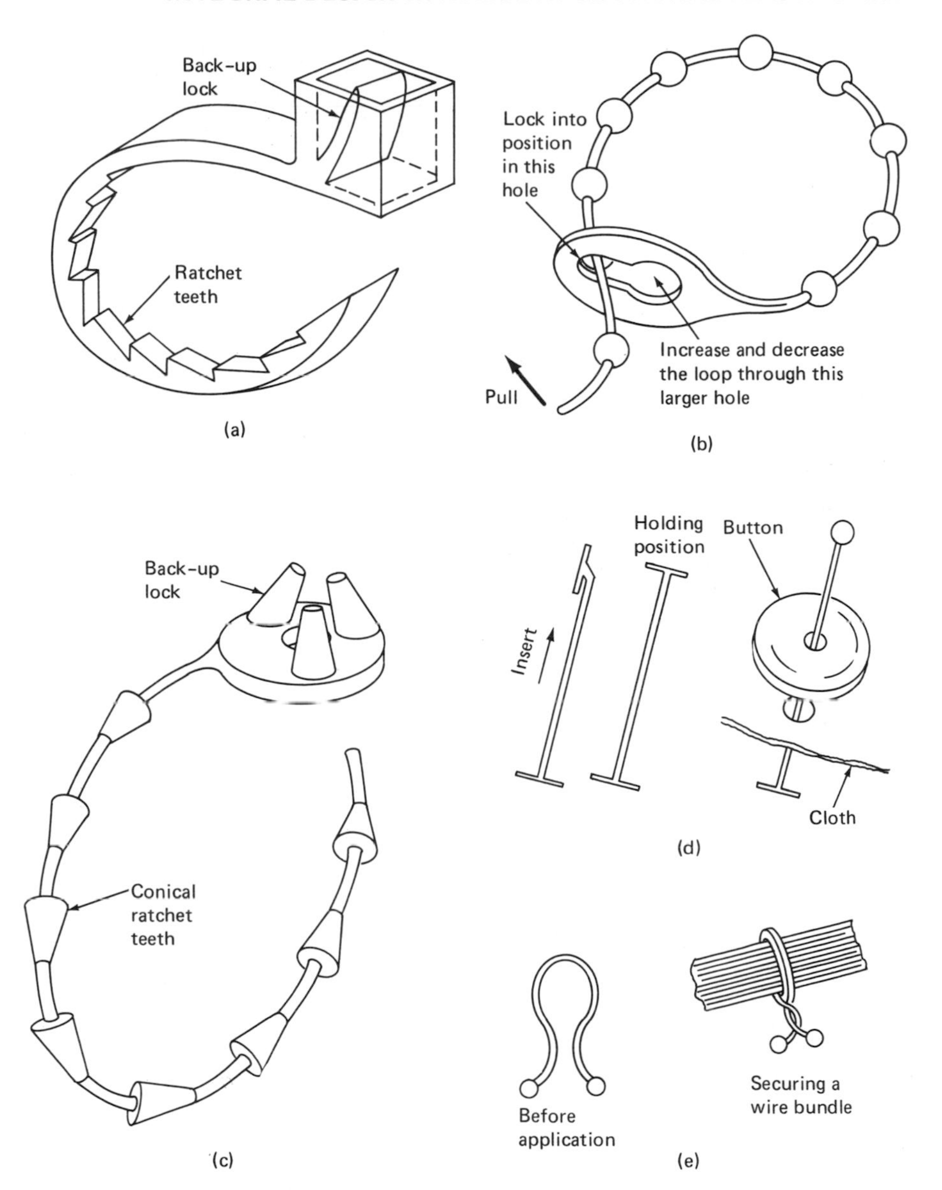

Figure 5.20. Plastic ties and twists. (a) A tie for electrical wire. (b) A tie that can be released and reused. (c) A nylon tie (by Dennison Co.). (d) Nylon ties for tags, labels, and a button (by Dennison Co.). (e) Nylon twist (by ITW, UK).

Inserts (Figure 5.19). Inserts are used to join and align pieces of extruded or formed metal. Plastic fittings are used in wood furniture, aluminum window frames, and picture frames.

Ties and twists (Figure 5.20). These fasteners are made for disposable and reusable applications. They are widely used for attaching tags, labels, buttons, and electrical wires. Automatic feed machines, built for these fasteners, enable fastening at a high speed. Plastic ties are often stretched after molding to align the molecules. After the mechanical deformation, the length of a tie is increased with no loss of strength. A nylon loop is also used as a "twist." The material can endure a large inelastic deformation, and will not unwind itself.

Shrink wraps (Figure 5.21). Plastic shrink wraps are used for packaging and electrical insulation. A common example is the six-pack of beer. Fluorocarbon shrink tubes are used for coating steel rollers (for processing sheet materials) and tubing fittings [71]. The tubes or sheets are stretched to an expanded shape. Because the plastic has cross-linking chemical bonds, internal stresses will not be released through molecular flow. The stresses are frozen-in at room temperatures. When heat is applied, the material softens, shrinks and returns to its original shape. This characteristic is called plastic memory. After application, a shrink wrap becomes a tension-skin structure. The materials for shrink wraps are: polyolefins, neoprenes, silicones, and fluorocarbons. Plastic memory can produce amazing engineering

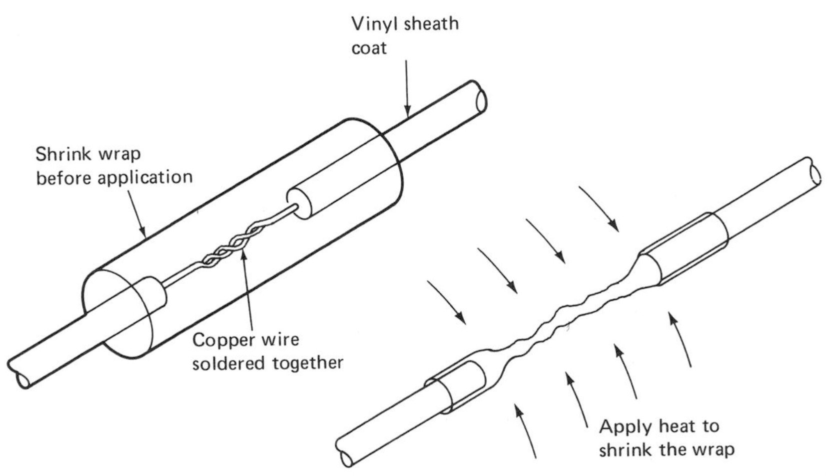

Figure 5.21. Shrink wrap for electrical insulation [71].

applications and apparently impossible assemblies. An automobile fender can be made such that it will return to its original shape after a traffic accident with just a hot-air blower. Seals and continuous rings can be made to shrink into piston grooves.

5.6. PLASTIC HINGES AND STRAPS

A hinge is a device that is compliant in rotation about a fixed axis, but strong in tension, bending, and torsion about any other axis. Since the front cover of a book can be flexed about the edge of the book, but resists motion in other directions, we can consider that the front cover is attached to the book by a paper hinge. Plastic hinges made from different polymers may endure from several to several thousand flexes. The most superior materials for hinges are polypropylene and polyallomers (copolymer of ethylene, propylene, and other olefins). Propylene and polyallomer hinges that are properly made can last for over one million flexes.

5.6.1. Technical Aspects of Integral Hinges and Straps

A strap offers a flexible connection with no fixed axis of rotation, and can usually flex in more than one direction. Figure 5.22 shows six applications of integral hinges, and Figure 5.23 shows four applications of integral straps. If a strap is used on a box, shoulders and mating grooves are needed to help alignment. However, straps are easier to make and they have higher tear strengths. The two major factors contributing to bending stresses are the thickness of the material and the radius of curvature of the bend. In a hinge, the material thickness at the hinge point is very thin (less than 0.005 in.) resulting in small bending stresses. A strap has a thickness of over 0.025 in. but also a larger radius of curvature (0.10 in.) at the bend. Straps can be made with any soft plastic: ethylene, vinyl, EVA, rubber, ionomers, etc.

Integral plastic hinges are also called *living hinges.* Propylene hinges can be manufactured by processes such as hot-stamping, extrusion, injection-molding, and blow-molding. Vinyl hinges are made by stamping or extrusion. Nylon, acetal, and even high-impact styrene can be molded or stamped into hinges. Some hinges are needed only in the assembly process, not in the finished product. There is just

Figure 5.22. Integral plastic hinges made of polypropylene or polyallomer.

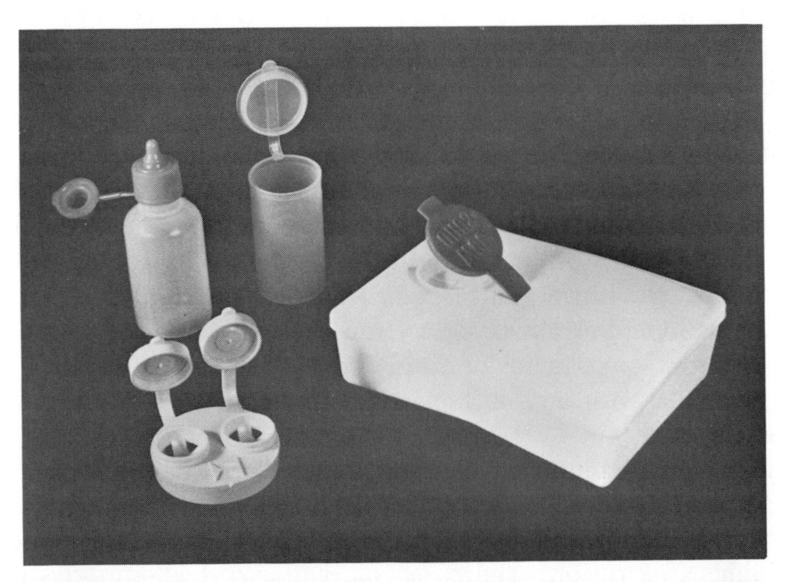

Figure 5.23. Integral plastic straps can be made with any soft plastics.

one cycle of flexing required of such hinges. Figure 5.24 shows an example of a 1-flex hinge.

The following are several useful hints about a molded living hinge. (1) The land of the hinge (Figure 5.25(a)) should be about 0.06 in. wide for a proper flow pattern of propylene melt. If the land is too wide, the pressure across the land will be too high. (2) The minimum plastic thickness at the center of the hinge should be 0.010 in. to 0.015 in. thick and 0.020 in. wide. (3) When the plastic melt flows across the small hinge gap, frictional heat will be generated. There should be sufficient cooling of the mold around the hinge area. (4) The hinge gap is a difficult area to flow across. Therefore, the gate should be placed so that the molten plastic can flow perpendicular to the hinge to ensure a good fill. If the melt flows along the length of the hinge, there is bound to be short-shot or cold-weld at the hinge (Figure 5.26). Extruded hinges are inferior in flex life because the material molecules flow parallel to the axis of the hinge. (5) Shoulders and lips should be included in the two mating parts to help alignment. (6) The finished piece should be flexed immediately upon ejection from the mold, while the heat from the mold is still present. A flex angle between 90° to 180° is recommended. The flexing action can stretch the hinge area by 200% or more; thus the initial 0.010 in. to 0.015 in. thickness will be

Figure 5.24. The polypropylene hinge on the box can be flexed thousands of cycles. The styrene hinge on the slide mount can be flexed only a few cycles.

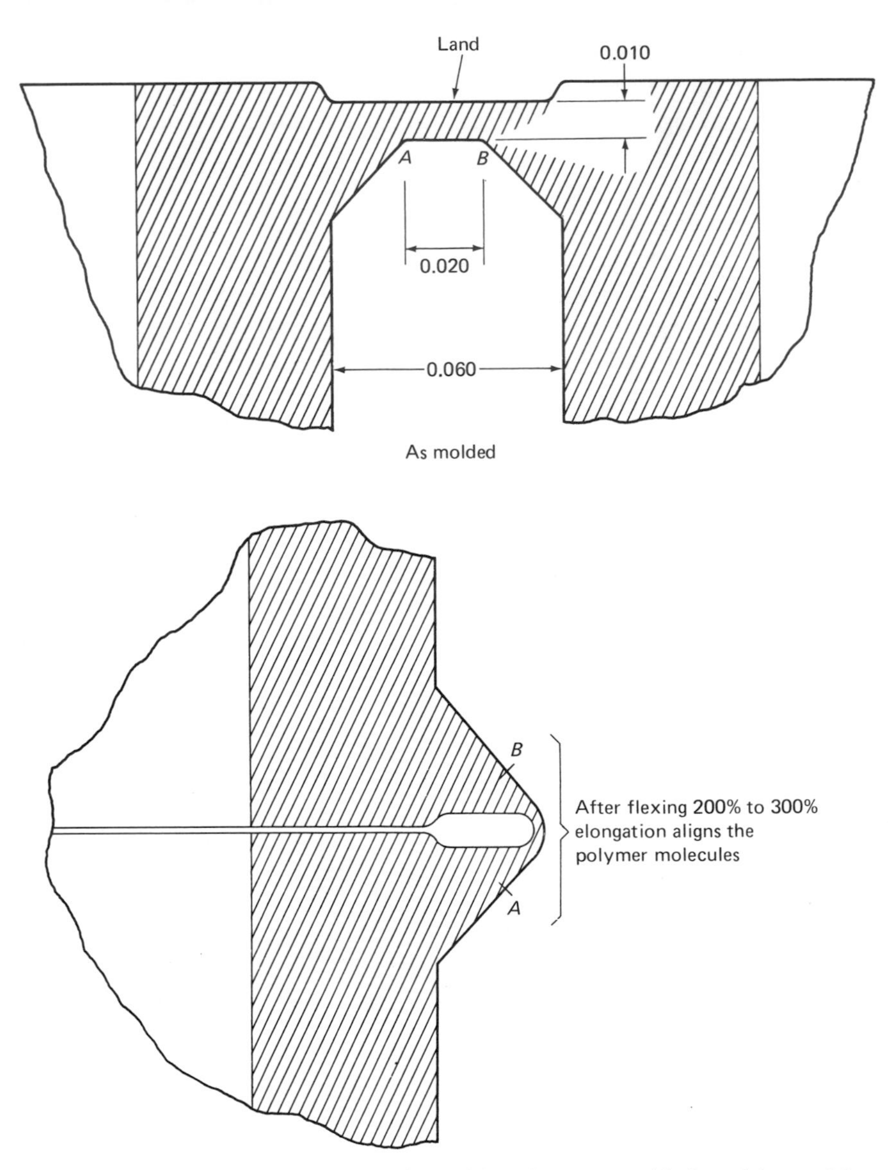

Figure 5.25. Details of a molded plastic hinge in an as-molded position and in a flexed position.

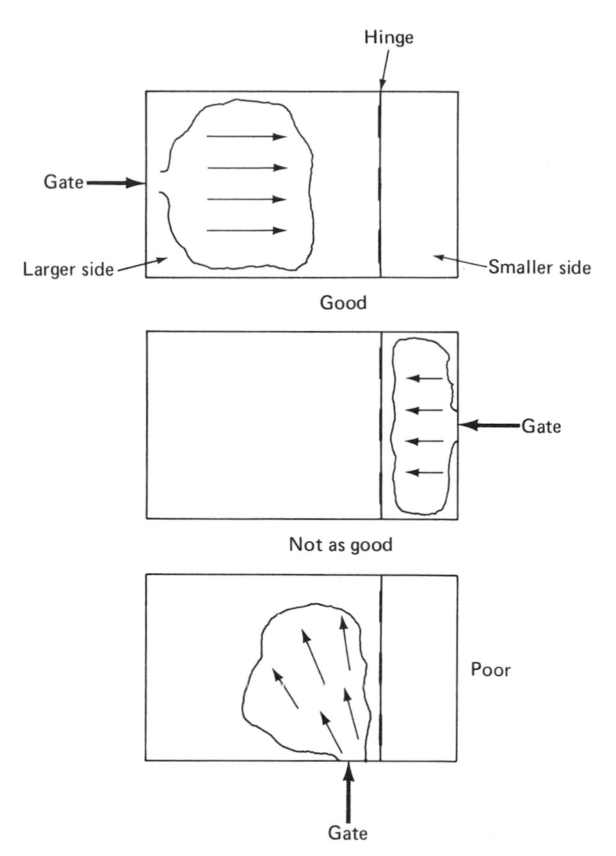

Figure 5.26. The position of a gate with respect to the hinge.

thinned down to less than 0.005 in. (Figure 5.25(b)). This elongation aligns the plastic molecules and increases the tensile strength from 5000 psi to 80,000 psi.

Hot- and cold-stamped hinges are made by compressing a sheet of material down to the desired thickness (about 0.010 in.). Stamped hinges are less durable in flexing, but are stronger in tear. Hinges in blow-molded parts are made by leaving a 0.010 in. gap at the pinch-off.

5.6.2. Design Ideas with Living Hinges

Hinges can be used effectively to reduce the number of parts in an assembly. A small toolbox can be molded with a hinge connecting the lid and the box. A 4-piece automobile accelerator pedal can

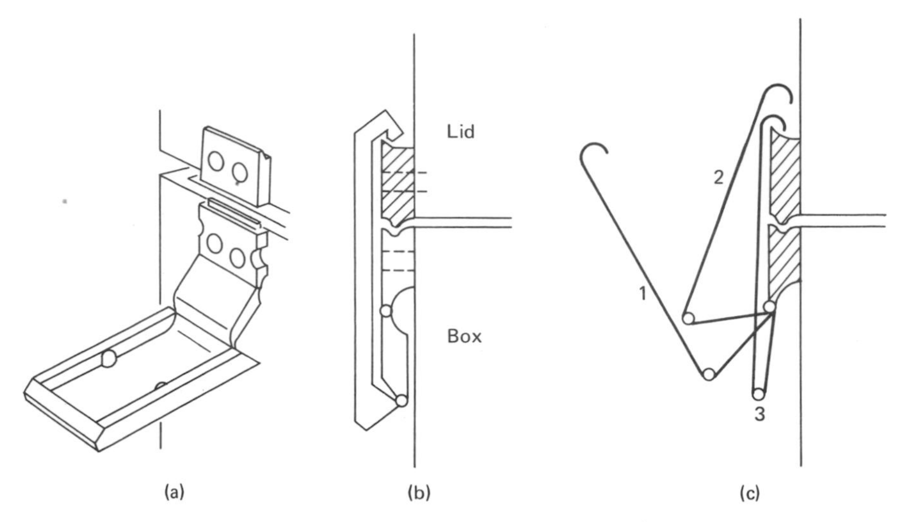

Figure 5.27. A propylene draw catch. (a) is an isometric view of the catch when open. (b) is a sectional view of the catch when closed. (c) is a simplified sketch showing the action during closing. (Courtesy Southco Inc.)

be reduced to 1-piece. The slide mount in Figure 5.24 shows how a hinge can improve alignment and provide a free joint. Since side cores in molds are expensive, boxes with detailed side patterns can be made by folding hinged panels together. In these designs, it is important to keep the plastic melt flowing perpendicular to the hinges.

The ability to produce many articulated parts in one shot opens new design possibilities. Figures 5.27 and 5.28 show a draw catch and a disposable forceps. Since the hinge is practically free, the cost of these products is just the cost of the material and the molding.

Living hinges combined with plastic spring components can produce preset position hinges (detent snaps). Figure 5.29 shows a bistable door closure made from extruded propylene. Three other patented preset position hinges are shown in Figures 5.30, 5.31, and 5.32 [75]. These gadgets act like spring clips and loose-leaf binders. Detent snaps are used in swing doors, hinged can lids, articulated dolls, jewelry boxes, eyeglass boxes, food containers, etc. Most of these detent snaps require 3 to 4 hinges and 1 spring. Although very inexpensive to manufacture, propylene is not the best material for

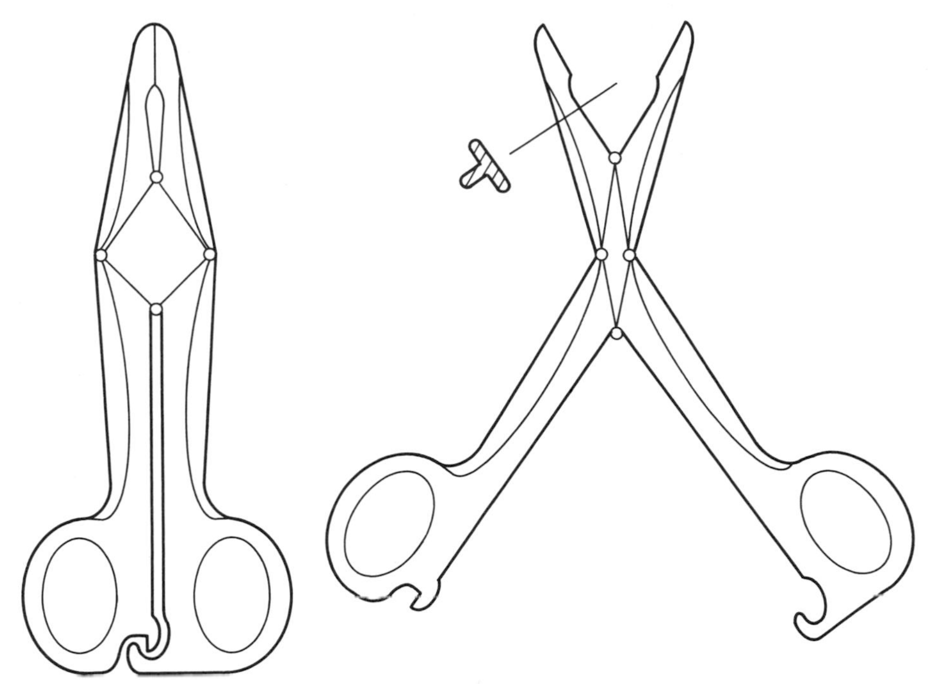

Figure 5.28. One-piece, molded, polypropylene forceps developed by Imperial Chemical Industries, Ltd, Great Britain [72].

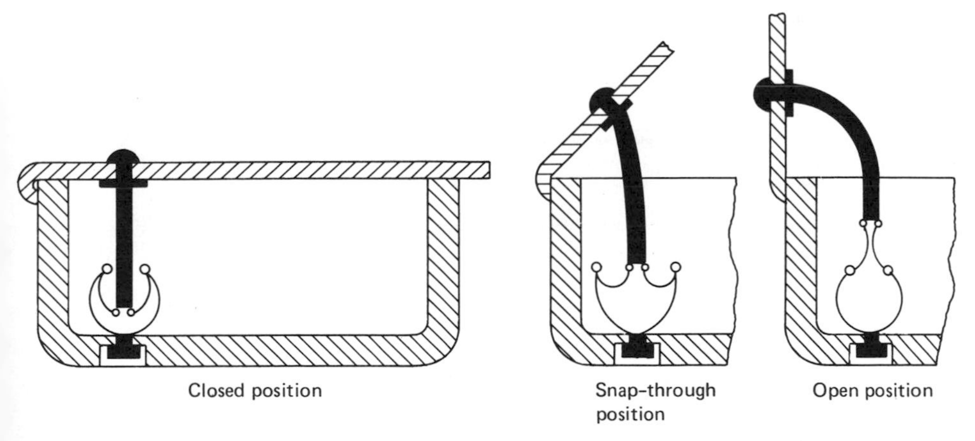

Figure 5.29. A bistable door closure designed by General Motors Corporation (U.S. Patent No. 3,541,370). The spring-loaded device is made of extruded polypropylene [73].

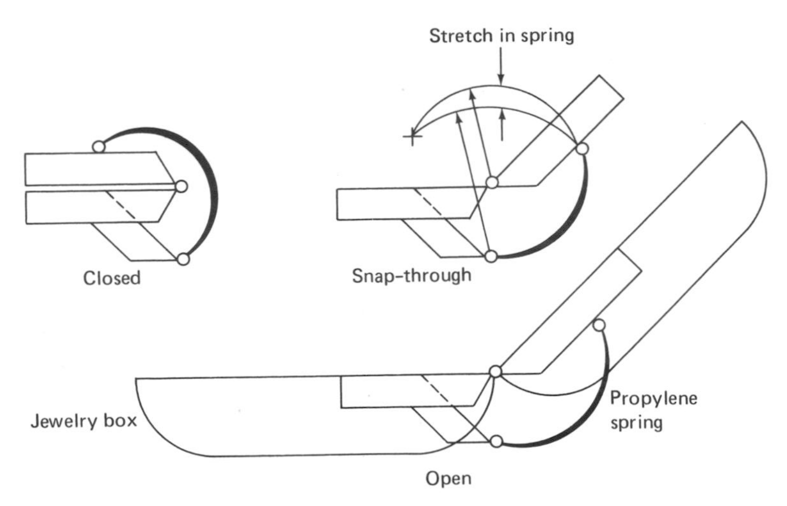

Figure 5.30. An injection-molded, polypropylene, spring-loaded hinge. In the open or closed position the polypropylene spring is stress-free. In the intermediate snap-through position the spring is stretched.

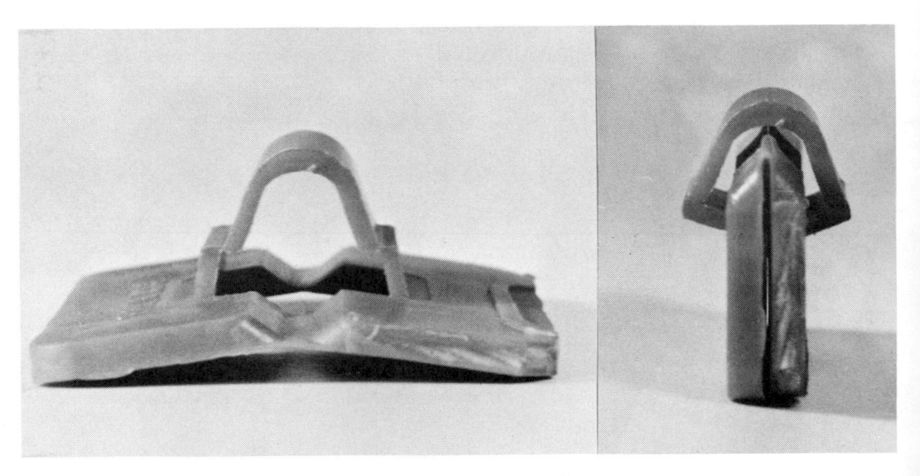

Figure 5.31. A spring-loaded hinge that opens 180°. (U.S. Patent No. 3,720,979, Aggogle Inc., 80 Fifth Ave., New York.)

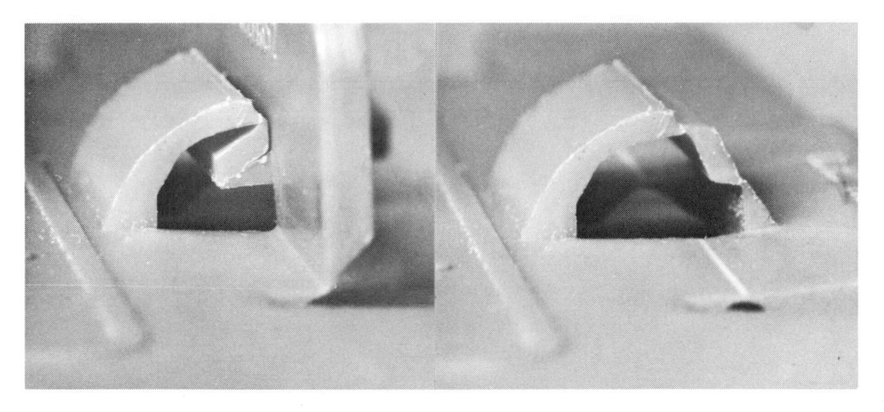

Figure 5.32. A spring-loaded hinge that opens 90° [75]. (U.S. Patent No. 3,594,852, Aggogle Inc., 80 Fifth Ave., New York.)

plastic springs. It works well for temporary spring actions with moderate forces. For longer periods of time and larger forces, it tends to relax and lose part of the spring force.

5.6.3. Some Other Plastic Hinges

Plastic hinges can be inexpensive and practical. The central idea behind their design is to make the part as simple as possible and to avoid using metal hardware. There is a trade-off between the intricacy of a mold and the cost of assembly and finishing. For production volumes of more than 100,000 pieces, it is definitely worthwhile to build an intricate mold that eliminates all the finishing processes. For small-volume productions, trial runs, and prototypes, the mold should be simple. The pieces can be drilled, heat-staked, and hand-fitted together. Since there is sliding between mating parts, the hinge acts like a bearing. If the hinge is to be truly durable, it is desirable to make at least one side of it of nylon, acetal, or olefin materials. The following methods of hinge formation do not require metal hardware in the hinge, and are suitable for mass-productions.

Insert-molded posts. A post made of nylon can be insert-molded into a different plastic to form a strong and precise hinge. The nylon part must be perfectly round and very smooth. The hinge will have enough friction to stay in any open position. With the proper selection of material, a hinge with an insert-molded post can last practically forever.

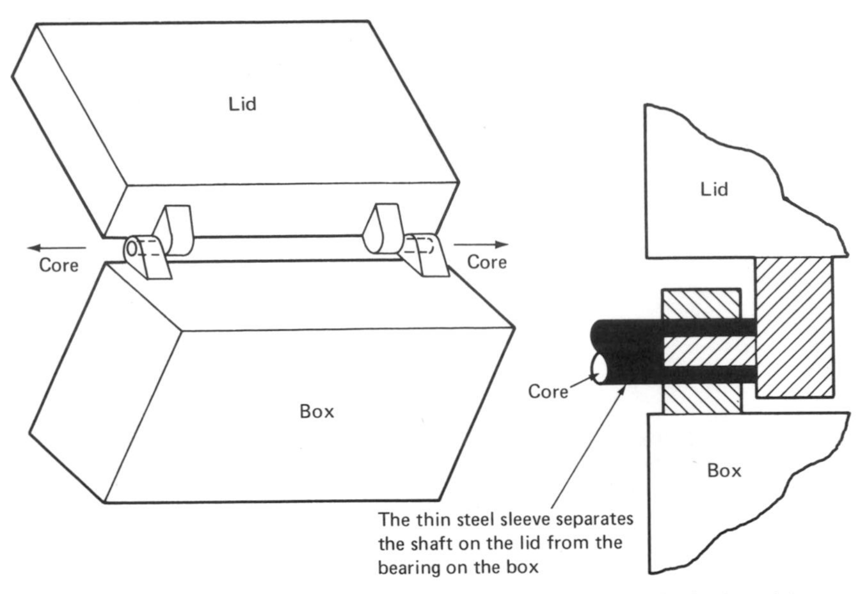

Figure 5.33. The lid and the box are molded in an interlocked position.

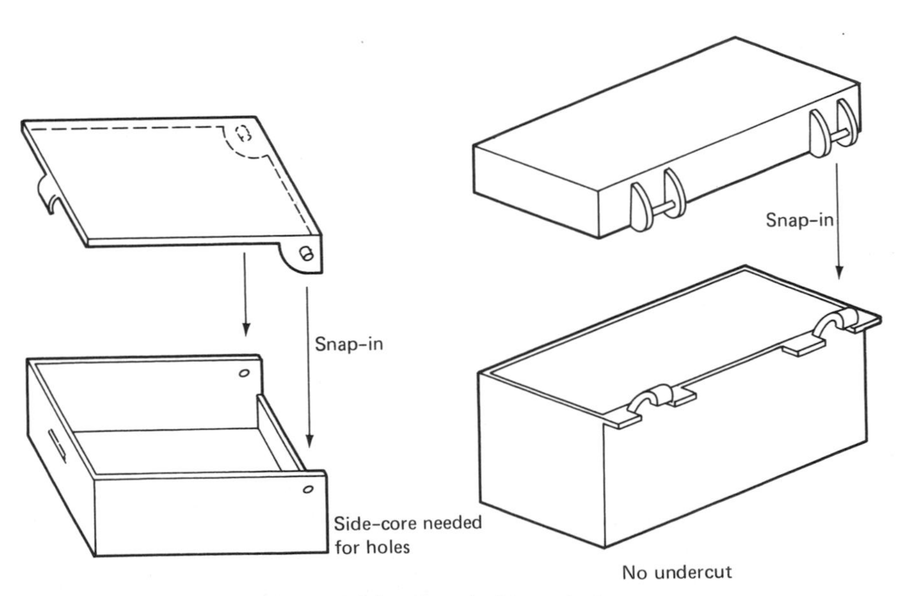

Figure 5.34. Snap-in hinge designs.

Side cores in the mold. If a small box has only 2 post-and-socket hinges, the box can be molded with the posts interlocked in the sockets by using side-cores in the mold (Figure 5.33).

Snap-in posts (Figure 5.34). In many injection-molded and blow-molded parts, snap-in hinges work quite well. Some manufacturers of toolboxes claim that snap-in hinges can be cheaper than living hinges because a lower grade of ethylene can be used.

Ball-and-grip hinge (Figure 3.18 (a)). This is the standard design for crystal styrene boxes. The 0.125 in. diameter balls snap on to spherical sockets 0.020 in. deep. The sockets create undercuts in the mold, but they can be ejected without any problems.

6. SIMPLIFYING ASSEMBLY

It has been suggested that ultimately well-designed components could self-assemble in a tumbler. We know from the second law of thermodynamics that there is a tendency for things to become disorganized and disassembled. In order to have things that can self-organize, it is necessary for them to have very low potential energy in the final assembled state. Such energy levels may be derived from magnetic fields, surface tension, fluid buoyancy, etc.

The easiest way to save assembly is to reduce the number of components and to use integral design. Section 6.1 covers general considerations of cost and design in assembly. Then, the following six topics are individually discussed: (1) snap-fits and press-fits, (2) fasteners, (3) electrical and electronic assembly, (4) adhesives, (5) soldering, brazing, and welding, and (6) welding plastics.

6.1. DESIGN FOR EASIER AND FASTER ASSEMBLY

It is a mistake to assume that the cost of fastening is the cost of the fastener plus a small amount of labor. In many actual situations, the fastener cost is very small (maybe 10%) compared to the total in place cost, but such other factors as equipment, joint preparation, labor, personnel, training, setting time, and operational cost contribute to the cost. It is rather difficult to evaluate the operational costs, which combine the purchasing, storage, inventory-control, and handling costs associated with the item. The handling cost alone often exceeds the cost of the item.

To install threaded fasteners, the joint preparation includes drilling a hole, cleaning the surface, and tapping the threads. If precise preload is required, torque wrenches are needed for assembly. Joint preparation may be reduced by the use of better fasteners. For example, high-strength bolts can reduce the number of bolts and cut down the number of holes needed. Self-tapping and self-drilling screws can eliminate a part of the joint preparation. For cement joints, some sort of clamping device is usually needed. The application of the cement and the handling of the workpiece usually costs more than the cement itself. The setting time of the cement can be translated into equipment, floor space, and labor costs. Ideally, the cement sets immediately after it is applied to the mated parts, but lasts forever in storage. Automated assembly is substitution of a machine for labor. The cost of equipment has to be evaluated against the savings in labor. There are many levels of automation. Each incremental investment has to be studied separately to find the degree of automation best suited for the situation. The cost of assembly labor can be reduced by time-and-motion study of the job. Proper setup of the work area, convenient arrangement of components, and proper workflow are essential. The design engineer can help by reducing the number of parts and avoiding the use of small parts. The assembly effort can be reduced by the following features: easy line-up, easy grip, low assembly force, and low assembly energy. Snap-fitting is a proven method that can save labor because of the assembly speed, the low energy needed, and the absence of small and loose parts.

A sure way to lower operational cost is standardization: reducing the variety of fasteners, increasing the quantity of the standard items, and using the easily available types. Of course, wherever special fasteners are functionally superior, they should still be used. Fastener reduction effort should begin by alerting the design engineers to the standard items available. Guidelines, which list fasteners according to function and recommend selected items as "preferred" among several alternatives, can be helpful.

6.1.1. Manual Assembly

Manual assembly combines the sensations of sight, sound and touch with hand movement to accomplish the task. Sensing is the control function. The information sensed may be color, orientation, size, roughness, alignment, etc. Humans have enough flexibility to deal with irregular situations and to make decisions. However, an assembly operation requiring less judgment and less in-process planning will be faster. Self-aligning parts, color coding, and identification tabs can facilitate identification and minimize error. To speed up assembly, hand movement should also be simplified. Some of the ways to simplify hand movement are: reducing the number of separate motions, reducing the distance of hand movements, reducing the required force, and using ballistic motions.

The speed of response to different external stimuli depends on the person and the type of stimuli involved. The typical response time to visual, sound, and touch stimuli are 0.23, 0.19, and 0.18 sec., respectively. Color-coding is one of the visual signals that require less concentration. Color-coding is used in electrical wiring to distinguish apparently similar wires and terminals. With metallic terminals and wires, a silver color is used for the negative terminal, and copper color is used for the positive terminal. Color coding is often applied to the packaging of chemicals, photographic materials, and machine components (though it can fail in its purpose when used under colored illumination or when used by a colorblind person). With components that have special orientation, markers can be used. A tapered tube with a diametric difference of a few percent between the two ends is difficult to orient. Some markings, such as a notch or groove will

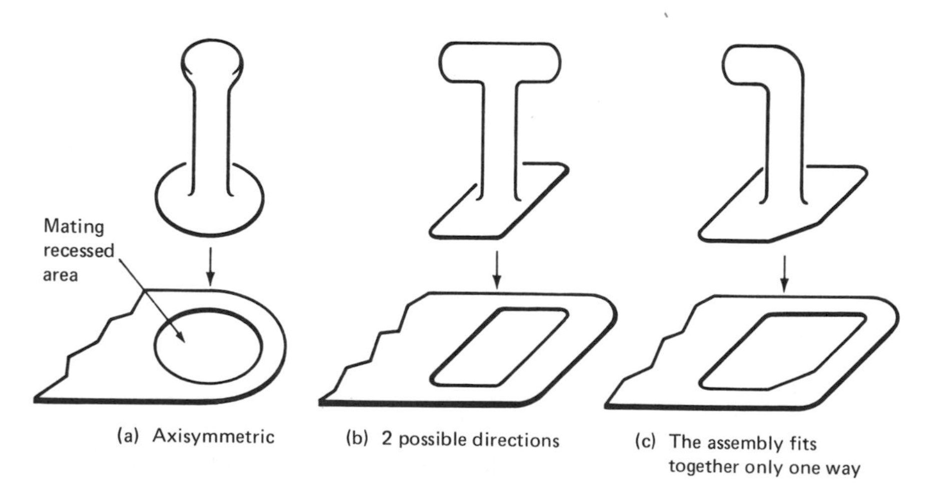

Figure 6.1. Three typical types of orientation. (a) is best, (b) is next, and (c) requires the most attention.

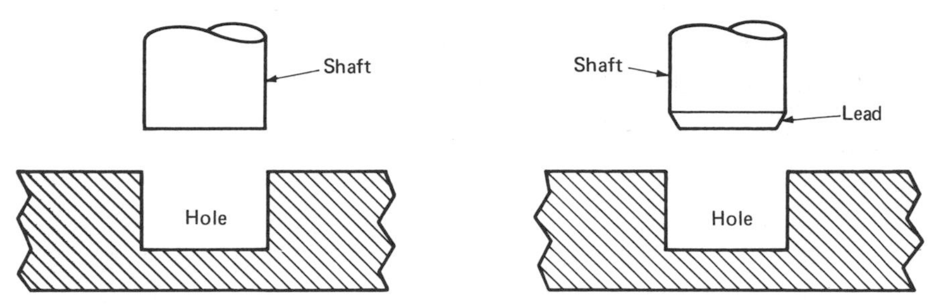

Figure 6.2. A lead helps inserting. The lead may be placed on the shaft, on the hole, or on both.

help identification. Figure 6.1 shows three typical types of orientation: complete symmetry, symmetrical in one axis, and no symmetry. The two common ways to facilitate orientation are complete symmetry and gross asymmetry. When a piece is completely symmetrical, orientation is eliminated. Gross asymmetry is helpful because the orientation is unmistakable. Electronic components have color dots, irregular tabs, and asymmetrical leg patterns, so that they can match the patterns on the printed circuit board.

Inserting a shaft into a hole or a ring onto a stud is a common operation. The effort required can be classified as dropping-in, rough positioning, and careful positioning with tilt and twist. A lead is very helpful in correcting for slight misalignment (Fig. 6.2). Alignment can often be simplified with fixtures (Fig. 6.3). Screws with coarse threads are easier to align and start. Fine threads may rip if the screw is not properly started.

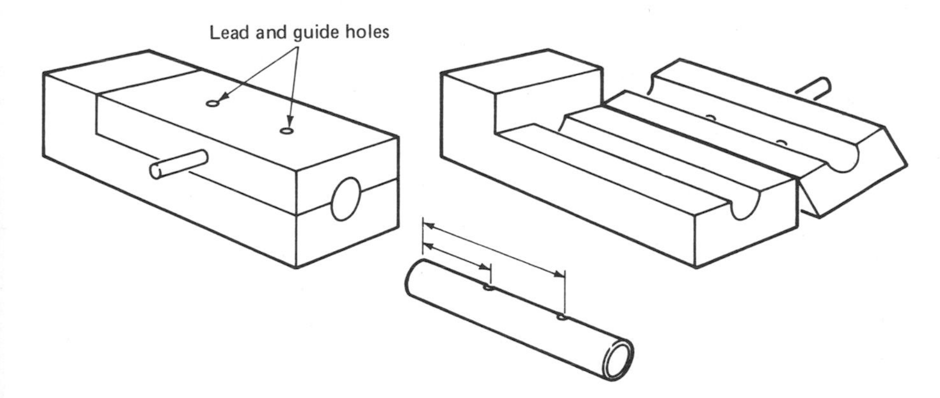

Figure 6.3. A fixture for drilling holes precisely on a tube.

Pieces should be designed so that they can be separated easily without tangling. When cups, trays, and boxes are stored or shipped in a nested position, the 3° to 10° draft angle often causes jamming. To prevent jamming, stops made of webs can be incorporated to separate the pieces (Figure 6.4).

The effort in picking up an object depends on its size, shape, rigidity, weight, and surface texture: a coin is often slid to the edge of a table to be picked up; a pencil is picked up by a pressure grasp; a tennis ball is picked up by a hook grasp; a small screw is often picked up and transferred on the tip of a magnetized screwdriver. Most of the problems with picking up and transferring parts can be solved by a proper design of bins, chutes, and containers.

The removal of finished parts from a fixture or platform can usually be mechanized rather easily. A hole on the table or a chute next to the platform is an easy way of removing parts. If the part is on a fixture, ejection by springs may be appropriate. The spring is compressed when the fixture is closed, and ejects as soon as the fixture is opened. If there are many scrap parts from the operation, a jet of air may be the best method.

The human energy required in an assembly operation can be reduced by power tools. Electric or pneumatic screwdrivers and socket wrenches are widely used. However, even with the help of power

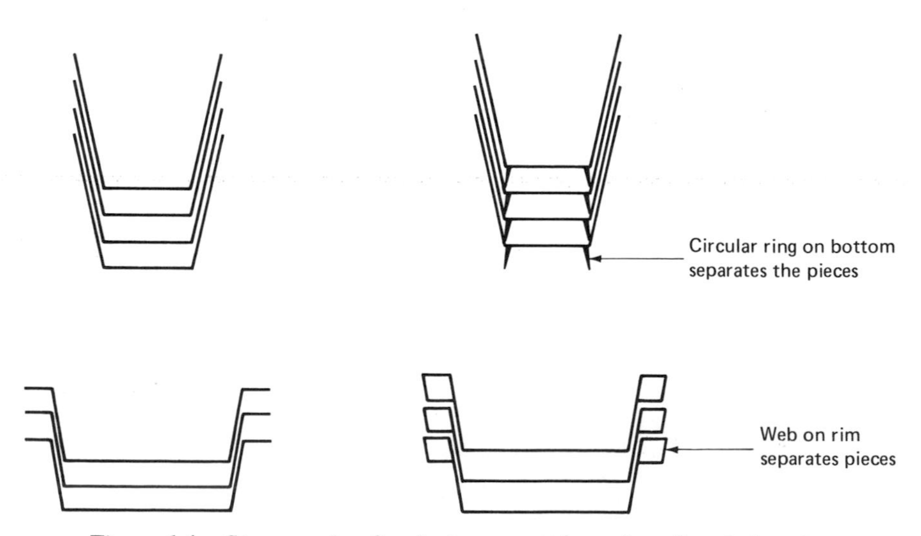

Figure 6.4. Stops made of webs to prevent jamming of nested parts.

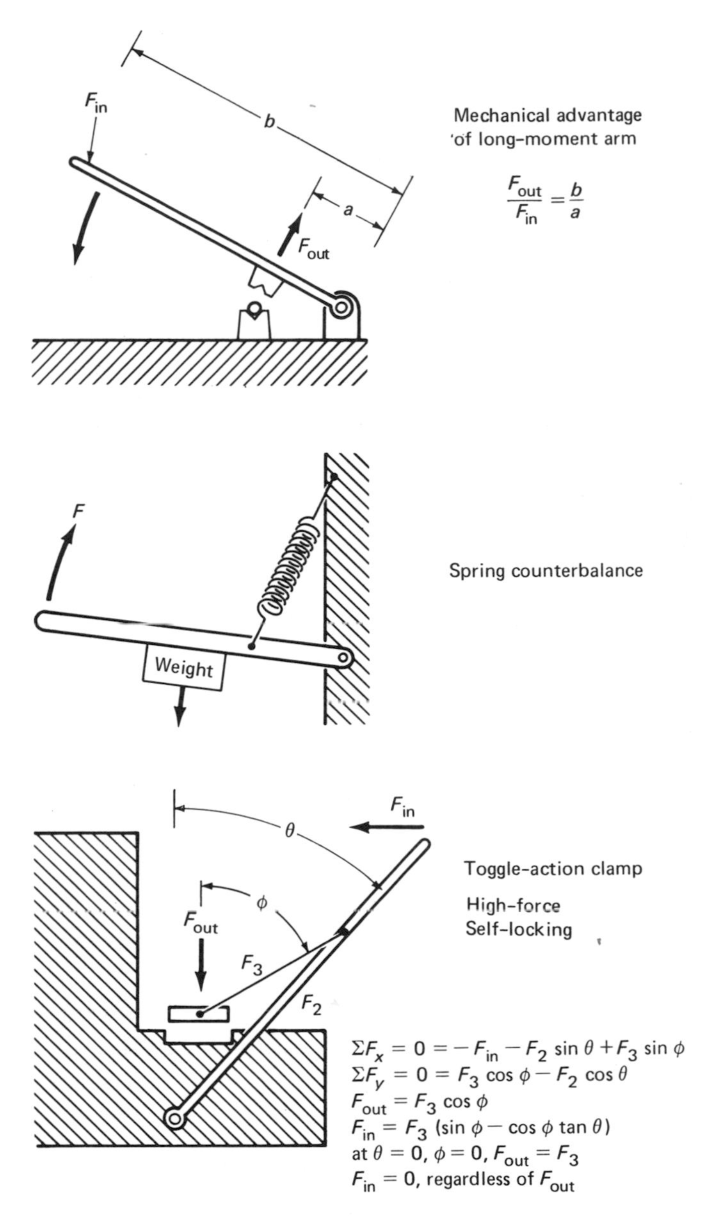

Figure 6.5. Methods of force reduction by counterforce and mechanical advantage.

tools, people can still get tired from the force and torque needed in holding the tools. Therefore, when possible, the power tool should be fixed to the ground (such as in a drill press). It is so easy for a design engineer to overlook the burden and difficulty of a worker in an assembly line. Exerting 1 lb of force with a thumb for a snap-fit may not seem to be difficult at all. However, a person doing this five times every minute for the whole day will have severely sore thumbs. Assemblies should be made as effortless as possible. Any assembly that has the potential to strain muscles should be made to fit on a fixture or arbor press. The force required can also be reduced by having long lever arms, counterbalance springs and weights, or toggle linkages (Fig. 6.5). Another step toward mechanization is the use of an air cylinder to substitute for manual force.

Automatic feed is a common way to speed up assemblies involving wires (welding), nails, screws, and plastic ties. Much of the trouble of automatic feed lies in the orientation of the fastener. Essentially, the fastener has to be arranged in coils or magazines. In one commercial machine, nails are lined up and fed on a paper strip. The paper strip is disposable after the nails are used.

6.1.2. Automated Assembly

An automatic-assembly machine can be considered a series of mechanized devices built to perform a predetermined job. A typical sequence of actions is "feed, position, assemble, and eject." In order to cope with an expected amount of disturbance and irregularities, some inspection and rejection detour may be needed. Additional intelligence in the device is needed for quality control and precision work. The design for automated assembly centers around product features that would help feed and position. It is better to assume that the assembly machine has no intelligence, and that things have to happen exactly as planned millions of times. When compared to manual assembly, there is a loss of flexibility, but an improvement in consistency, speed, force, and energy. Many design rules for manual assembly are equally useful for automatic assembly.

The feeding of components usually include orienting the components. Besides the coil-and-cartridge method, loose, small parts are also fed by vibration-feeders, chutes, magnetic devices, reciprocating devices, and centrifugal devices. Success in feeding depends on the

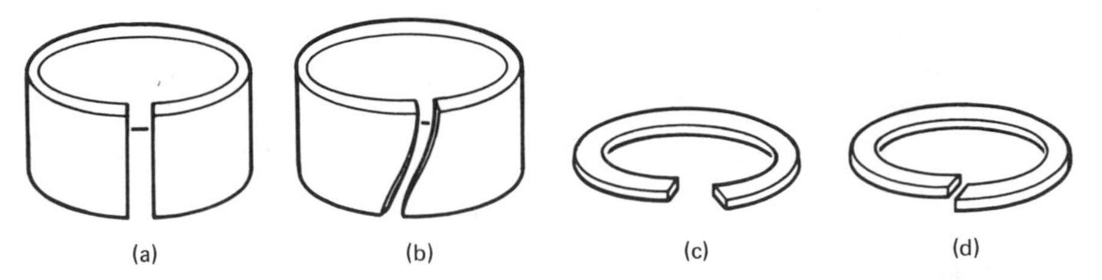

Figure 6.6. Ways to prevent component tangling. A straight gap (a) tangles easily. A curved gap (b) is unlikely to tangle. Regular lockwashers (c) hook together easily, but those with narrow gaps (d) do not.

physical characteristics of the components. Design features not needed for the function of the products may be added to facilitate feeding. A discussion of several design considerations follows.

Tangling. This problem is more serious in automatic assembly than in manual assembly. Figure 6.6 shows two ways to prevent tangling of two particular parts. With springs, the end-loop can be closed to lessen the chance of tangling.

Rough handling. Mechanized feed and transfer often subject parts to abuse and impact. Years ago, when Del Monte first mechanized the picking and processing of tomatoes, the interesting problem of how to grow tough-skinned tomatoes arose. Delicate components may have to be strengthened to endure the mechanized feed.

Tolerance and consistency. The required consistency and tolerance in mechanized feed may be more stringent than those needed for the product's function. As an example, there is a strong tendency for mold-makers to put a gate on the rim of a round part (Fig. 6.7). Such an irregularity could impede the rolling of this round part down

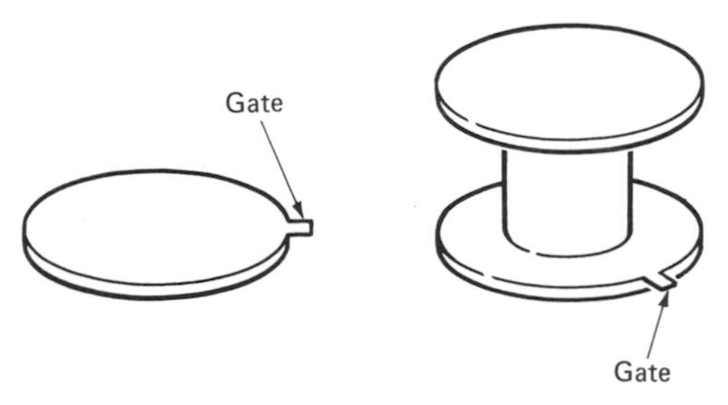

Figure 6.7. A gate on the circumference prevents the part from rolling down a chute.

a chute. The tolerance in mechanized devices may necessitate presorting and screening to eliminate foreign materials, faulty parts, and irregular parts.

Orientation features for use on a vibration-feeder. Small parts are loaded in bulk into a bowl. Then by vibration, the parts will move slowly along an upward spiral to the assembly machine. Along the path, there are cutoff areas, wiper blades, and special rails to reorient or reject parts that are not oriented as expected (Fig. 6.9). The output feed-rate is equal to the feed-rate of the feeder times the probability of a correct orientation. Figure 6.8 shows how the feed-rate can be increased by incorporating more symmetry and increasing the probability of correct orientation. If a fair die is tossed, the chance of getting any specific number is ⅙. By weighting the die, we can increase the probability of a specific orientation. Therefore, parts with gross asymmetry can be oriented because of the location of their center of gravity. When going over a specially-designed cutoff area, all the parts with an unstable location of their centers of gravity would fall off, leaving the properly oriented parts (Fig. 6.9). Small holes, internal features, and minute features on a part are difficult to feel. Prominent external features can permit detection of orientation by simple mechanical means. Square edges prevent pieces from climbing and wedging. Fig. 6.10 shows two possible behaviors of pieces in a track and in a chute.

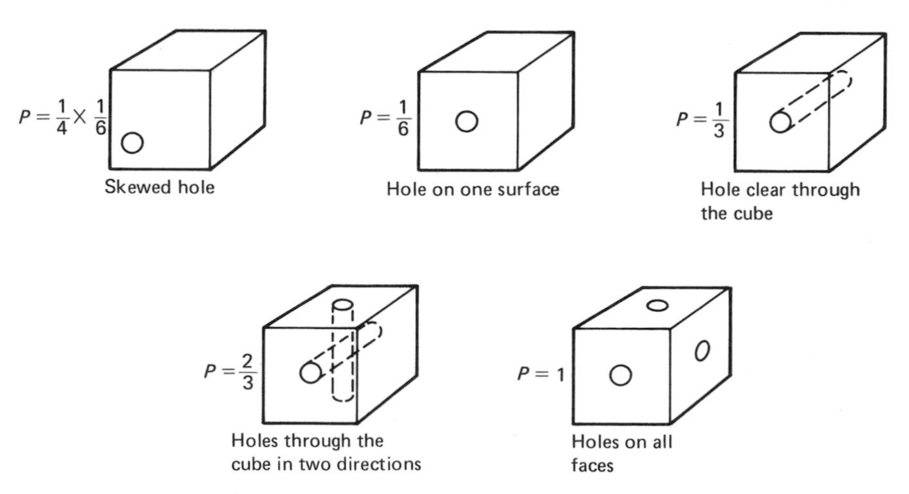

Figure 6.8. The probability of proper alignment from a random feeder.

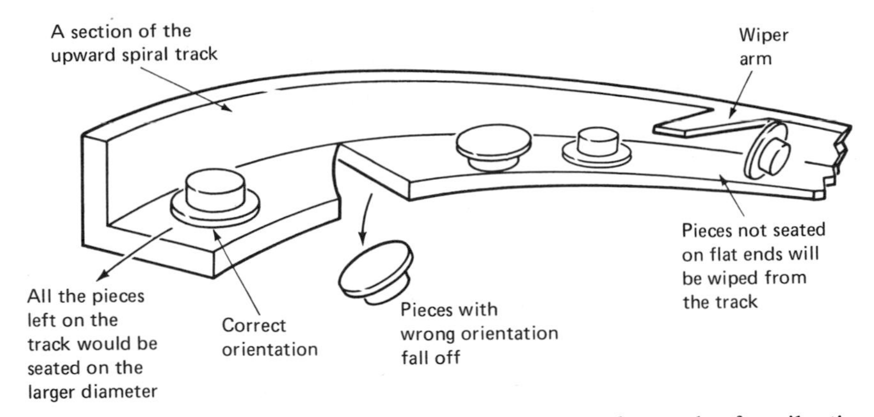

Figure 6.9. Component feeding and orientation on the track of a vibration-feeder.

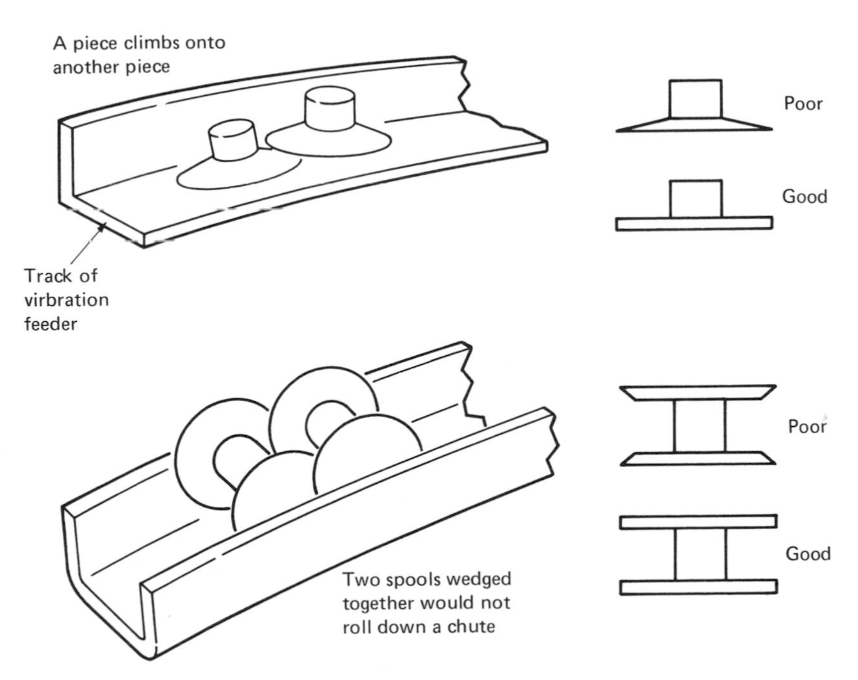

Figure 6.10. Square edges prevent pieces from climbing and wedging.

Other feed and orientation methods. There are many other feeding methods used less frequently than the vibration-feeder. These methods utilize magnetic characteristics, rolling abilities, and any special characteristics of the parts. If a piece is very delicate or very difficult to orient, it might be best to fabricate the piece in-line. For example, a stamping machine may be set up near an assembly machine. Springs are often wound on the spot and fed directly into the assembly process. Maintaining part orientation from the previous processing is a widely used method. As an example, primitive robots, called pick-and-place devices, may be used for transferring parts from a press onto a conveyor belt.

After the feeding is accomplished, the part will be transferred, positioned, indexed, and assembled. Frequently, technical assistance is available from manufacturers of such equipment as riveting and welding machines. It is desirable to start assembly with a base plate onto which other components may be added. The assembly process may consist of stacking up pieces, inserting parts into holes, or mating independent parts. The use of chamfers, radii, tapers, fillets, or leads helps insertion. Self-locating receptacles help to hold the pieces in place while transferring and indexing. If the piece is held together with several screws, the tightening of the first screw may distort it and prevent the installation of other screws. Using one screw at the center would prevent such distortion problem. Increasing the rigidity of the part, providing clamping features, or tightening the screws all at once are alternative ways to solving the problem.

The strength of the part must be consistent. Parts must be made with enough safety factors to prevent bending and collapsing under the force of assembly. It is important for the assembly action to be simple. Insertion of a piece involving tilt and twist is a very difficult task. It is almost impossible for a machine to thread wires. Design the product so that assembly can be carried out in straight-line motions, circular motions, or motions in simple geometric loci.

In the design of automated systems, all the subassemblies must be faster than the main assembly. Holding-capacity is needed to soothe the slightly varying speeds of different segments of the assembly line. Alternate routes and manual back-up methods should be considered in case of equipment breakdown. An operator is usually needed to load the raw material or the basic components. The operator may also check and correct the system's malfunction.

6.1.3. Joint Configurations from Molding

Some joint preparation can be obtained by carefully designing the mold. We will start with a brief discussion of product geometry and mold design.

A shut-off is an area of the mold at the parting line that confines the plastic melt in the cavity. The easiest shut-off is perpendicular to the direction of the mold movement (Figure 6.11). If a slight gap develops at the shut-off, clamping forces can be applied to the mold halves to close the gap. A shut-off parallel to the movement of the mold is impossible because the slightest mold mismatch will cause gaps that cannot be remedied by any means. A gap will result in flash in the molded part. It is generally accepted that 7° is the smallest angle a shut-off can have with the direction of the mold movement. The larger the angle, the more dependable the shut-off is.

Consider molding a hook as shown in Fig. 6.12. There are three ways to put the parting plane of the mold: in the *xy* plane, the *yz* plane, and the *zx* plane. The first two are easier for getting the hook. However, they necessitate drafting on the flat surface or side-cores in the mold. An engineer may choose to put the parting line in the *xz* plane in order to avoid these difficulties. To get the hook, a hole has to be provided for a core to protrude through the flat surface. All the sides of the core involving shut-offs would need 7° or more slope. Every attempt should be made to have hooks on the circumference of the object and facing outward (Figure 6.13).

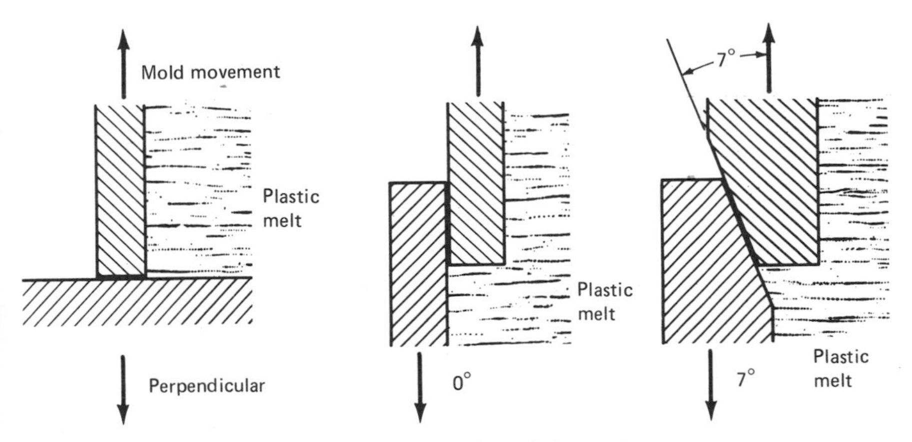

Figure 6.11. Angles of shut-off.

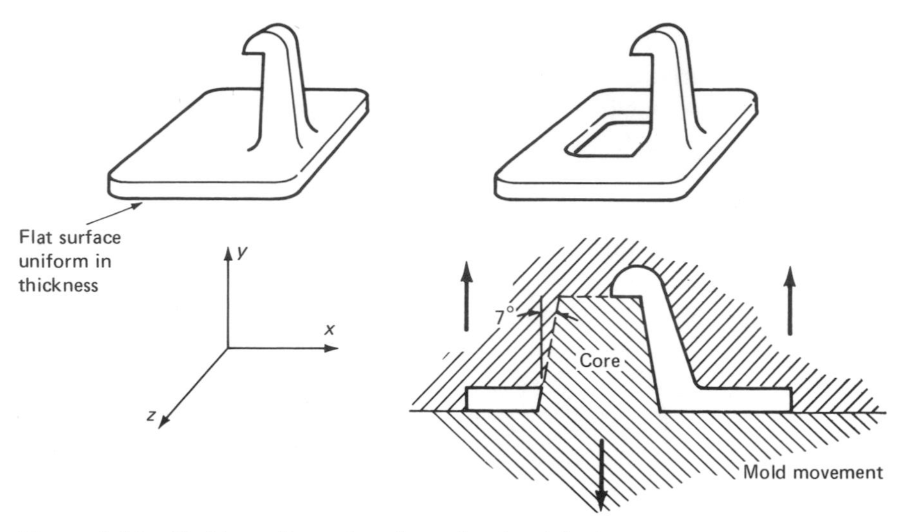

Figure 6.12. Mold configuration for a hook with the parting plane in the *xz* plane.

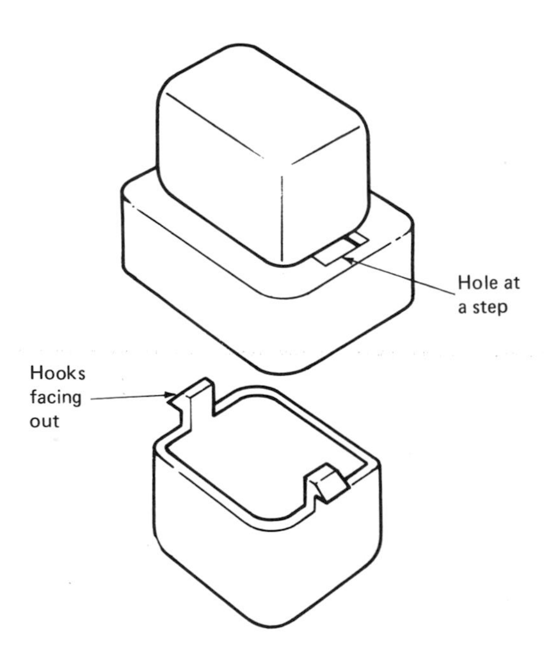

Figure 6.13. The components of a snap-fit. Holes on vertical walls are placed on a step. Hooks should be on the circumference of the object and facing outward.

A hole on a surface perpendicular to the mold movement is easy to obtain. To mold a hole in a surface parallel to the mold movement is rather tricky. Usually it requires either a very large slope in the surface or a step where the hole is needed (Fig. 6.13). One typical problem is to provide a bearing hole for a shaft of some sort (Fig. 6.14). The solution is an over-and-under design. The flat surface is segmented into pieces going over and under the shaft alternately. A core is a weak part of the mold. It is generally a good idea to provide large cut-out areas for a core, and to limit the height of a core. The alignment of two mold halves and the presence of alien matter impart side forces on the cores. Weak cores may fail during production.

To get an undercut off a mold, the part must be flexible enough. A closed-box shape will result in tears at the corners (Fig. 6.15). The battery box in Figure 6.15, with free hanging webs, can easily flex over a core. Polyvinylchloride, PVC, has incredible stretch and recov-

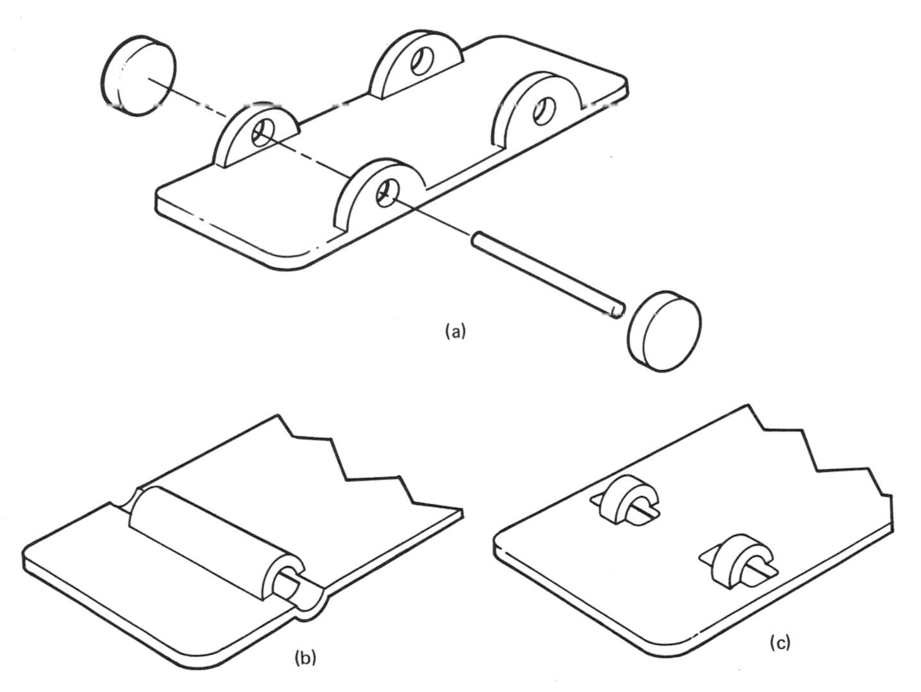

Figure 6.14. Over-and-under method for holes in vertical walls. (a) is a very costly configuration to make. (b) and (c) are two alternative configurations that simplify the mold.

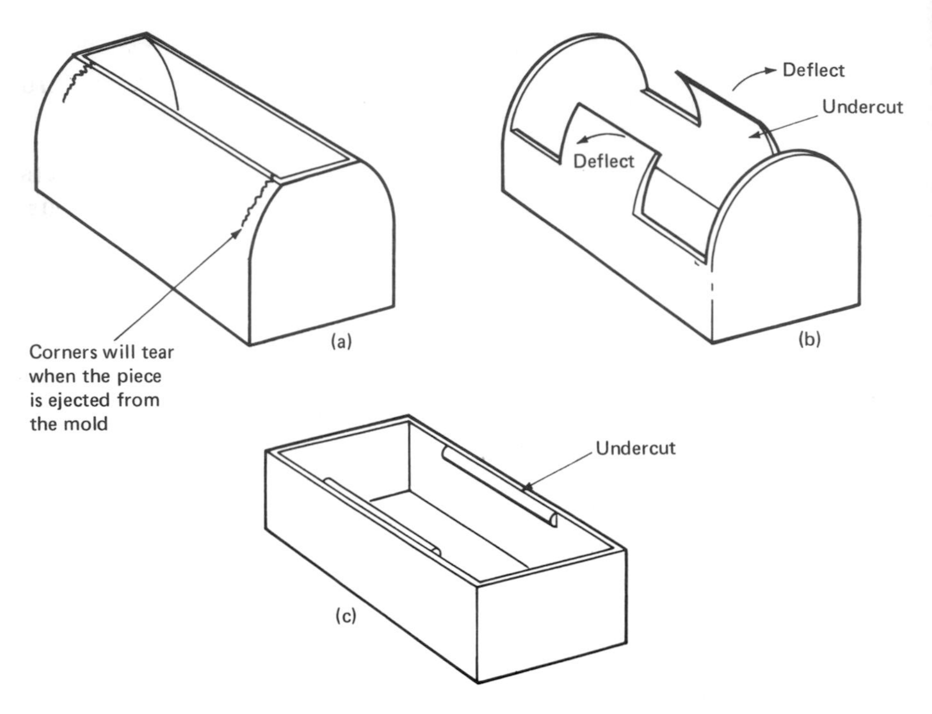

Figure 6.15. Ejection of boxes with undercuts. Undercuts should be placed at places that are capable of flexing over the mold core as in (b) and (c).

ery ability while hot. Sometimes rather striking undercuts can be overcome by using polyvinylchloride.

One could argue that a side-slide in the mold would make the best hooks and holes. A side-slide with cams or air cylinders is not only costly to make, but costly to run. The mold would be much larger, and would require a larger tonnage molding machine. The cycle time for molds with side-slides is also longer. It is best to do without complicated moving parts in the mold.

6.1.4. Joint Configurations from Forming

Formed pieces start from a blank. It is possible to prepunch the joint details in the blank to eliminate postforming machining.

The design engineer can start by designing joints such that the shape of the holes and notches is not critical. Figure 6.16 shows examples of formed pieces with prepunched details (Strasser [86]).

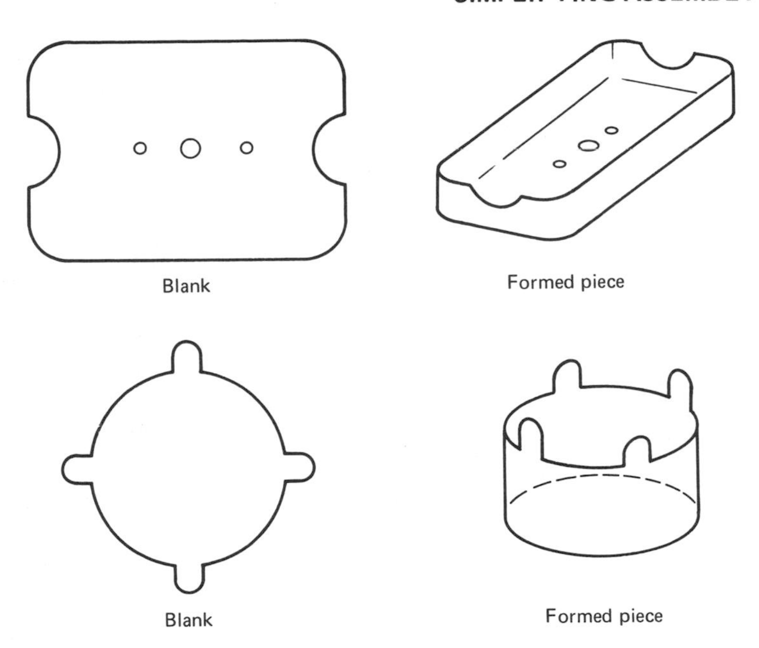

Figure 6.16. Formed pieces with prepunched details. (*Courtesy Machine Design.*)

Note that slots and holes in a blank will improve forming because they relieve stresses and reduce forming forces. Plane anisotropy of sheet material may cause rather odd deformation patterns. The material and direction of grain should be carefully controlled.

6.1.5. Insert Molding and Cast-In

Molding one component into another eliminates the assembly of the two components (Figure 6.17). Shoes, for example, are commonly made with the rubber sole molded onto the shaped leather tops. This process is called *cast-in* if the enclosing material is metal. The final product has a very intimate fit because the cast-on material shrink-fits around the insert. The process also produces exceptionally good joint strength because notches on the insert lock mechanically with the enclosing material.

With insert molding, there may be a shut-off between the two halves of the mold and between the mold and the insert. A proper shut off-configuration is essential for reducing flash (Figure 6.18). Figures 6.19 and 6.20 show five different designs utilizing zinc and lead die-

Figure 6.17. Examples of insert-molded products.

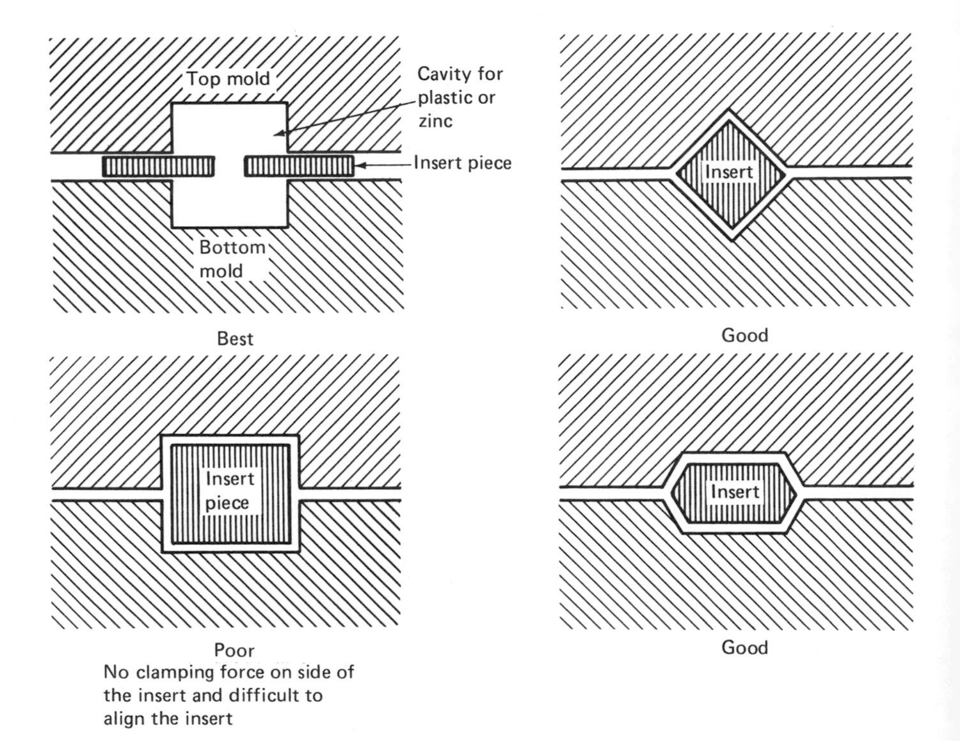

Figure 6.18. Shut-off between the mold and the insert.

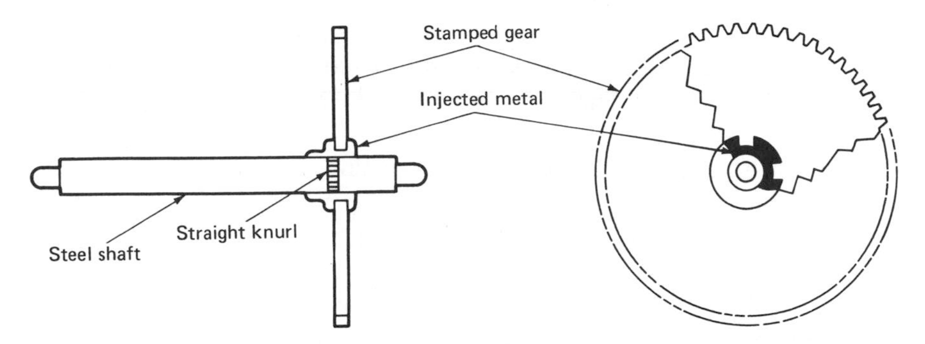

MAXIMUM TORQUE FOR DIE-CAST GEAR/SHAFT ASSEMBLY

Steel shaft dia. (in.)	Length of straight knurl on shaft (in.)	Length of cast metal (in.)	Maximum torque (oz-in.) Lead alloy	Zinc alloy
0.062	0.062	0.120	55	Fractured shaft[a]
0.062	0.125	0.120	80	Fractured shaft
0.125	0.125	0.250	440	Fractured shaft
0.125	0.250	0.250	530	Fractured shaft
0.250	0.250	0.500	—	Fractured shaft
0.312	0.250	0.250	—	Fractured shaft

[a]Shaft fractured before breaking torque of joint could be measured.

Figure 6.19. Zinc and lead die-casts are used to join a steel shaft and a stamped gear. With zinc alloy, the shaft is often the weakest component. (*Courtesy Machine Design.*)

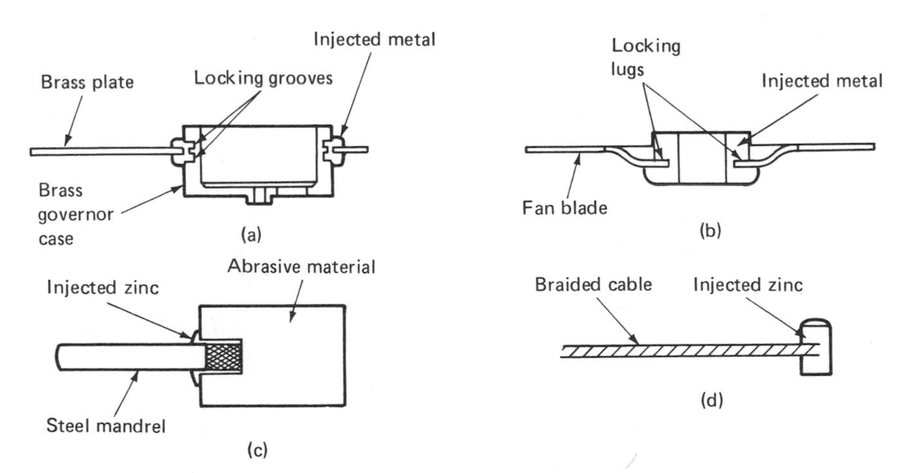

Figure 6.20. Several examples of joining with die-casting. (a) a small governor housing; (b) a small fan with die-cast hub; (c) an abrasive grinding tool; and (d) a control cable. The components have grooves, knurls, and lugs to facilitate interlocking. (*Courtesy Machine Design.*)

cast (Jay [89]). The flash problem with metal die-cast is eliminated by twisting the piece before ejecting it from the mold cavity. For this reason, at least one cylindrical or spherical surface is needed in the piece.

The speed of metal die-casting is amazing—about 750 units per hr. The metal freezes almost instantly in the mold cavity and does not burn the insert even if it is plastic or paper. The volume of die-cast metal is usually less than 0.3 $in.^3$

Because the joint strength is derived from the shrinkage of metal around the interlocking features of the insert, surface cleaning and preparation are not needed. The use of grooves, notches, lugs, and knurls is shown in Figures 6.19 and 6.20. The joint's strength depends on the strengths of the materials, the locking features, and the size of the insert. For the brake cable in Figure 6.20:

$$\text{pull-out strength} = \text{shear area} \times S_{ys}$$

$$= \text{cable diameter} \times \pi \times \text{insert length} \times S_{ys}$$

where S_{ys} is the yield strength in shear of zinc alloy.

If the end of the cable is spread to a larger diameter, the pull-out strength will improve. The torsional strength of a die-cast hub is shown in Figure 6.19.

6.1.6. One-Screw Assembly

The biggest objection to weld joints and adhesive joints is that they cannot be disassembled for repair. Screws are desirable because everybody knows how to tinker with them. The location of screws usually suggests the disassembly scheme. There are many assembly situations where one screw is sufficient. Auxiliary holding strength may be obtained by using keys, guides, mold-in pins, shoulders, and hooks. Figure 6.21 shows four situations where one screw is used for aligning and holding the pieces together. Too often, several screws are used just because the engineer has not considered the one-screw assembly method. Sometimes several pieces can be assembled by one screw between the base plate and the cover plate. The other pieces are trapped in holes, in grooves, between pins, or between slots.

applied to any combination of materials, such as metal and plastics,

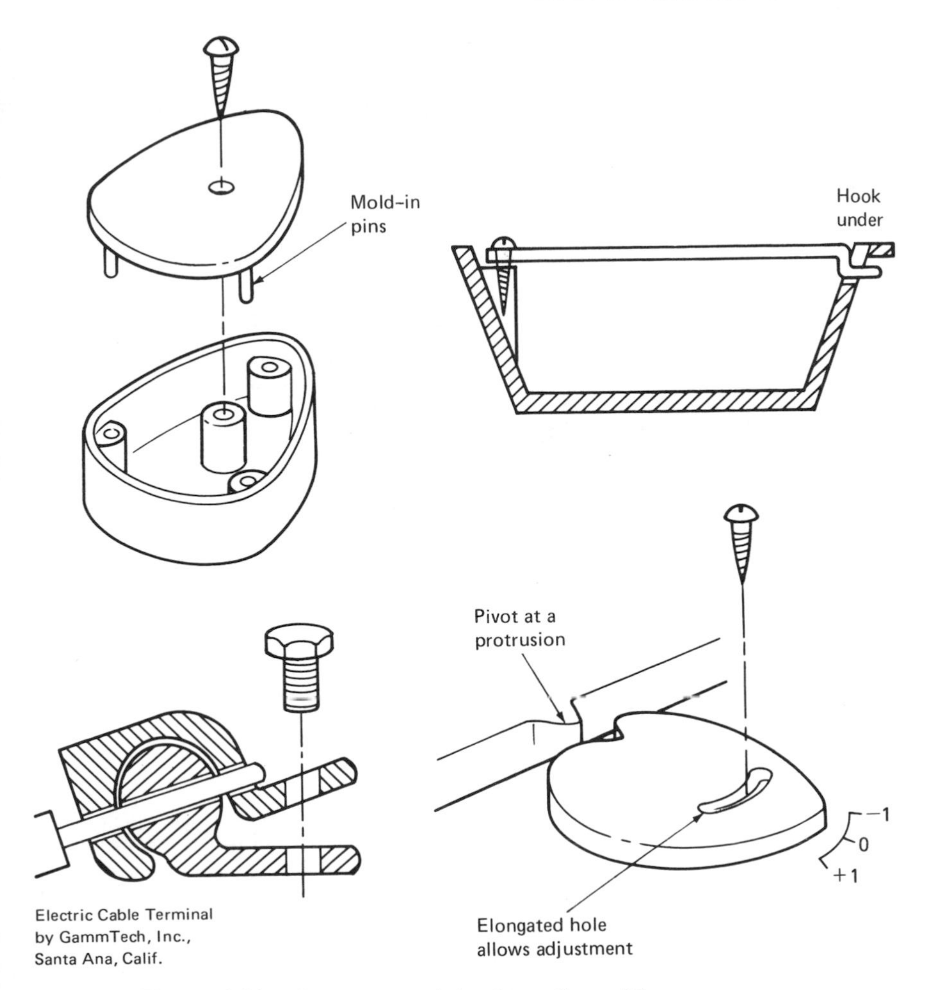

Figure 6.21. Components joined together with one screw.

6.2. SNAP-FITS AND PRESS-FITS*

Snap-fits are widely used today for temporary and permanent assemblies. Besides being simple and inexpensive, snap-fits have many superior qualities.

As a method of permanent assembly, snap-fitting can be compared to sonic-welding, spin-welding, and press-fitting. Sonic-welding and spin-welding require that the plastics to be joined are at least compatible, preferably of the same material. A polyethylene lid, for example, cannot be welded to a polystyrene bottle. Snap-fitting can be

*Reprint from Ref. [91].

glass and metal, and others. Snap-fits are stronger and more dependable than press-fits. In a press-fit, the pull-out strength is derived from friction, and is about the same as the press-in force. The strength of a snap-fit comes from mechanical interlocking, as well as from friction. The pull-out strength in a snap-fit can be made hundreds of times larger than the snap-in force. In the assembly process, a snap-fit undergoes an energy exchange with a *click* sound. Once assembled, the components are not under load. Unlike the snap-fit, a press-fit component is constantly under the stress resulting from the assembly process. Therefore, over a long period of time, stress relaxation and creep may cause a press-fit to fail. The strength of a snap-fit will not decrease with time.

When used as demountable assemblies, snap-fits can compete very well with screw joints. Again, the loss of friction under vibration can loosen bolts and screws. A snap-fit is vibration-proof because the assembled parts are in a low state of potential energy. A screw-type light-bulb socket is seldom used in machines where there is vibration, which is why snap-type light-bulb sockets are widely used in cars, motorcycles, etc. There are fewer parts in a snap-fit, which means savings in component costs, inventory costs, and bookkeeping costs.

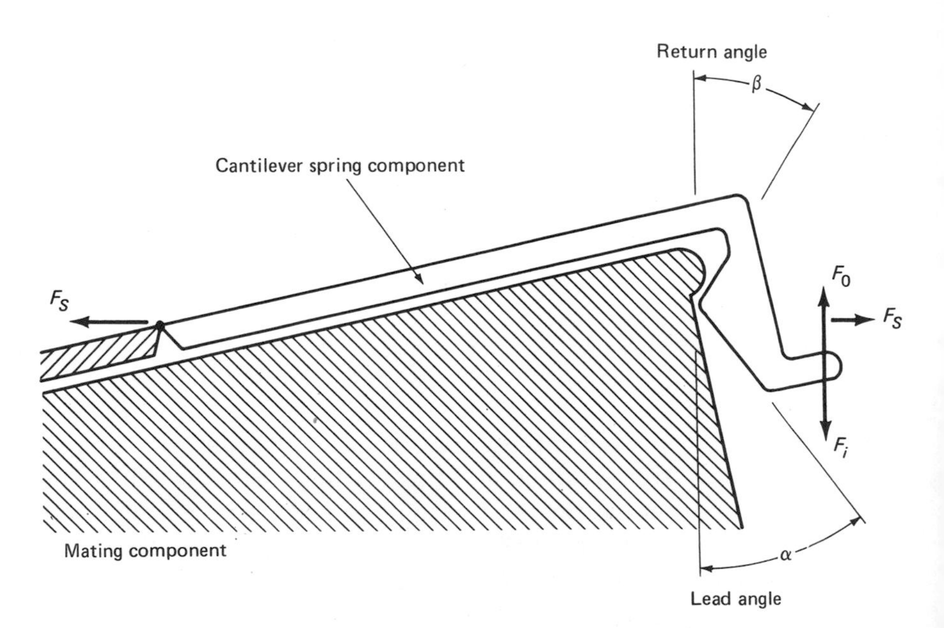

Figure 6.22. Analysis of a cantilever type snap.

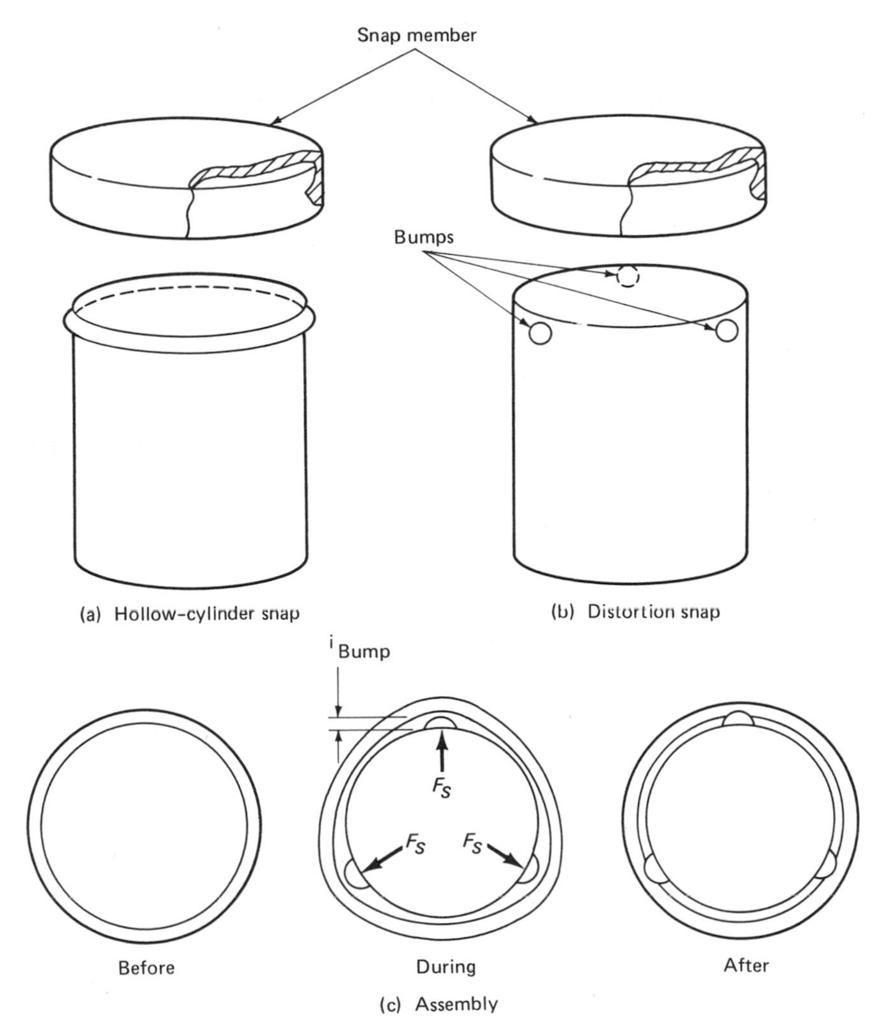

Figure 6.23. Hollow-cylinder and distortion snaps. The mating parts of a hollow-cylinder snap (a) have the same shape. In the distortion type snap (b), the shapes of the mating parts are different. During assembly (c), the lid of the distortion snap is deformed to a triangular shape to pass over the bumps.

The snap-fit is also considerably easier to assemble than all other joining methods.

The successful design of snap-fits depends on observing a set of rules governing the shapes, dimensions, materials, and the interaction between mating parts. The interference in a snap-fit is the total deflection in the two mating members in the assembly process. Too much interference creates difficulty in assembly, and too little interference causes low pull-out strength. A snap-fit can also fail from permanent deformation or breakage of its spring component. A drastic change of friction due to abrasion or oil contamination may ruin the snap.

A snap-fit can be characterized by the geometry of its spring component. The most common snaps are: (1) cantilever type as shown in Figures 6.22, 6.24, and 6.25, (the name *cantilever* is used rather loosely to include snaps with leaf-spring components); (2) hollow-cylinder type, as in the lid of a pill bottle (Figure 6.23(a)), (the name *cylinder* is also used loosely to include noncircular section tubes); (3) distortion type (Figure 6.23(b)). Distortion type snaps include any shape that is deformed or deflected to pass over an interference. While the shapes of the mating parts in a hollow-cylinder snap are the same, the shapes of the mating parts in a distortion snap are different by definition.

6.2.1. Predicting Snap-In and Snap-Out Forces

It is sometimes thought that snap-fits are not dependable because they might "pop-out." The following analysis will show that a well-designed snap-fit cannot be pulled out easily. First, we will examine the simplest case of a cantilever type snap with zero friction. Next, the same snap will be analyzed with the effect of friction. The results will then be generalized to cover all kinds of snaps.

Assuming zero friction, the snap-in force is related to the spring rate of the snap by the tangent of the lead angles and the interference (Figure 6.22).

Assuming zero friction, $\mu = 0$,

$$\text{snap-in force } F_i = F_s \cdot \tan \alpha = K \cdot i \cdot \tan \alpha \qquad (6.1)$$

$$\text{snap-out force } F_o = F_s \cdot \tan \beta = K \cdot i \cdot \tan \beta \qquad (6.2)$$

where

F_s = snap force, the force that reduces i to 0,
i = interference,
K = spring rate of the snap,
α = lead angle, and
β = return angle.

The smaller the lead angle is, the easier it is to assemble the snap. The return angle affects the snap-out force in the same way. The smaller the return angle is, the less force is needed to disassemble the snap. When the return angle is 90°, the snap is self-locking. When the return angle is greater than 90°, the interference is a barb.

When friction is considered, the tangent term becomes a more complicated function of the lead angle α and the coefficient of friction μ. With friction, $\mu \neq 0$,

$$\text{snap-in force} \quad F_i = F_s \cdot \left(\frac{\sin \alpha + \mu \cos \alpha}{\cos \alpha - \mu \sin \alpha}\right)$$

$$= K \cdot i \cdot \left(\frac{\sin \alpha + \mu \cos \alpha}{\cos \alpha - \mu \sin \alpha}\right) \tag{6.3}$$

$$\text{snap-out force } F_0 = F_s \cdot \left(\frac{\sin \beta + \mu \cos \beta}{\cos \beta - \mu \sin \beta}\right)$$

$$= K \cdot i \cdot \left(\frac{\sin \beta + \mu \cos \beta}{\cos \beta - \mu \sin \beta}\right) \tag{6.4}$$

With friction, the force required to snap-in and snap-out is larger. If the friction coefficient is set to zero in Equations 6.3 and 6.4 the sine and cosine functions reduce to a tangent function as in Equations 6.1 and 6.2. The minimum return angle that can cause locking ($F_0 = \infty$) is $(90° - \phi)$ instead of 90° as in the frictionless case. ϕ is the friction angle and is equal to $\tan^{-1} \mu$. The detailed derivations of Equations 6.1 to 6.4 are given in Appendix 5.

With an appropriate redefinition of the spring rates, the above analysis can be applied to hollow-cylinder snaps and distortion snaps. The spring rate of a snap is defined as the snap force F_s that is required to reduce the interference to zero divided by the interference. The spring rate for a hollow-cylinder snap K_{cyl} can be expressed as the total force in the radial direction F_r divided by the increase in radius i_{rad}

$$K_{cyl} = \frac{F_r}{i_{rad}}$$

Figure 6.23(b) shows an example of a distortion snap. The lid is the same as that of the hollow-cylinder snap (Figure 6.23(a)). The mating part, the bottle, is a smaller cylinder with 3 bumps. In the

TABLE 6.1. THE DIMENSIONS AND CHARACTERISTICS OF THREE SIMPLIFIED SPRING COMPONENTS.

	K	σ^a	ϵ^a	i^b
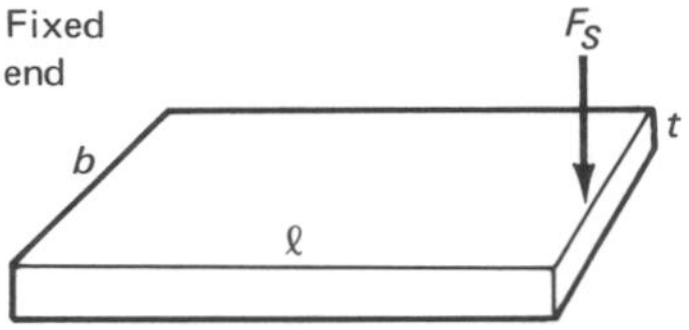 Uniform cantilever (Fixed end; F_S; b; ℓ; t)	$\frac{Eb}{4}\left(\frac{t}{\ell}\right)^3$	$\frac{3Eit}{2\ell^2}$	$\frac{3it}{2\ell^2}$	$\frac{2S_y\ell^2}{3Et}$
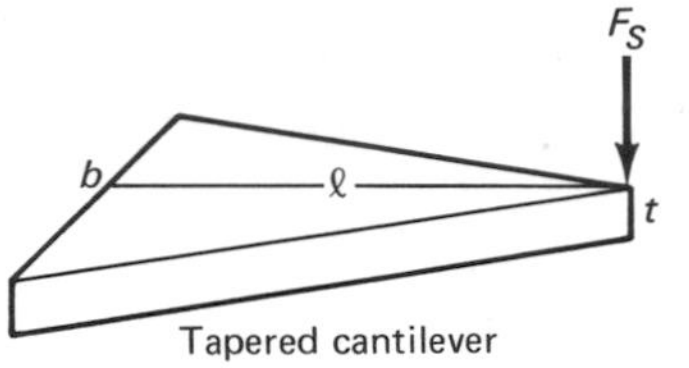 Tapered cantilever (F_S; b; ℓ; t)	$\frac{Eb}{6}\left(\frac{t}{\ell}\right)^3$	$\frac{Eit}{\ell^2}$	$\frac{it}{\ell^2}$	$\frac{S_y\ell^2}{Et}$
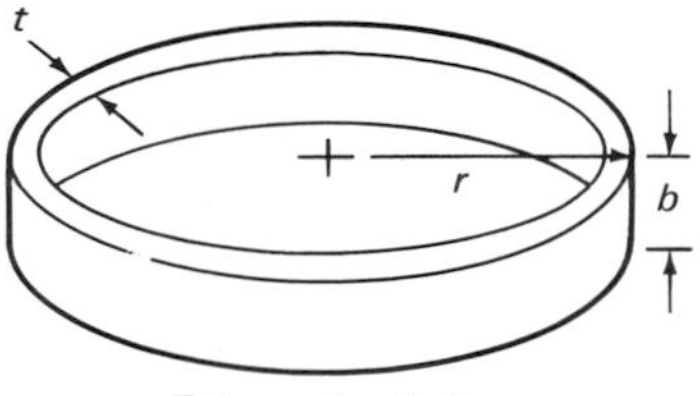 Thin-wall cylinder (t; r; b)	$2\pi Eb\left(\frac{t}{r}\right)$	$\frac{i}{r}E$	$\frac{i}{r}$	$\frac{S_y r}{E}$

[a] σ and ϵ are the maximum stress and strain due to the interference i.
[b] i is the largest value of design interference allowed for a given strength S_y.

assembly process, the lid has to be distorted from a circular to a triangular shape to pass over the bumps (Figure 6.23(c)). Here the spring rate K_{dist} could be defined as three times the force at each bump F_b divided by the interference at the bump i_{bump}

$$K_{dist} = \frac{3F_b}{i_{bump}}$$

The spring rates are so defined that the product of K and i gives the snap force, and the product of the snap force and the tangent* of the lead angles gives the snap-in and snap-out forces. Typical snap-in and snap-out forces are the range of 2.5 to 250 Newtons (about 0.5 to 55 lbs). Interferences are usually between 0.25 mm to 2.5 mm (about 0.010 to 0.100 in.). If the coefficient of friction is 0.15, the lead angle is 30°, the interference is 1 mm, and the spring rate is 5000 N/m, Equation 6.3 shows that the snap-in force would be 3.98 N. Table 6.1 shows some approximate spring rates for three simplified spring geometries: a uniform cantilever, a tapered cantilever and a thin-walled hollow-cylinder. The calculation of spring rates may be a formidable problem for many practical designs. Computer methods, such as the finite element method, and experimental techniques with scale models can give useful results.

Special keys can be incorporated in a hollow-cylinder snap such that the snap-out force is dependent on alignment. When the index is aligned, the interference decreases to a minimum. Such snaps are widely used in childproof medicine bottles.

6.2.2. Designing the Spring and the Interference

The geometry of the cantilever (also called a leaf spring or a finger) determines the spring rate, K. Besides providing the designed spring rate, the spring component has to be designed against breakage and permanent deformation. In this section, we will examine the stress and strain induced in the spring component during assembly. The stress and strain have to be kept below the material strength S_y (about 62×10^6 N/m^2 or 9 ksi for plastics) and the elastic limit (from 2% to 9% for plastics), respectively.

The force F_s induced during assembly equals the product of the K and the interference i: $F_s = K \cdot i$. When this force is acting at the tip of the cantilever, the maximum stress at the root of the cantilever has to be lower than the strength of the plastic material (Figure 6.24). Refer to Table 6.1:

$$\text{bending moment } M = F_s \cdot l$$

*Or sine, cosine, μ function if $\mu \neq 0$ (Equations 6.3 and 6.4).

$$\text{stress } \sigma = \frac{M(t/2)}{I} = \frac{F_s l \cdot t/2}{I} = \frac{Kil(t/2)}{I}$$

$$= \frac{3Eit}{2 \cdot l^2} \leqslant S_y \text{ yield strength of material}$$

$$\text{design interference } i \leqslant \frac{S_y 2l^2}{3 \cdot Et}$$

Where E is the material elastic modulus, and I is the moment of inertia of the cross-section.

Similar analysis can be applied to other types of springs. The spring rate, K, the stress, σ, the strain, ϵ, and the designed interference, i, are listed in Table 6.1 for three simplified spring components. For a given interference, a lower spring rate means less stress. The spring rate of a cantilever can be decreased by increasing the length of the cantilever or by tapering the cantilever toward the tip. The tapered cantilever, with 1/3 lower spring rate, can flex pass the interference with 33% less force and stress (Table 6.1). Using a smaller thickness t to decrease the stress may cause difficulties in stability, pull-out strength, and molding. If the return angle is larger than 90° minus the friction angle ϕ, the snap-fit would not snap-out (see the explanation following Equation 6.4). Pull-out disassembly would cause shear failure around the interference or tensile fracture along the length of the cantilever (Figure 6.24). In Figure 6.25, the spring component is stronger because it experiences only compressive stresses in the pull-out process.

Referring to the spring rates listed in Table 6.1, the ratio (t/l) and (t/r) are of the order of 1/10. Therefore, for snaps of the same mate-

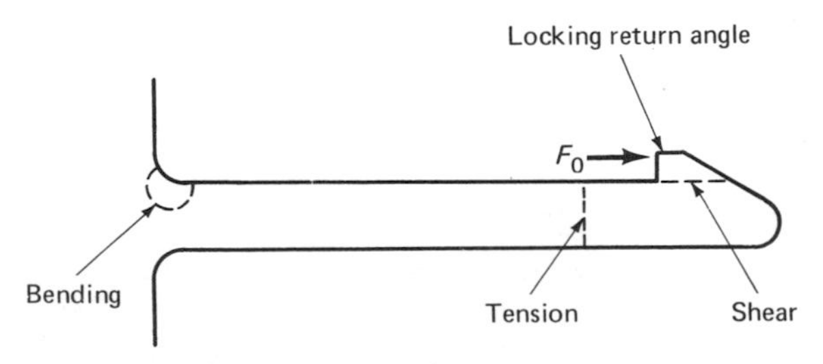

Figure 6.24. Locations of potential failure. The cantilever snap has a locking return angle and cannot be snapped out.

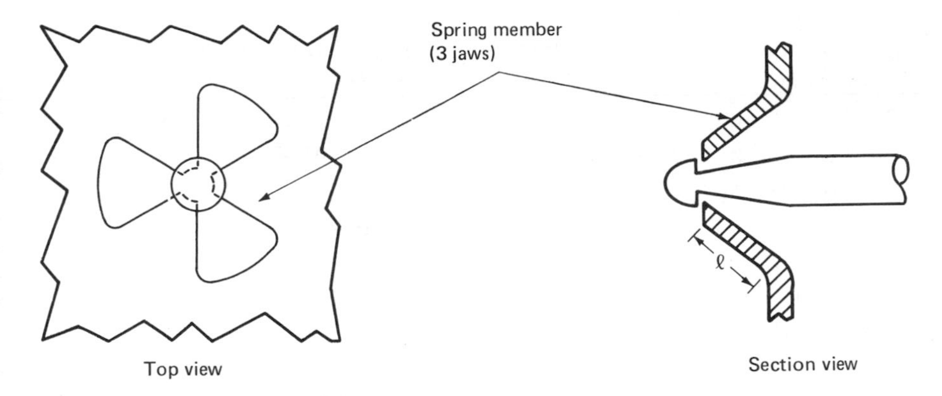

Figure 6.25. A permanent snap design.

rial, the spring rates of hollow-cylinder snaps are about 1000 times higher than those of cantilever snaps. For high pull-out strengths, hollow-cylinder snaps of rigid plastics can be used. The interference should be considered as a percentage of the radius. An interference of 1 mm in a 10-mm radius is the same as a 2-mm interference in a 20-mm radius cylinder.

Figure 6.25 shows a snap design that was satisfactory until the components were vacuum-metalized to improve appearance. The spring components then started to break during assembly. Investigation showed that the vacuum-plating caused the following changes: (1) the

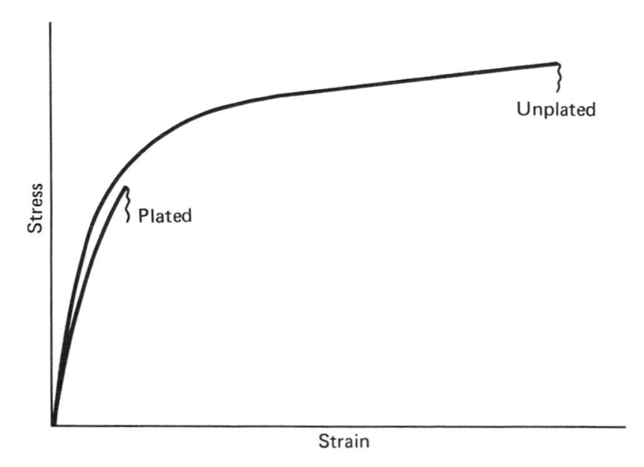

Figure 6.26. Stress-versus-strain plot for plated and unplated plastic parts. The lacquer used in vacuum-plating makes the plastic material brittle.

interference was increased because of 0.05 mm coating on all surfaces, (2) the friction coefficient was increased from 0.15 to 0.3, and (3) the plastic material became brittle (Figure 6.26). The solution to the breaking of the spring components is to control both the interference and the spring rate to decrease stress. The interference could be cut back slightly and the length of the spring can be increased to lower the spring rate.

One of the challenges in the design of snap-fits is the undercut problem in the mold caused by the interference. The hollow-cylinder snap design in Figure 6.23(a) would lead to undercuts in both members. The undercut in the lid can be overcome simply by forcibly ejecting the lid, since the plastic is soft enough. The undercut in the bottle can be handled by side-slides in the mold or by blow-molding. For practical mold-making, the interference is the factor that can be increased with the least amount of trouble. Initially, the mold would be made with the interference on the low side. Sample parts would be molded from this semifinal mold. The snap forces and the strength of the spring components in these samples would be tested. If everything behaves as expected, the snap-in and snap-out forces would be slightly less than the desired ones. A final cut in the mold would increase the amount of interference and perfect the snap-fit. If the snap fails in tension in pull-out tests, the thickness t can be increased while the interference is kept the same.

6.2.3. Friction and Lead Angles

Friction and lead angles influence the snap-in and snap-out forces as described by Equations 6.1 to 6.4, but they are unrelated to the stress and failure of the spring component. Small lead angles and low friction will provide smooth assembly. The coefficient of friction between plastic parts is usually in the range 0.15 to 0.25 [93]. The actual value depends on the materials and the surface textures of both members. The presence of dirt or oil would, of course, affect the friction coefficient. Lubrication might be a necessary step in the assembly of high-friction components made of rubber, vinyl, or polyethylene.

The kinematics of the spring component can greatly influence the strength of the snap-fit. Figure 6.27(a) shows a plastic toy car with integral bearings to accept snap-in axles. The two snap designs shown

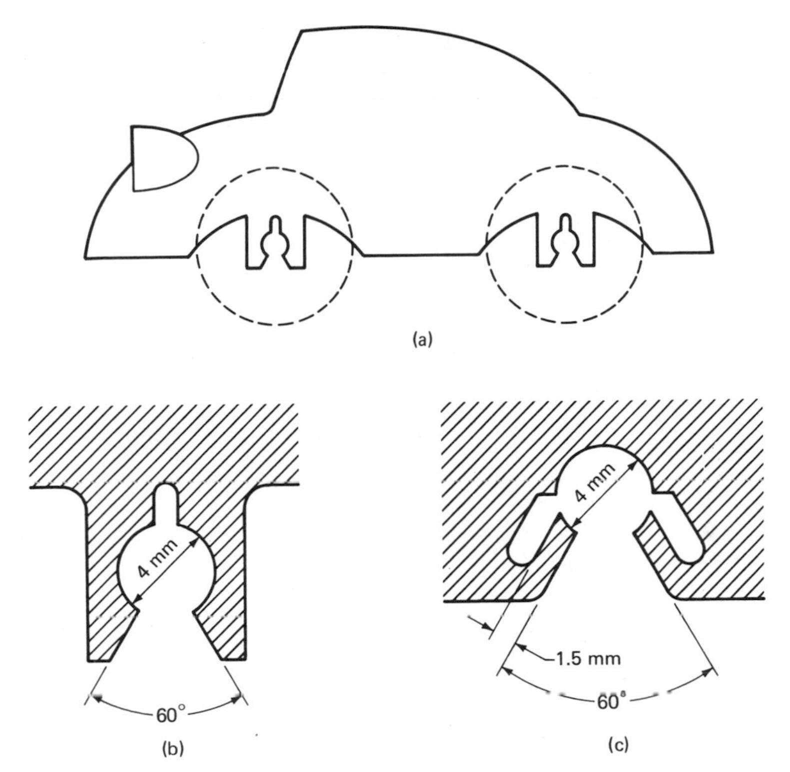

Figure 6.27. Molded bearings to accept snap-in axles. The configuration in (b) has a 45° return angle. That in (c) has a locking return angle.

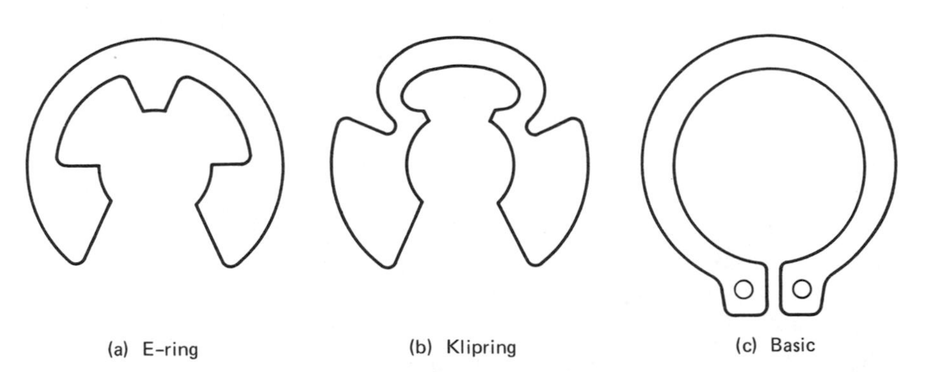

Figure 6.28. Three kinds of snap rings. The basic ring shown in (c) has locking lead and return angles. Such rings have to be assembled with special pliers.

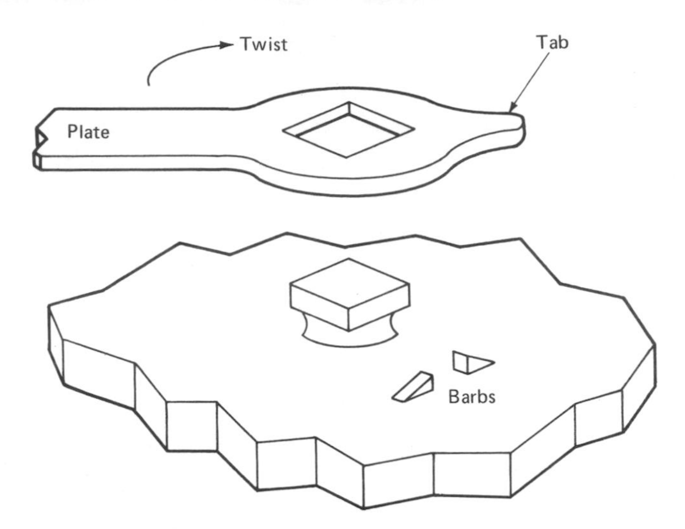

Figure 6.29. A twist snap. The plate is keyed onto the square stud and twisted through 45° until the tab is trapped between the two barbs.

in Figures 6.27(b) and (c) have equal lead angles but different return angles. The snap (b) has a 45° return angle whereas the snap (c) has a 90° return angle. The kinematics in Figure 6.27 can be applied to other designs such as the snap rings in Figures 6.28(a) and (b).

Many snap-fits are designed with locking lead and return angles. To assemble and disassemble these, auxiliary devices such as a pair of pliers are required. The snap ring in Figure 6.28(c) has to be pulled open at the holes with special pliers. Similarly, slotted rod snaps with locking return angles (Figure 6.31(c)) can be disassembled if the two cantilevers are pushed toward their common center by auxiliary devices.

Twist snaps have been successfully applied to different products. In Figure 6.29, the plate with the square hole is first keyed onto the square stud. The plate is then twisted through 45° until the tip of the plate (the tab) is engaged between the barbs. The tab is a cantilever spring. The lead angles are on the ramps of the barbs. The light-bulb sockets commonly used in automobiles are twist snaps, but they do not have lead angles to facilitate assembly or disassembly. Many single-lens reflex cameras employ a bayonet mount for interchangeable lenses. The basic principles are illustrated in Figure 6.29.

The kinematics of lead angles are further complicated by the deflection of parts under load. A rigid ball to be inserted into a plate of soft material can start with a 90°lead angle. Upon loading, the soft material can deflect to form favorable (smaller) lead angles. Most lead angles calculated from unloaded configurations change upon loading. If a snap with locking lead or return angles deflects in such a way as to increase the angles, the snap is a true locking snap. The configurations in Figures 6.25 and 6.27(c) exhibit such characteristics.

6.2.4. Choosing Material for Rigidity and Dimensional Stability

That snap-fits are so widely used in the plastic industry is because plastics can tolerate a large strain ϵ at their elastic limits and they have low elastic moduli E. Materials, such as cast iron and glass, that have poor resilience* are not suitable for use in the spring component, which is essential for a snap.

In the following discussion, "soft" plastics refers to those with an elastic modulus E of 0.41×10^9 N/m^2 (60 ksi) or less, and "hard" plastics refers to those with an elastic modulus of 2.07×10^9 N/m^2 (300 ksi) or more. "Soft" plastics such as polyethylene, ionomers, EVA, and soft vinyls are suitable for hollow-cylinder snaps. Hollow-cylinder snaps experience circumferential tension, which demands an exceptionally low modulus of elasticity and a high elastic limit (Table 6.1). Soft plastics also have good sealing properties that enable an airtight assembly. Since soft plastics lack dimensional stability, they are not suitable for cantilever or distortion snaps. Polyethylene parts will warp enough to completely miss the design interferences in cantilever or distortion types. Hard plastics such as polystyrene, ABS, nylon, acetal, polycarbonate, and rigid vinyl are candidates for cantilever and distortion snaps. Dimensional stability and consistent shrinkage of these plastics permit a precise control over the designed interference.

The hollow-cylinder snap and the distortion snap in Figure 6.23 are similar in shape but different in material requirements. The snap member and the mating member (bottle) in Figure 6.23(a) have the same shape. To pass over the interference, the snap member expands and experiences hoop stresses. In Figure 6.23(b), the two mating

*The modulus of resilience is the elastic energy-storage capacity of a material. It is the area under the stress-strain curve below the elastic limit.

members have different geometries and the snap member is distorted as it passes over the interference. The stress experienced by the distortion snap member is due to bending. In Figure 6.23, the material for snap (a) should be stretchy, and the material for snap (b) should be rigid. Table 6.1 shows that the strain in a hollow-cylinder snap is about 10 times that in a tapered cantilever snap. The higher strain in the hollow-cylinder snap should be compensated for by the softness of the material (the E of soft plastics is 20 times lower than that of hard plastics), so that the stress in both types are of the same order of magnitude. The stress state experienced by these snaps is analogous to those in a rubber band and a paper clip. Dimensional stability, which is trivial for a rubber band, is critical for a paper clip.

Cantilever snaps are less stable than hollow-cylinder and distortion snaps, especially in torsion and impact. Figure 6.30(a) shows a regular dual cantilever snap and a modified version with thin, curved webs to improve stability. This mixed design between cantilever and distortion is used in automobile bumper guards with success. Another design has the tip of the cantilevers joined together to improve the strength of the individual spring components (Figure 6.30(b)).

In the choice of materials, other factors that should be considered are: (1) resistance to corrosion, solvents, abrasion, and creep; (2) material stability at extreme temperatures; (3) consistence of friction and molding tolerances; (4) consistence of strength with different molding cycle times, colorants, additives, and regrind materials; and (5) ease in ejection from molds with undercuts. All the above factors are important for the production and reliability of plastic products. As an example, vacuum-metalizing often destroys a snap completely because the lacquer solvent attacks some plastics.

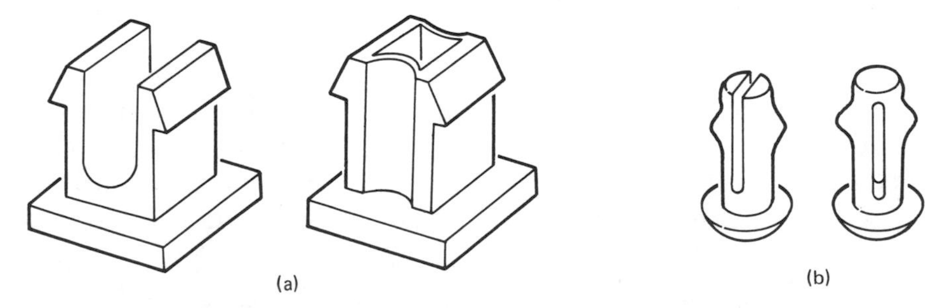

Figure 6.30. Modified dual cantilever. The cantilevers are modified for better stability and higher strength: in (a), two thin webs are added along both sides of the dual cantilevers; in (b), a thin web is put across the tip of the cantilevers.

The change of material properties under adverse temperatures can be used to a designer's advantage. Two mating parts that are difficult to assemble may be warmed up to increase the elastic limit and to ease assembly. A plastic part can also be chilled to increase rigidity and facilitate assembly.

6.2.5. Snap-Fit versus Press-Fit

It takes less force to assemble a snap-fit than a press-fit. A snap-fit also has higher strength, and it can be assembled and disassembled through many more cycles. It does not lose its strength with time and is less demanding on tolerances. The advantages of a press-fit over a snap-fit are simpler geometry and no undercuts. A cylindrical press-fit can also support torsion, which makes it ideal for assembling wheels onto axles.

An unwise attempt to use a rigid plastic (polystyrene) in a hollow-cylinder snap was made (Figure 6.31(b)). The amount of interference had to be made very small because of the high spring rate of the styrene hollow cylinder. The tight tolerance resulted in molding difficulties and rejects. The remedies would be: (1) changing the hollow-cylinder material from styrene to vinyl, or (2) cutting a slot in the hollow cylinder or the rod. A slot in the rod would in effect

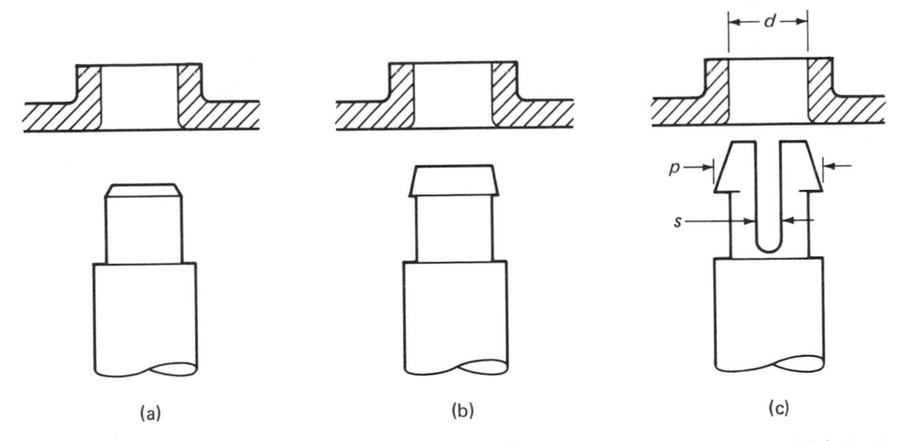

Figure 6.31. Comparison of (a) press-fit, (b) hollow-cylinder snap, and (c) dual cantilever snap. (a) In a press-fit, the interference stress is very high and the tolerance is most critical. (b) The high spring rate in this assembly dictates small interference and tight tolerance. (c) Since the spring rate is lower, the interference can be increased and the tolerance becomes less critical.

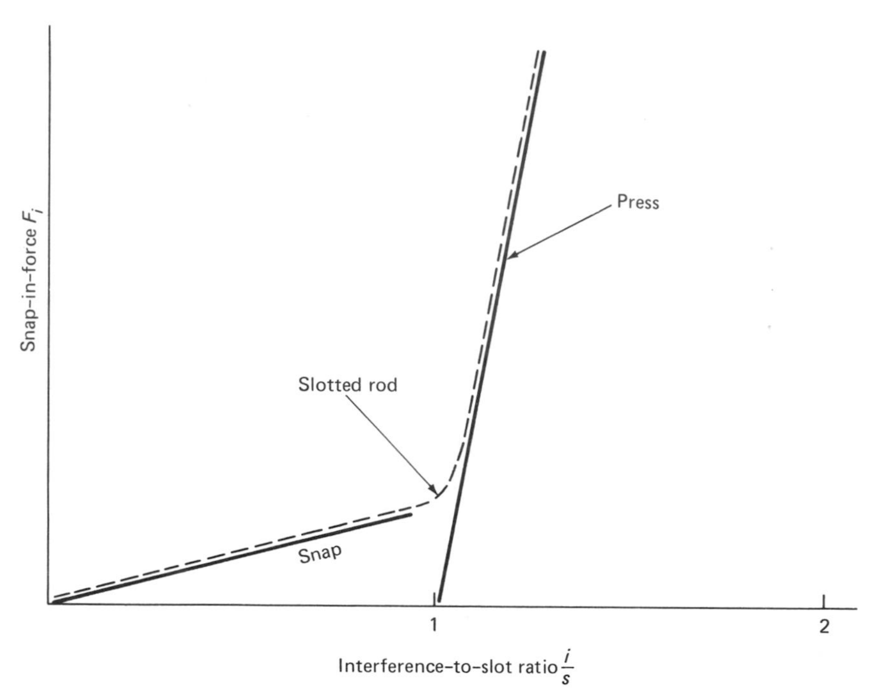

Figure 6.32. The snap-in force versus interference-to-slot-width ratio. When the interference exceeds the width of the slot, the snap-in force increases very drastically.

change the hollow-cylinder snap to a cantilever snap (Figure 6.31(c)). The interference could then be increased and the tolerance at the interference would be less critical. A press-fit requires extremely tight tolerance (Figure 6.31(a)). In Figure 6.31, the spring rate of the rods decreases from (a) to (c), while the interference increases. For the cantilever snap in Figure 6.31(c), the spring rate is rather low until the slot is closed by the snap force F_s. Therefore, the interference ($i = p - d$) must be made smaller than the slot width, s. If i/s is larger than 1, the slot is not wide enough, and the snap-fit behaves somewhat like a press-fit in the assembly process (Figure 6.32).

If $i/s < 1$, snap-fit behavior:

$$K = K_{\text{rod}}$$

$$K_{\text{cyl}} = \infty$$

The plate can be assumed rigid.

If $i/s > 1$, press-fit behavior:

$$\frac{1}{K} = \frac{1}{K_{rod}} + \frac{1}{K_{cyl}}$$

Both the plate and the rod deflect in a press-fit.

In the snap-fit in Figure 6.31, the stress in the cantilever is released as soon as the assembly is completed. In press-fits, the contact stress experienced during assembly remains between the two members. This strain energy may be released over a long period of time through the viscoelastic properties of plastics (stress relaxation and creep). When the interference stress disappears in the press-fit, the pull-out force also goes to zero, which means the loosening of the press-fitted members.

One logical way to establish a design would be: (1) select the type of snap and the general shape from overall considerations, (2) find the spring rate of the spring component, (3) select an interference based on the allowable stress and the elastic limit of the spring material, (4) specify the lead angles to obtain desirable snap-in and snap-out forces.

6.2.6. Designing the Press-Fit

A press-fit consists of an outer member that is hollow and an inner member that is either hollow or solid. Because the members are cylindrical in shape and have no undercuts, they are rather easy to fabricate. The outside diameter of the inner member is larger than the inside diameter of the outer member. There is an interference, i_{rad}, between the radii of the members. During assembly, the outer member is stretched and experiences a tensile circumferential stress (hoop stress). The inner member is compressed and experiences a compressive stress in the circumferential and radial directions. The sum of the increase in the inside diameter of the outer member and the decrease in the outside diameter of the inner member equals the interference. The ratio of the two deformations can be predicted by a model with two springs in series.

The contact pressure p between the two members is assumed to be uniform. The contact force F equals the product of the pressure p and the area of contact $2\pi r_b h$ (see Figure 6.33 for notations). The

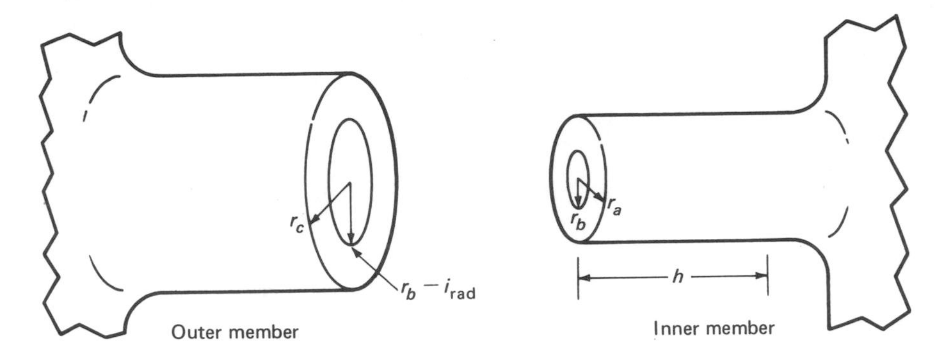

Figure 6.33. The geometry of a press-fit.
r_a = inside radius of inner member,
r_b = inside radius of outer member and outside radius of the inner member,
r_c = outside radius of outer member,
i_{rad} = interference between radii, and
h = length of engagement.

push-on and pull-out forces both equal the contact force times the coefficient of friction μ. The torsional strength equals the contact force times the coefficient of friction times the radius r_b. In a press-fit design, the material properties are usually given (E, ν, S_y, μ). The objective is to design a sufficient pull-out force and torque strength while keeping the stresses in the members below the yield strengths.

$$\text{spring rate of outer member, } k_o = \frac{F}{\Delta r_b}$$

$$= \frac{E_o 2\pi h}{\left(\dfrac{r_c^2 + r_b^2}{r_c^2 - r_b^2} + \nu\right)}$$

$$\text{spring rate of inner member, } k_i = \frac{F}{\Delta r_b}$$

$$= E_i 2\pi h \frac{1}{\left(\dfrac{r_b^2 + r_a^2}{r_b^2 - r_a^2} - \nu\right)}$$

contact force, $F = i_{rad} \cdot$ (combined spring rate of the two members)

$$= \frac{i_{rad}}{\dfrac{1}{k_i} + \dfrac{1}{k_o}}$$

$$\text{push-in force} = \text{pull-out force} = F \cdot \mu$$

$$\text{torque strength} = F \cdot \mu \cdot r_b$$

$$\text{contact pressure, } p = \frac{F}{2\pi r_b h}$$

The maximum compressive stress on the inner member is at r_a:

$$\sigma_{\text{HOOP}} = -\frac{2pr_b^2}{r_b^2 - r_a^2}$$

The maximum tensile stress on the outer member is at r_b:

$$\sigma_{\text{HOOP}} = p\frac{(r_c^2 + r_b^2)}{(r_c^2 - r_b^2)}$$

Because of the higher compressive strength of many materials, the inner member is less likely to fail. The two common types of failures are: (1) the splitting of the outer member when the tensile hoop stress exceeds the strength of the material, (2) the interference is so large that the members are difficult to assemble. The designer has to find a set of dimensions that satisfies the above equations. Sometimes one or two of the dimensions are fixed by other design considerations.

Lead angles are often used on the mating members to help alignment and start the fit. The torque strength can be increased by fluting the shaft (i.e., putting longitudinal grooves on the surface of the shaft as shown in Figure 6.34). The pull-out strength can be increased by broach configurations (Figure 6.35). Figure 6.36 shows several ways to increase tolerance and decrease stress. The principle behind all these modifications is to decrease the spring rates of the members. The spring rates are decreased because the external member is subjected to distortion instead of uniform radial expansion. The decreased spring rate increases the tolerances in the diameter and in the alignment.

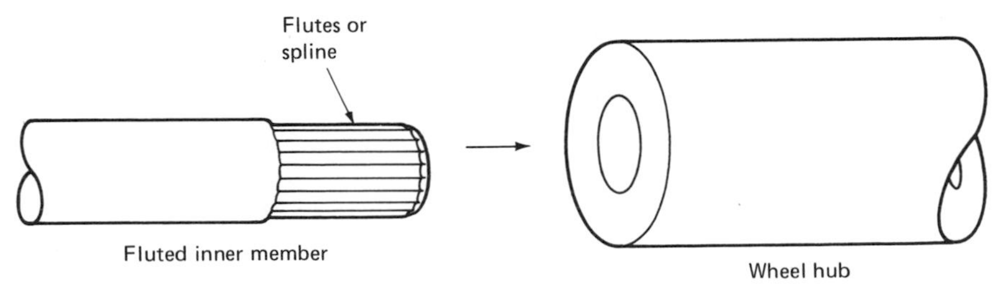

Figure 6.34. A press-fit with high torque strength.

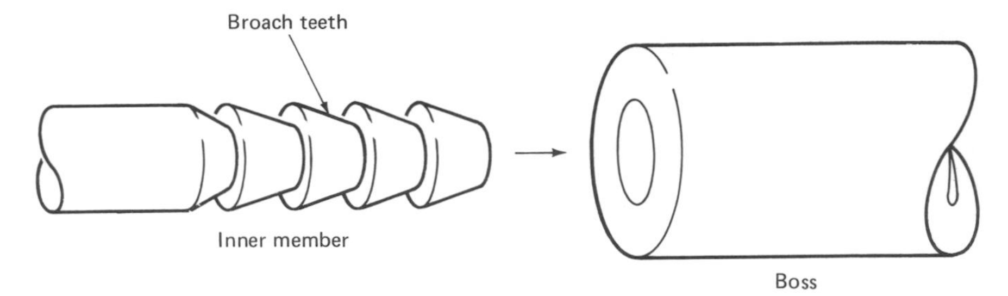

Figure 6.35. A press-fit with high pull-out strength.

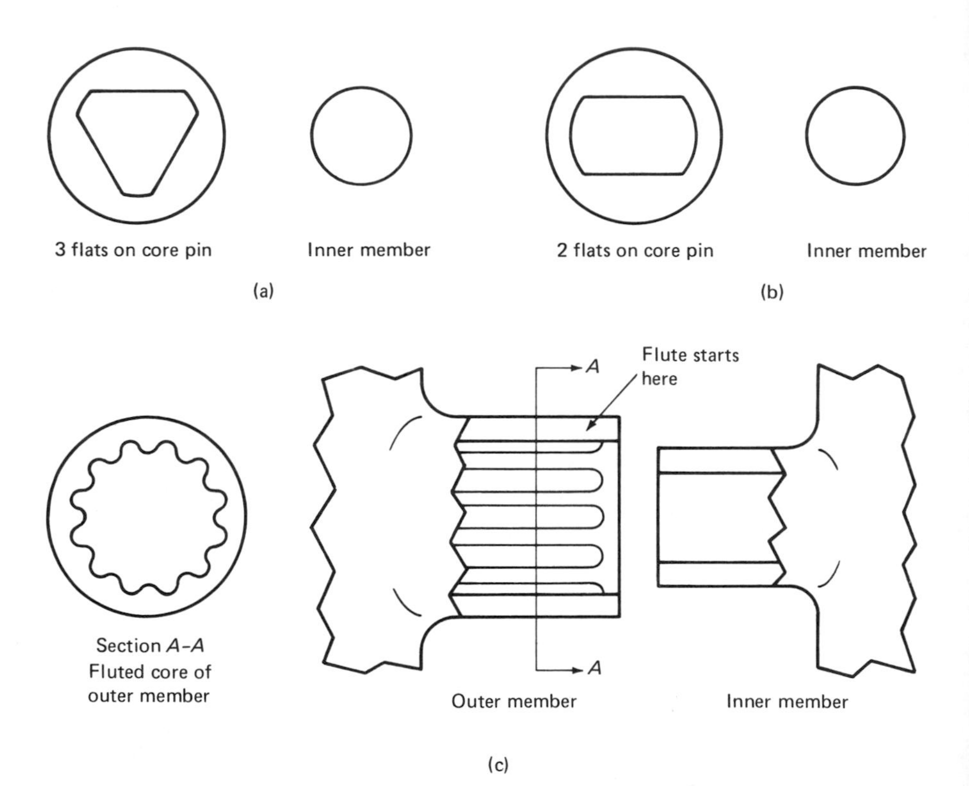

Figure 6.36. Improved press-fits by noncircular external members. (a) is a distortion press-fit. (b) is an alignment-tolerant press-fit. (c) is a design with better tolerance and added strength. Because of the more complicated joint configurations, these designs are suitable for molded or die-cast components only.

6.3. FASTENERS

Fasteners perform the crucial function of holding parts of a product together. They come in different types and details. A significant amount of labor is usually required in the use of a fastener. A suitable choice among the numerous varieties of fasteners can save labor and time, even though fasteners sometimes make up less than 2% of the weight of a product.

Special fasteners are often designed to speed up assembly or disassembly. Fastener thread forms are often specially tailored to the joint materials or the load conditions. Fasteners are usually designed and manufactured by special companies. The job of a designer is to employ the proper types of fasteners that are functionally sufficient, easy to install, and decent in cost. Also, fasteners should not be detrimental to the appearance of a product.

6.3.1. Cost Reduction with Fasteners

The in-place cost of a joint can easily be 4 to 10 times the cost of the fastener itself. The in-place cost includes the costs of purchasing, storage, inventory control, installation, and inspection. To save handling costs and equipment costs, fasteners are often standardized. The principles of standardization are discussed in Section 3.3.1 and Section 6.1.

Extending the use of standardization, one is tempted to add washers, spacers, and other accessories to adapt a standard bolt to a particular application. This is the proper way for models and prototypes. However, once the application is frequent enough, savings can be realized by combining these build-up members into an integral piece. Cold-heading technology is capable of producing special fasteners at low cost. The high speed and no-scrap features of this room temperature process make it very economically competitive. A careful functional analysis of each feature in the fastener may simplify the design.

Many modern fasteners are aimed at reducing or eliminating joint preparation. Selftapping and selfdrilling are two common examples. Selftapping fasteners can be removed and reinstalled in the same location several times if the threads are carefully aligned. Selfdrilling screws often suffer from decreased holding strength if an old hole is reused. Though these fasteners are more expensive, they offer several advantages: (1) savings in drilling, punching, or tapping, (2) prevention of hole blockage by paint or misalignment, (3) increase in production rate, and (4) convenience for visual inspection. The material of selftapping and selfdrilling fasteners must be harder than the materials to be fastened. For a metal-piercing rivet, the column strength and ductility of the fastener also need to be considered. The reduction of joint preparation is sometimes necessitated by poor accessibility. Such fasteners can reduce cost substantially if they replace a time-consuming or difficult assembly operation.

Automation is an attractive strategy for high-volume products. If the production rate is slow because of material movement or intensive inspection and hand fitting, automatic fastening will probably help very little.

For some rivet assemblies, two workers are needed if one person cannot reach both sides simultaneously. A blind rivet can complete the whole assembly from one side and save half the labor. Service requirements should also be considered in a design. The service personnel often have to tinker here and there to find ways to disassemble a product. Components that need frequent inspection or lubrication should be accessible without tearing open the whole assembly. Several kinds of screws should be used to distinguish those that hold together the housing from those that hold together internal structures. The housing screws should have slotted heads, while the internal screws should have Phillips heads. This is to avoid internal structures being loosened unintentionally. Permanent assemblies should be assembled by rivets. Several fasteners used on the same product should either be the same or grossly different. Indistinctive fasteners with slight differences in torque rating, diameter, or length often cause misassembly and endless trouble. Furthermore, it is better to have fasteners weaker than the structures because broken fasteners are easier to replace than broken structures.

6.3.2. Improving Strength of Bolts and Screws

From the strength standpoint, it is advantageous to tighten a bolt as much as possible, even to the point that yielding occurs in the bolt. The stress in a bolt during installation is a combination of tension and torsion. Since the torsional stress will relax after the bolt is in place, a bolt that does not fail during installation is unlikely to fail afterward. Special wrenches are sometimes used to insure that the bolt is not pulled into two pieces by careless tightening. The material for the nut should be soft and ductile in order to get more uniform force transfer from the bolt to the nut. Sometimes the nut material can be five times weaker than the bolt material without degrading the strength of the whole system. Initially, the pitches on the screw and the nut are the same. As the assembly is tightened, the bolt is under tension and the nut is under compression. The pitch on the bolt will decrease and that on the nut will increase. This will cause a high stress concentration at the first thread. A relief groove at the first few threads on the nut will soften the nut and spread the stress (Figure 4.56). The part of the nut that is relieved is under tension as is the bolt.

After the bolt is preloaded, it acts as a redundant spring with the plate being clamped. External forces on the joint cause disturbances in the bolt stress and in the plate stress. The disturbances are proportional to the stiffnesses of the bolt and the plate. It is desirable to have the plate infinitely stiffer than the bolt so that all external load fluctuations are absorbed by the plate. If the bolt is stiffer than the plate, the stress variations in the bolt may eventually cause fatigue failure of the bolt. Figure 6.37 shows how a gasket is recessed to increase the plate stiffness. A common way to increase plate stiffness is to have large areas of steel in contact. In a shear joint, preloading can cause the clamped plates to develop very high frictional forces. The shear strength of the joint can be increased by increasing the coefficient of friction between the joint surfaces. Since the hold-

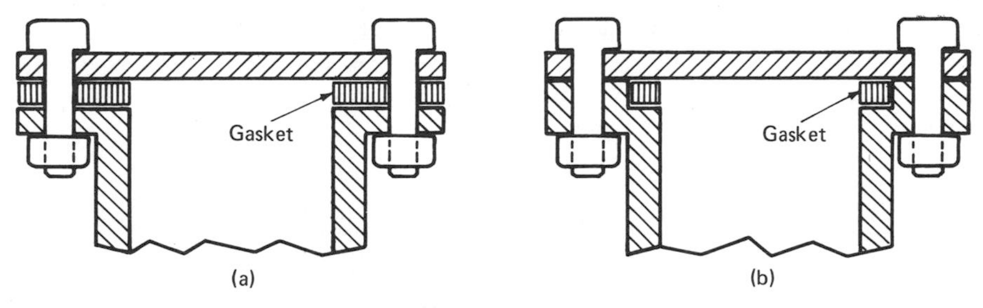

Figure 6.37. Fatigue failure of fasteners can be prevented by stiffening the plates. In (a), soft plates cause large stress fluctuation in bolts, but in (b), the gasket is recessed to stiffen the plates and reduce stress fluctuation in bolts.

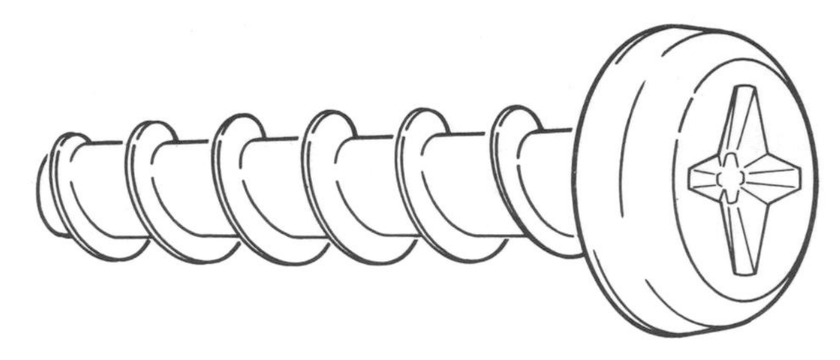

Figure 6.38. A plastic screw with sharp, deep, and widely-spaced threads.

ing force is derived from friction, external load variations will only vary the frictional force. The fastener is under constant tension and free from fatigue.

For a given pitch diameter, the coarse thread is stronger than the fine thread. If the bolt fails by ripping the threads, a cylinder of material between the major and minor diameter is sheared off from the bolt. A coarse thread has more material between the major and minor diameters and is, therefore, stronger. Coarse threads are also used with power wrenches because power wrenches tend to rip up fine threads when they are not carefully aligned. A metal screw is usually much stronger than plastic boss or sheet metal. The threads designed to hold in plastics or sheet metals are usually sharp, deep, and widely spaced (Fig. 6.38). A plastic molded boss should have an ID (internal diameter) the same as the pitch diameter of the screw and an OD (outside diameter) about 2.5 times the pitch diameter. Normally, the thickness of a nut is between one to two times the diameter of the screw. For sheet-metal assembly, the length of engagement can be increased by extruding the pilot holes (Figure 6.39). Other features to increase the thread strength are: spot-welded nuts, rolled-on inserts, ultransonic inserts (in plastics), spin inserts (in plastics), selftapping inserts, knurled press-fit inserts, and screw-expanded inserts. A selftapping screw with lobular threads

Figure 6.39. An extruded hole can increase the length of engagement in sheet metal.

(threads on a somewhat triangular prism instead of a cylinder) can also improve the holding strength (Figure 6.40). As the screw taps into a plastic boss, no chips are formed because the hollow boss continuously distorts into a triangular shape to conform to the screw. The threads are gently rolled onto the mating member. The triangular shape also resists loosening caused by vibrations.

Besides the strength of fasteners, the configuration of fasteners also strongly influences the strength of a connection. Figure 1.12 shows some ways to arrange fasteners for higher strength. Basically, the fastener cluster should be spread as far away as possible to gain bending moment arm and bending strength. Fasteners should not be placed at locations of small deflections, as discussed in Section 4.4.3. A fastener (or fastener location) is important if its absence can cause drastic decrease in the connection strength (see Section 4.4.4). In a cluster of fasteners, the one that causes the smallest decrease of connection strength upon its removal is the one that has the lowest value.

There are many tricks to make bolts vibration-proof. A fine thread has a shallow helix angle and is more difficult to shake loose than a coarse thread. If a bolt with right-hand threads loosens in a particular application, a left-hand thread may cure the problem. However, the confusion between left-hand and right-hand threads can cause difficulty in maintenance. Nylon inserts and adhesives are used with success. Various ways of deforming the nut or screw thread also increase friction and produce a locknut. Several of the oldest methods are still among the best. A lockwasher is one example; another is to use a wire to seal through a hole in the bolt and slot in the nut.

Corrosion protection should also be considered when using fasteners. In many applications, a zinc coating lasts longer and costs less than a cadmium coating. Aluminum corrodes rapidly in the presence of noncompatible materials. Care must be taken in using aluminum fasteners, as well as in choosing fasteners for aluminum structures.

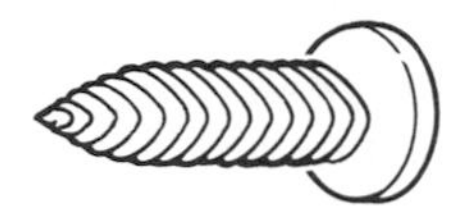
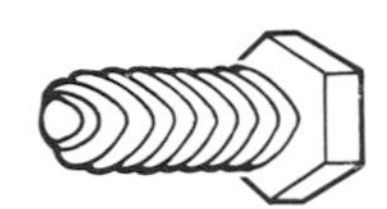
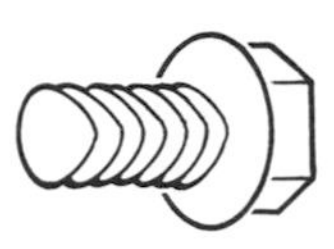

Figure 6.40. Threads on triangular lobes. (Elco Industries, Inc.)

6.3.3. Eyelets and Rivets

Eyelets and rivets are strong, dependable, and inexpensive fasteners. Eyelets have tubular bodies with flanges on both ends. Rivets have solid shanks bearing a head on one end and a flange or a head on the other. Eyelets are drawn on progressive dies and are even cheaper than rivets. Because of the hole, an eyelet is often used as a bearing to protect plastics, leather, and fabrics from frictional wear. For example, an eyelet can be used on a styrene hole bearing a steel shaft. Before assembly, one side of the fastener is straight without flange or head. After the eyelet or rivet is inserted in place, a flange is formed by rolling, clinching, hammering, or staking. Eyelets are usually used for light-duty applications. The setting of an eyelet does not require much force, and is often done by a hand-operated arbor press.

The design rules for rivets and eyelets are quite similar. The length of the rivet or eyelet must be long enough for proper forming of the flange. The fastener must protrude from the plates by 50% to 70% of its own diameter. The existing flange or head should be placed on the weaker sheet. The thicker or harder sheet can bear the forming force better. If both sheet materials are soft, a washer should be used underneath the flange that is formed during assembly. A rivet should be loaded in shear (in a lap joint). An eyelet should be under very light load. The flange or head should be seated flatly on the plate. Some rivets are not set tightly so that the plates can rotate around the rivets. The axial space should be less then 0.020 in. and uniform around the flange. Three common modes of failure are buckling of the shank, cracking of the flange, and distortion of the fastened plates. The shank may buckle because of insufficient material hardness, excessive clearance in the punched hole, or an improper tool. The hole should be 5% to 7% larger than the diameter of the shank. A tight hole can provide lateral support and prevent buckling. The flange may crack because of insufficient ductility of the material or improper setting. If a special rivet is machined on a screw machine, there is a tendency to use materials that are easy to machine but not really good for riveting. Distortion of fastened sheets can be caused by insufficient clamping, improper riveting sequence, or excessive impact force. When many rivets have to be installed in a row, it is better to first install the two ends, then the middle, and then a few more at wide spacings. The gaps can be filled in later. If the

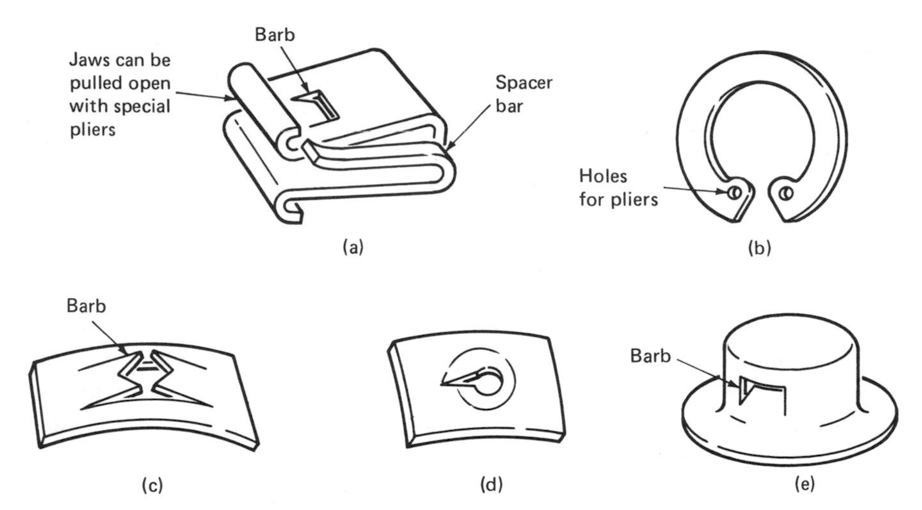

Figure 6.41. Sheet-metal fasteners: (a) clip with spacer, (b) retaining ring, (c) speed clip, (d) single-thread nut, and (e) push cap.

rivets are installed continuously from one end, there may be enough cumulative distortion to prevent alignment at the other end. In a small assembly, it may be possible to set a row of rivets in one stroke.

6.3.4. Sheet-Metal Fasteners

Although fasteners made from sheet metal sometimes have an unappealing appearance, they are very efficient and strong for many applications. Figure 6.41 shows several types of sheet-metal fasteners: spring clips, snap rings, retaining rings, single-thread nuts, and push nuts. Barbs are easily incorporated into these fasteners to resist pullout. When used on aluminum or soft steel, the barbs can actually chisle into the metal if pulling is attempted. A spherical or conical shell can also be incorporated into these fasteners to resist buckling. Those types with poor appearance are often concealed in a product.

6.4. ELECTRICAL AND ELECTRONIC ASSEMBLY

There are seven steps for an electrical wiring assembly: (1) wires and cables are cut to length, (2) the insulation is stripped on both ends, (3) terminals are installed, (4) the prepared wires are threaded through openings in housings, fittings, conduits, etc., (5) the appropriate connections are made, (6) the circuit is checked visually and electrically, and (7) the wires are secured into bundles and labeled. The wires should be protected from heat, moisture, chemicals, mechanical abuse, and electrical interference, and should also be readily accessible for maintenance. Modern developments in electrical wiring have produced improved ways to accomplish these objectives.

Electronic assemblies used to be very bulky and tedious. Now, IC circuits, PC boards, and electronic components are made very compact. The interfacing usually occupies significant space. The present technology is such that the microprocessors may be less expensive than the housing and wiring in a product. The competition regarding the design and manufacture of these assemblies is often the critical factor in product design. Electronic assembly usually comprises the following steps: (1) the circuit is drawn to 4× scale with the use of tapes and decals, (2) the circuit is photoreduced and etched onto a phenolic-copper laminated board, (3) holes are drilled or punched in the board, (4) components are inserted on the board, (5) the connections between the components and the board are made by wave-soldering, and (6) the board is inspected visually and electronically.

Standardization is especially important in electrical and electronic assemblies. With automatic assembly, standard modules are much cheaper than customized packages. For mechanical insertion of components onto PC boards, the shape of components, the spacing of terminals, and the orientation of components should all be as uniform as possible.

It is also desirable to have designs with as much electrical tolerance as possible. Components with wider tolerances are less expensive. Wide tolerances permit a choice among components with different mechanical configurations and dimensional tolerances. Close cooperation between design and manufacturing is necessary to find the best overall cost.

Material-engineering is an area that is always changing. Injection-molded polyester with glass filler is used for fittings and housings previously made with thermosetting plastics. New materials for PC boards allow smaller holes to be punched with greater ease. Copper foils may be stamped and cemented to boards, instead of etching away costly copper materials. Plate-on processes can deposit copper where it is needed. Tin contacts are often good enough to prevent oxidation. Where gold contacts are needed, there are processes that will apply gold locally instead of over the whole surface. A tiny gold-plated disk can be spot-welded to the contact point. The thickness of gold should be held very close to the specification in order to conserve the precious metal.

6.4.1. Wiring Tools and Hardware

Automatic wire-processing machines can measure, cut, and count wires, as well as strip insulation, install terminals, and print wire labels. There are also hand tools that can strip insulation and cripple terminals with calibrated forces.

Terminals are used to insure a good contact between the wire strands and the electrical device. The bare wire is inserted into a barrel. The compression of the barrel insures a good contact and prevents oxidation and corrosion. Because of the different thermal and chemical properties of copper and aluminum, terminals designed for one type of cable may not be suitable for another. Fig. 6.42 shows four ways to increase reliability and speed up assembly: (a) an insulated terminal can save the manual installation of a plastic sleeve, which is often needed on the outside of a terminal; (b) a flared opening on the barrel will help the insertion of wires; (c) tabs on the prongs of a terminal make a snap-fit with the screw post so that the terminal need not be held by hand while the screw is tightened; (d) a stop at the end of the barrel helps proper insertion. Even though each feature adds cost to a terminal, choosing the appropriate features will save time and reduce the number of rejects.

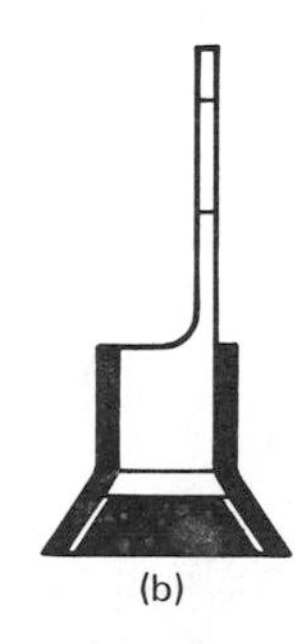

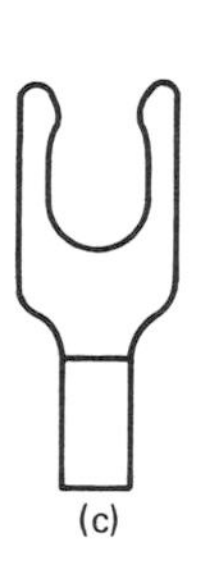

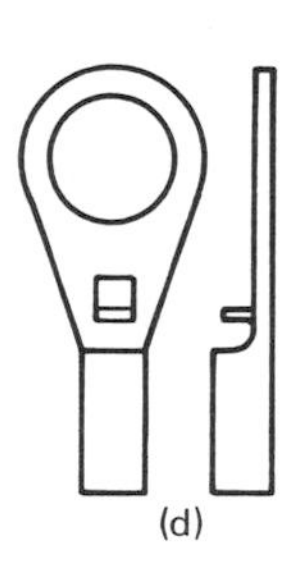

Figure 6.42. Special features on terminals [100]. (a) Integral insulation eliminates the hand-installed plastic sleeve. (b) A flare helps insertion of wires. (c) Bumps on tip of prongs provide snap action. (d) A stop helps to control the insertion depth of copper wire. (*Courtesy of Assembly Engineering*)

Wires are collected into bundles and tied together by plastic ties. Power tools are available to do the tying in one or two seconds. Some of the ties are designed to fracture at a calibrated tension at a point just outside the lock point. Other ties are designed to be released and reused if necessary. If a company always uses wire bundles from 10 to 12 strands, it may be more economical to buy 12-strand cables and leave one or two strands unused. By wasting a few wires, a lot of hand-tying can be eliminated. Many wire harnesses and fittings are designed with snap-fits. These are push-in rivets for attachment to the housing, and V-grooves for accepting the wires and cables. Junction boxes often have many perforations. The user can pick an appropriate perforation and knock out a hole for use. The unused perforations will be left closed for protection against dirt.

6.4.2. Printed Circuit Boards and Wave-Soldering

There are many ways to make a PC board. Some methods are obsolete; others are suitable only for large productions. With the changing technology and availability of new materials, old methods may be revived and new techniques may be added. The electrical current rating of copper foils is surprisingly high due to the large cooling surface area. The circuitry on a PC board tends to be very involved and difficult to visualize. Computers are often used to aid design. More and more components are packed onto fewer boards. Cross-over wiring can be achieved by drilling through the board, plating through the hole, and making the connection on the other side. Punching holes in a PC board is less expensive than drilling holes. Since the terminals on a DIP (dual-in-line package) IC chip are spaced at 0.100 in. apart, most holes are about 0.030 in. in diameter. Punched holes are easily obtained for diameters of 0.050 in. or greater, and the cost of punching goes up as the diameter gets smaller. The size of the holes has to include the alignment tolerances for the punching machine and the inserting machine. It is desirable to punch the pilot holes and the component lead holes in one operation. Also, since phenolic boards warp in the direction of the grain, the boards should be arranged to have the grain running in the shorter dimension, and the components oriented transverse to the grain.

For small productions, the components are manually inserted onto the boards. It is important to let the worker sit comfortably and to have the components placed for easy access. Silk-screen printing of component outlines on the board may help prevent insertion error [101]. Some workers learn quickly and can memorize the pattern. A teaching aid should be available if needed, but it should be non-imposing when the worker resents it. Mechanical insertion is appropriate for high-volume productions. The saving increases as the complexity of the board increases. Mechanization can be accomplished by an indexing machine with a turntable, or a modular machine with various stations each handling one component. The x-y table insertion machine uses one insertion head that handles many different components. (The components are arranged on a tape in a prescribed order by a numerical-control machine.) As the inserting head manipulates the component, the table transverses and travels until the place of insertion is directly under the head. Placing the components close together reduces the transverse time. A uniform orientation and a uniform part size reduce head movements. After the part is inserted, the leads are clinched to secure the part. When all the parts are in place, the board is ready for wave-soldering.

Wave-soldering works rather like the stream of water in a drinking fountain. A continuously recirculating stream of molten solder is forced to flow under the pressure of a pump. The board and the inserted components are chemically cleaned, fluxed, dried, and preheated. Then they are conveyed past the crest of the solder wave with half of the board thickness submerged in molten solder. As the board exits from the pool of solder, the solder peels back from areas that are not wet by the solder. Sometimes a solder bridge short-circuits two terminals that should not be joined. Solder bridges have to be visually inspected and manually removed, since electrical testing may ruin the components. The lump of solder at a bridge is under three forces: gravity, surface tension, and wetting force (cohesion). The first force acts against the formation of the solder bridge, while the latter two help to form the bridge. The gravitational force is proportional to the volume of solder; the surface tension is proportional to the curvature of the solder, and the wetting force is proportional to the area of contact. Many factors contribute to the formation of solder bridges. Among them are temperature, flux, and component orientation.

In a study done at the University of Illinois and Micro Switch divi-

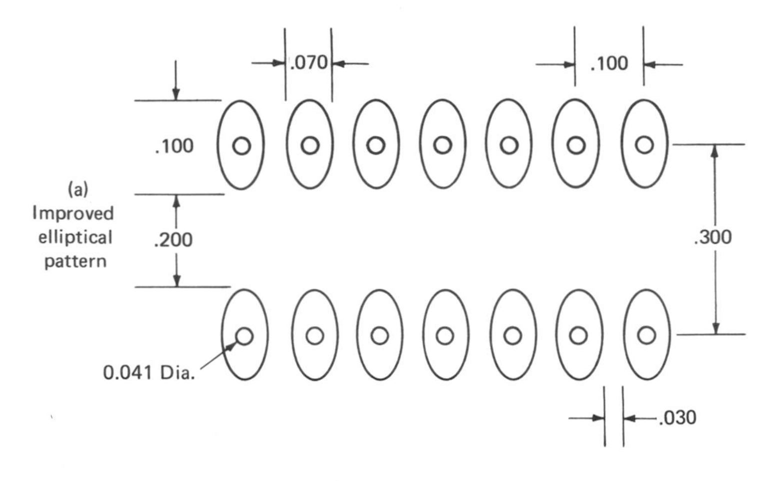

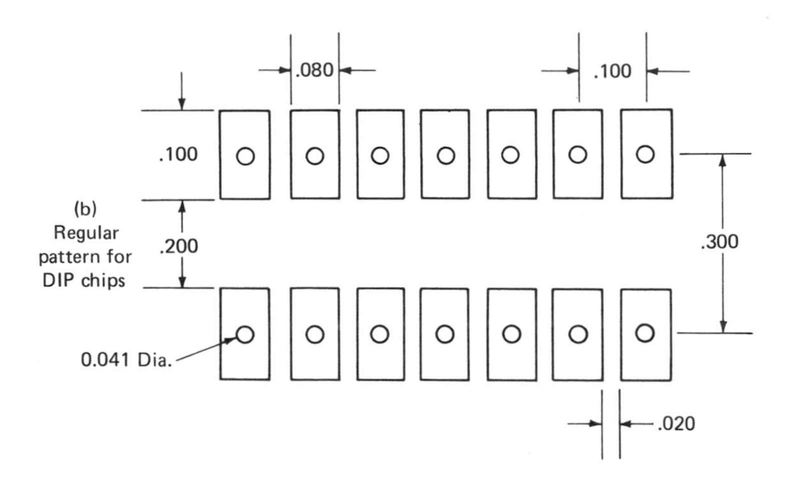

Figure 6.43. DIP (dual-in-line package) IC chips are locations of frequent solder bridges. The elliptical solder pad shape, (a), can eliminate over 90% of the solder bridges with the rectangular pads, (b). All dimensions are inches [102].

sion of Honeywell, the incidence of solder bridge was 1% to 3% [102]. It was found that shortening the leads and changing the shape of solder pads (Figure 6.43) eliminated more than 90% of the solder bridges. The lead length below the board surface was changed from 0.080″ to 0.050″. The wetting force was decreased because both changes reduced the area of contact. If the minor diameter of the elliptical solder pads can be further reduced to 0.055″, the incidence of solder bridge would be further decreased.

6.5. ADHESIVE JOINING

There are many reasons for using adhesives. Sometimes adhesive joining is used to provide simple joint configurations, smooth contours, or airtight seals. In automobiles, adhesives are used to bond trunk covers to structural members because the bond offers wide tolerances and allows for thermal distortions. In aircraft, adhesives offer a high strength-to-weight ratio. In consumer products, cost reduction is often possible through simplified joint configurations and lower fastener costs. Cement joints often have larger surface areas and lower stress concentrations. Cement joints are more versatile than welding because they can handle dissimilar materials.

6.5.1. Joint Configuration and Strength

The reliability of adhesive joints depends on the surface preparation, the curing pressures, the surrounding temperature, moisture, the nature of stress, aging, and other factors, some of which are not encountered in other assembly methods. The following is a general review of the factors affecting the strength of a joint.

The strength of a joint is highest with a specified thickness of the adhesive. If the thickness were above or below the optimum, the joint would be weaker. The proper thickness is accomplished by a proper joint configuration and by applying a suitable curing pressure to the joint. The thin film of adhesive between two joint surfaces may be designed to withstand one or more of the kinds of loading discussed below and depicted in Figure 6.44.

Compression. Under compression, the bond is usually very strong. The joint would stay together even if the cement were cracked or absent. A compression load can generate lateral strain in the members due to Poisson's Effect.

Shear. A typical example of a shear joint is a sleeve-and-tubing configuration as in PVC pipes and fittings. Even materials with very poor compatibility can develop high strength. The shear stress distribution in a plain lap joint is nonuniform. By tapering the components at the bond area, we can change the stiffness and obtain better joint compatibility (Figure 6.45).

Tension. Usually, tension is accompanied by bending when the load is off-center.

Cleavage. Cleavage occurs when bending moments are transmitted through the bonded surface. Such stress is very nonuniform. The severity of the stress depends on the bending moment and the I/c of the cement film area. Failure usually starts at the point of the maximum tensile stress.

Peel. When one or both of the bonded bodies is flexible, an extremely high stress can be generated by peeling. The forces would be concentrated on a line where separation occurs. Peel resistance of a joint depends on the flexibility of the cement. If the cement itself is elastic and soft, the area of stress would be small but tolerable. Rigid

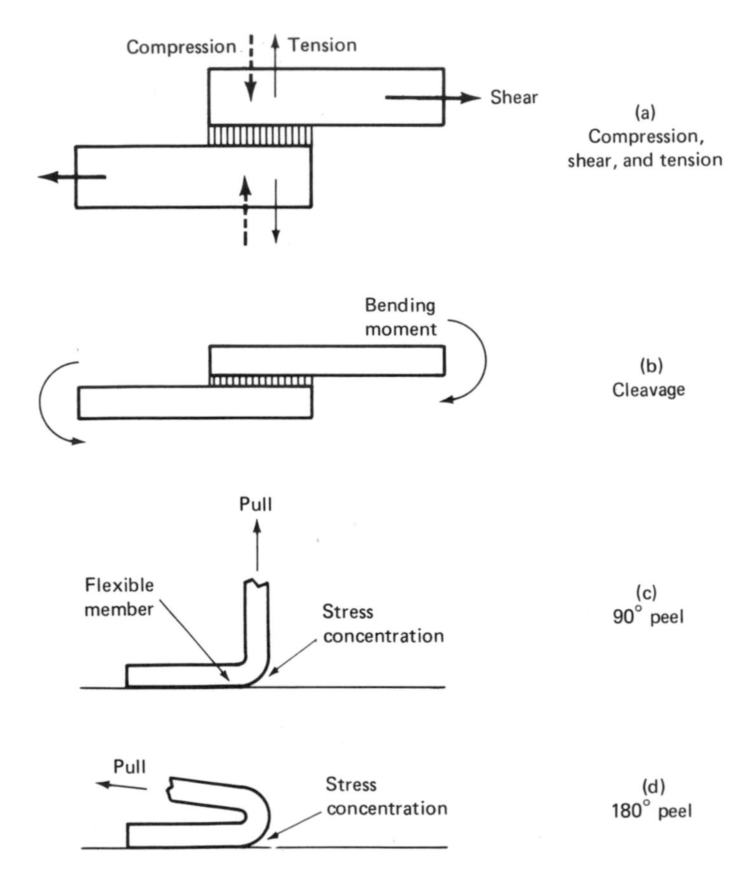

Figure 6.44. Load conditions for adhesive joints. The loads in order of severity are: compression, shear, tension, cleavage, and peel. The worst load is 180° peel.

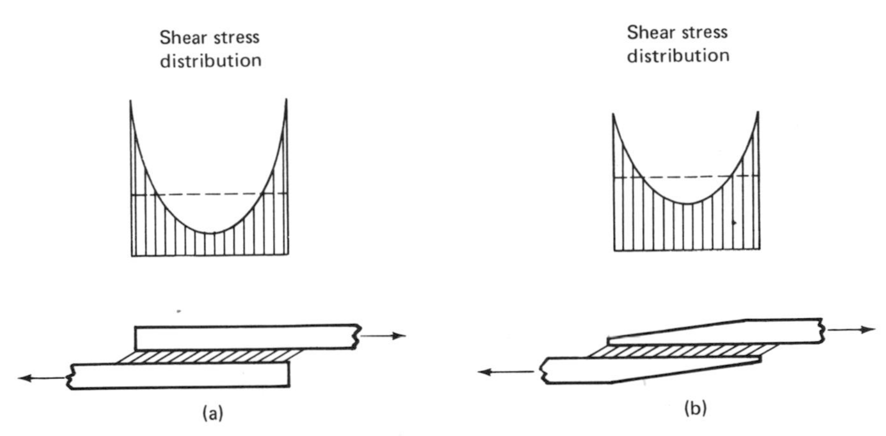

Figure 6.45. Joint improvement by adjusting the mating member stiffnesses. By changing the thickness of the lap material, the stiffness can be tailored. Where the stiffness is decreased, the stress is lowered. The tapered lap joint in (b), has a more uniform stress distribution and a higher strength than the simple lap joint in (a).

cements usually have very poor peel resistance due to the small stress area and the high stress concentration. The peel strength is proportional to the width of the cement film and independent of the length of the cement film. To avoid peeling, loose ends of flexible components should be covered, concealed, or secured with rivets. If a flexible sheet is bonded onto a rigid member, the flexible member may be extended over the edge and wrapped under the rigid member. The rigid member may also provide a recess area so that the corners of the flexible member do not protrude and invite peeling.

The load states above are listed in order of increasing severity. It is not difficult for an adhesive manufacturer to show a picture of a hoist picking up a car or an elephant with an adhesive joint in the link. As engineers, we must know if an adhesive will work in our particular product, in the real environment and time span. The strength of an adhesive may decrease if the material to be bonded is difficult to wet. The two surfaces should be flat and the thickness of the cement film should be uniform. An irregular thickness of cement film and the presence of air bubbles lead to nonuniform stress distribution. To increase the conforming characteristics and the area of contact, pressure-sensitive wall hangers are often made with a thin layer of soft, foam lining. Surface preparation is extremely important because the joint strength can be greatly reduced by traces of oil, water, dirt, or mold-release agents. If a cement were applied over a painted surface, the maximum strength would be limited by the bonding strength of the paint.

The environment and the stress history of a joint are also important. Heat, moisture, salt, and chemicals may weaken a bond. Differences in thermal expansion can cause very high interface shear. The strength of a joint may decrease with time when the adhesive is constantly under load (such as the weight of a picture frame), while an unstressed specimen may maintain its strength as it ages. Some adhesives are known to be vulnerable to cyclic loads. Elastic adhesives have better impact strengths. Increasing the thickness of the cement film increases the impact strength of an elastic adhesive. Thermosetting adhesives, such as phenol-formaldehyde, epoxy, and cyanoacrylate, have good creep resistance and high strength even at higher temperatures.

6.5.2. Labor and Cost Considerations

There are many hidden costs in making a cement joint: the mixing equipment, the dispensers, the clamp fixtures, the spoilage loss, for example. An adhesive may be spoiled by an expired shelf life or by preparing more than is used. An expensive adhesive may reduce the jointing cost if the amount per joint is very small, or if it cures quickly (as with cyanoacrylate adhesive). White glue (for paper and wood) is deliberately made extrathick (concentrated) to decrease the setting time.

One-part adhesives are more convenient than two-part mixtures. Whenever special cleaning or mixing is required, there is more chance for errors. Adhesive joints are often difficult to disassemble and rework. Therefore, alignment by built-in lids and shoulders is very desirable. Dispensing the correct amount of adhesive is quite difficult. When clamping pressure is applied, the adhesive should spread and cover the entire surface. Too much adhesive may spoil the product's appearance or contaminate the work area. Too little adhesive will give a weak joint. A worker ordinarily will apply an arbitrary glob of glue. The designer can design the product such that the glue line is concealed or recessed. Excess cement should be made to flow into a built-in reservoir or trap. The messy application of cements can be avoided by using pressure-sensitive adhesives or adhesive films. Solvent type adhesives can be applied by dipping if the bond contour is flat. Solvent adhesives should be applied to pieces that are porous or absorbent. Hot-melt adhesives are quite desirable because of their long shelf-life, easy application, and short curing time. Hot-melt glue is especially tolerant to joint mis-match and non-uniform glue thickness. The mating pieces should be easily accessed. Hot-melt joints can be separated by heating, if the bonded pieces can tolerate the temperature.

6.6. SOLDERING, BRAZING, AND WELDING

With these three methods, metals are melted, fused, and cooled to produce a joint. Soldering is performed with a low-temperature filler (below 800°F). Brazing is done with a high-temperature filler, but the temperature is still below the melting point of the base metal. Welding joins by melting the base metal, with or without a filler. The strength of a joint is related to the temperature of the jointing operation. High-temperature joints are stronger.

The three methods involve similar steps: joint preparation, cleaning, fluxing, heating, filler depositing, cooling, and defluxing. A flux is used to clean the joint surface, because it breaks away oxides and also enhances the wetting of the surface by the melt. During the heating process, the flux acts as a shield to prevent the hot surface from oxidizing. Some fluxes are corrosive and are cleaned thoroughly after the joint is completed. Vacuum-brazing, friction-welding, and resistance-welding do not require fluxes because the weld is isolated from atmospheric contamination.

Brazing and welding are strong and dependable assembly methods. The strength of a joint is proportional to its area. Stress concentration is minimal because the load can be transferred gradually from one member to another through a large area. The joint should be designed to facilitate the depositing of the filler and the fluxing and defluxing operations. Shrinkage distortion of the joints can be reduced by good joint designs and proper welding procedures. The balance of strength should also be considered here. The joints should not be stronger than any one of the original components.

Brazing and welding often reduce the cost of a product when they replace casting, forging, or machining operations. Figure 6.46 shows two components with a long cylindrical body and a fat end. Because

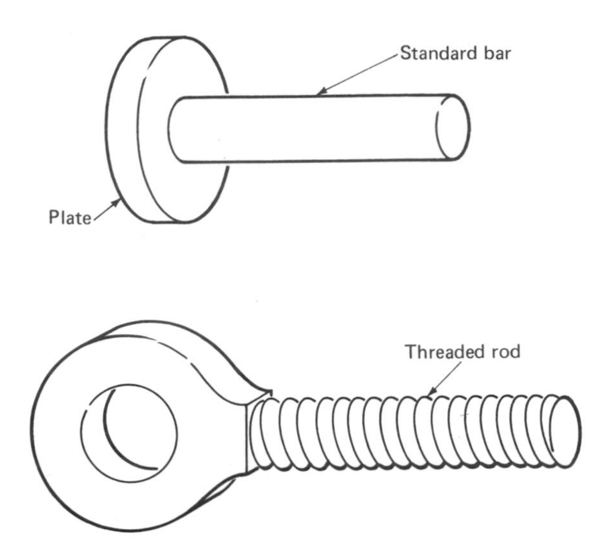

Figure 6.46. Manufacturing by inertia-welding. Components suitable for inertia-welding usually have a long bar of standard material.

of the two drastically different shapes, it is difficult to forge or machine the part. Fabrication from two pieces by friction-welding is the most economical method. Welded structures are usually lighter than comparable cast structures because the material thickness in welded structures is better controlled and better distributed. A welded structure can easily be stronger than a cast structure, but it is usually not as rigid.

Automation of these methods is widely used for high-volume productions. An automated process can increase reliability and lower material costs by dynamically and precisely monitoring the temperature, the duration, the location, and the deposited quantities of a welding operation. The solder may be preformed into sheets coated with flux and mechanically placed at the joints. The production rate from automation may be a few seconds per joint.

6.6.1. Soldering

A solder joint is a low-strength joint often used to maintain electrical contact or provide a seal at a seam. Wave-soldering (see Section 6.4.2) is the standard method for electronic manufacturing.

An operator is often in an awkward position when making manual electrical connections. He needs three hands to manipulate the soldering iron, the solder, and the lead wire. The first step toward mechanization is to set the workpiece on a fixture and use a foot pedal to control the soldering iron. The operation can be further simplified by pretinning the mating members. A closely fitted joint can absorb solder by capillary action, thus saving solder. The temperature of the surface has to be carefully controlled and matched to the weight of the solder tip and the power of the soldering iron. Solder will not wet a cold surface. A solid copper terminal may draw away heat so fast that a very high-power iron is needed. Since a heavier member is more difficult to heat, the soldering iron should be in direct contact with the heavier member.

To prevent a weak joint a solder joint should not be disturbed until cool. Any motion with the solidifying metal could cause a *cold joint.* A eutectic solder (63% tin and 37% lead) has the lowest melting point (361°F) and no pasty range. Since the melting and solidifying occur at precisely one temperature, the chance of a good joint is improved. The hole on a PC board can be plated through to increase the joint surface. Though a solder joint is relatively weak, it can be strengthened by a sleeve arrangement or by mechanically twisting the leads together.

6.6.2. Brazing

Brazing may be considered as high-temperature and high-strength soldering. In production, the brazing joints are often very closely matched (less than 0.003 in. gap) and the whole assembly is heated so that the solder flows into all the cracks quickly. Brazing is used for radiators, bicycle frames, precision valves, instruments, and tools.

One area of improvement is the material cost. A brazing alloy with a lower silver content may be used. The closer the joint fitting, the less alloy is needed. If the joint strength is more than needed, the lap area or joint configuration may be changed. Mechanized application of a brazing alloy and the use of preformed alloys can make the quantity of solder deposit very precise.

6.6.3. Welding

Unlike the overall heating of the brazing operation, welding requires a very intense and localized application of heat. As the pieces are exposed to the cycle of heating and cooling, distortion and residual stress become major problems. Many joint configurations and weld procedures are designed specially to minimize distortion. A long seam is first tacked at several locations by spot welds. Then the gaps are filled in, following a special sequence such that the shrinkage forces counteract one another. If a wrong procedure is used, the shrinkage can pull in one direction and cause a large distortion. If consistent distortion is observed, the parts may be arranged in an off-position, which will shrink into the desired position. The joint location can also be on an axis of symmetry of the structure (e.g., the neutral axis). Shrinkage at a symmetry point will induce symmetric reactions and will not cause a deflection. Shrinkage-induced stresses in a weld may be decreased by peening (striking the surface with a hammer).

Welding that melts the base metal may soften the metal by annealing it. Friction-welding just softens the base metal and squeezes it out of the way to get a good bond. The neighboring material will not lose its heat-treatment. Several capacitor discharge processes can complete a weld in less than 1/1000 sec, such that the neighboring material has no chance to heat. Processes that minimize the heating cycle of the base material will cause less metallurgical change and distortion.

The joint design should also include accessibility considerations for the welding operator. The location of a welded seam can greatly influence its strength. The moment of inertia of the weld area about a bending or torsion axis should be maximized.

6.6.4. Spot Welding

Spot welding is a very efficient method for production, as no shielding or filler is needed. The sheets or wires are clamped between two electrodes, and a large electric current is passed through them. The interface between the sheets becomes hot and the metal softens. The sheets are then pressed together. This process requires no cleaning beforehand or afterward. Though spots are sometimes visible, they are generally not concealed by further decoration.

6.7. WELDING PLASTICS

Many ways to weld plastics have been developed for different kinds of jobs. However, for large-scale production at low cost, ultrasonic welding and spin-welding are the best candidates. The welds are ac-

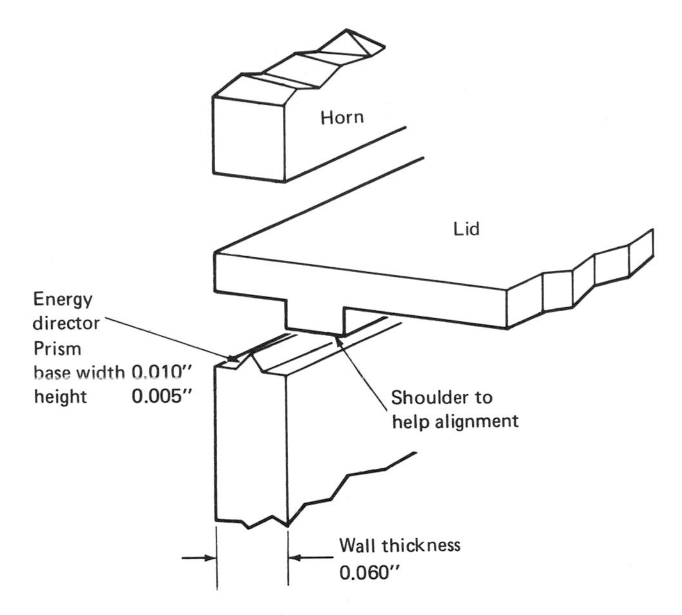

Figure 6.47. Joint configuration for ultrasonic welding.

complished in a rather short time (less than 3 sec), no additional material is needed, the process is clean, and the joint strength is very consistent. Ultrasonic welding and spin-welding can be set up at the molding press such that the press operator welds the pieces together while waiting for the next shot. The cost of the weld is just the rent of the welding equipment.

Sonic welding is a method utilizing the energy of high-frequency vi-

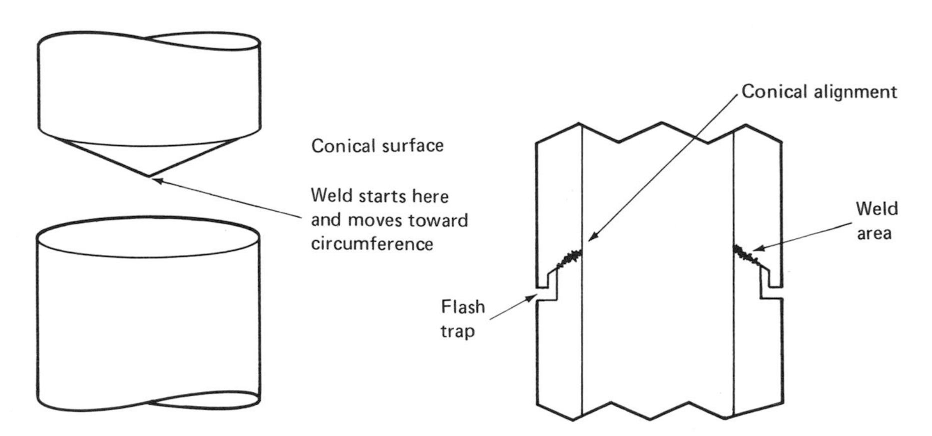

Figure 6.48. Joint configurations for spin-welding.

bration. The pieces are clamped under a horn. As the vibration is transmitted through the pieces to the fixture and to the ground, any loose surface will rub and develop surface friction and intermolecular friction. The weld line must be under pressure from the horn so that the melted plastic is pressed together and welded. It is preferable to have the weld surface less than ¼ in. from the horn to utilize *near field* welding. The energy of the 20-kHz vibration may cause cracks and burn at unexpected locations on the plastic part. The horn itself may crack under the vibration. The design and tuning of the horn requires technical expertise from the ultrasonic equipment manufacturer. The properties of plastics also affect the success of the joint. Soft plastics, such as polyethylene and polypropylene, usually get weak joints because of poor vibration transmission. Plastics with a high melting point, a low friction coefficient, or both require longer times or higher energy levels. Figure 6.47 shows a joint detail commonly used. The triangular prism is called an energy director. It gives a precise starting point of a weld. As the energy director melts, it is squeezed down to fill the complete surface. The energy director decreases the welding time and eliminates the possibility of air pockets in the weld.

Ultrasonic welding can be applied to studs between mating parts. Ultrasonic staking, sealing, spot welding, and inserting are also widely used.

Spin-welding was initially invented for metal joining. It is sometimes called inertia-welding or friction-welding. One of the joint pieces is spun at a high speed (up to 5000 rpm) and the circular mating contour is pressed together. After the material is softened by frictional heat, it is squeezed out of the interface by pressure. The spinning stops, and the pressure is maintained until the joint cools. Noncircular sections can be joined if the weld contour is circular, and if the parts are aligned after the spinning stops. Spin-welding can be used for any thermoplastic, including some rigid foams. The process is ideal for polyethylene and polypropylene. The joint design should facilitate the alignment of the joint pieces, contain the flash, and help to develop a solid weld (Figure 6.48). A weld should start at a fixed location and spread over the complete joint area so that no air pockets can remain in the joint.

7. IMPROVING MANUFACTURING METHODS AND MATERIALS

7.1. ALTERNATIVE METHODS AND SUBSTITUTE MATERIALS

The production of goods is the final step in realizing the fruit of engineering efforts. The design engineer, not in charge of manufacturing, often sees and feels differently than people on the production floor. Since product design influences manufacturing tremendously, the engineer should keep his knowledge of production problems and opportunities up to date. The design engineer is basically a creative person who turns ideas and needs into a physical product. He is likely to be excited by spectacular performance, outstanding features, and unusual specifications. A manufacturing engineer may be a very practical person who prefers normal products to unusual ones. Functional design requires that a product meet specifications and performance, whereas production design requires that the product be easily producible with a consistent quality throughout the entire production volume. The difference between laboratory conditions and production conditions cannot be overlooked. It is possible to convert other elements into gold, but such a process is hopelessly uneconomical in production scale. The gap between the two types of engineering activities should be closed as much as possible by cooperation in the early stages of product design.

Some components can be manufactured by several alternative methods. The product's features, the production method, and the material used may all hinge on economic decisions. The economic environment may favor one method now and another within a few months or years. In order to exploit all the advantages of alternative methods, the design engineer has to keep up with the developments in manufacturing technology. For example, replacing machining with blanking, forming, forging, or powder metallurgy very often results in a substantial reduction in cost. Sometimes a slight modification in the features of a product can permit the use of a cheaper method of production. Figure 7.1 shows two similar products, one of which is produced by welding two stock pieces, whereas the other one is cast. Figure 7.2 shows that bending and drawing can produce the same shape.

Historically, substitution of material has always taken place in response to economic, political, and ecological pressures. The earliest forms of iron and steel were developed because the unavailability of tin made bronze scarce. Artificial rubber was invented because there was a shortage of natural rubber during World War II. Man has proved to be resourceful enough to substitute a common material for a scarce one. In the near past, because of their economic advantages, synthetic materials have replaced natural and metallic materials in many applications. Engineering plastics have found many applications that formerly used die-cast metals. Plastics are not as rigid as metals and the deflection of plastic parts is not tolerable in some applications. On the other hand, the ability to deflect means better impact strength and improved toughness. Some plastics surpass metals

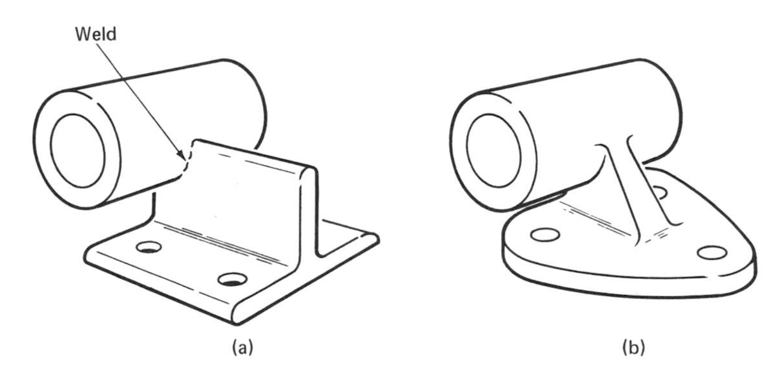

Figure 7.1. Two similar pieces produced by different methods. (a) is a piece welded together from two sawn, stock pieces. A cast piece (b) is likely to have thicker walls and sections. The flat bottom and bearing hole have to be machined.

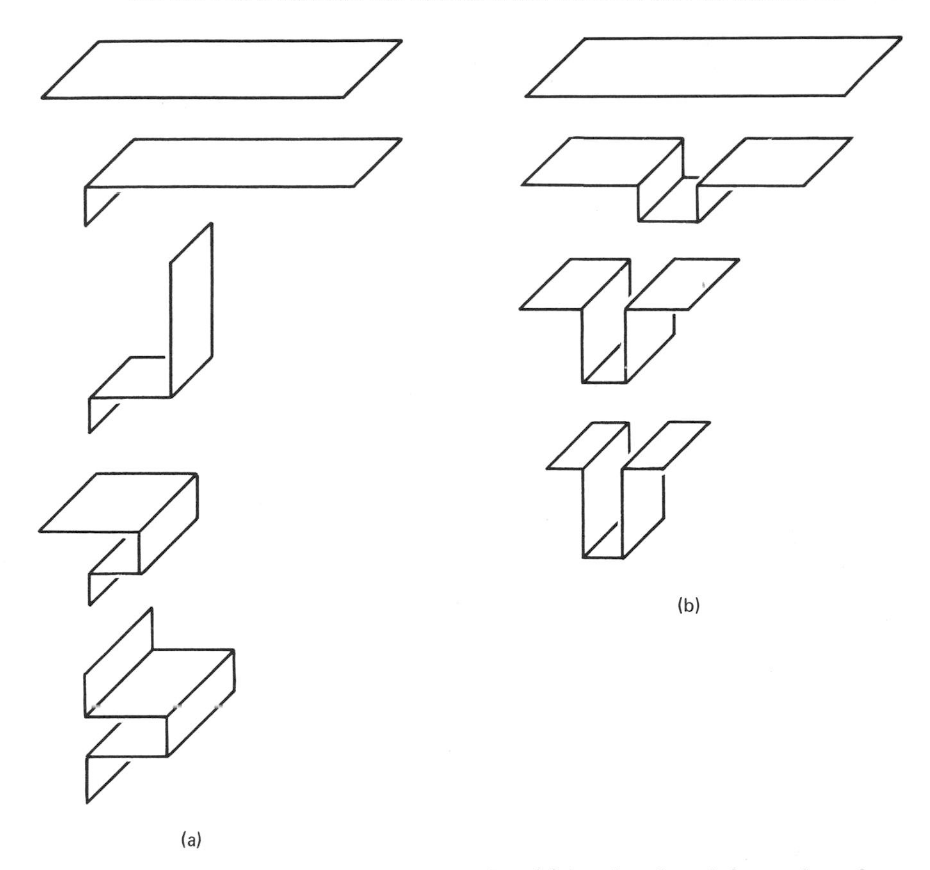

Figure 7.2. Bending and drawing. Bending (a) involves less deformation of material, more reorientation and realignment of the piece. Drawing (b) is often a one-stroke process, which requires a hold-down system and causes a large deformation of material.

in resistance to wear and corrosion. A design engineer should be aware of the possibility of using substitute materials that are functionally sufficient.

To save metal in a stamping process, pieces can be nested and fitted on a blank (Figure 7.3). Thin-wall die-casting, now being explored, may reduce the cost of die-cast pieces. Powder metallurgy is now producing many parts that used to be machined or die-cast. If conservation of raw material is the main concern, one might consider using more processing and assembly efforts to reduce the material consumption. For example, chemical processing can stretch the utility of petroleum by producing high-strength plastics. Mechanical

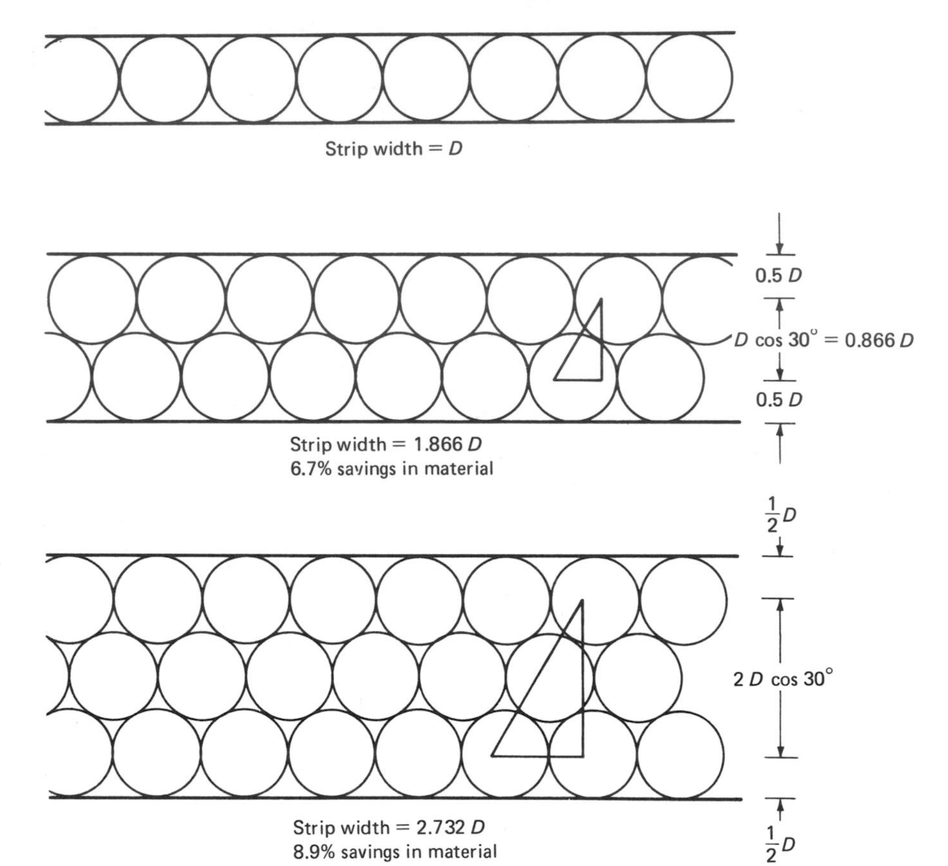

Figure 7.3. Nesting and fitting shapes onto a blank can result in material savings. For a strip n wide, the strip width = $(0.866n + 0.134)D$. Savings compared to a strip with $n = 1$ is

$$1 - \frac{0.866n + 0.134}{n}$$

For n very large, savings = 13.4%. The 2-wide arrangement can produce half the savings of the infinite-wide strip. The difficulty of handling wide strips of material has to be considered. In the model, the minimum part separation is assumed to be 0.

deformation can reorient nylon molecules so that the same strength can be obtained by using only half the nylon material (Figure 7.4). Optical fiber technology can realize incredible savings of copper (Figure 7.5). In electronic transmission, a multiplex system is used to reduce the number of transmission wires. In terms of material conservation, increasing the service-life of a product is equivalent to

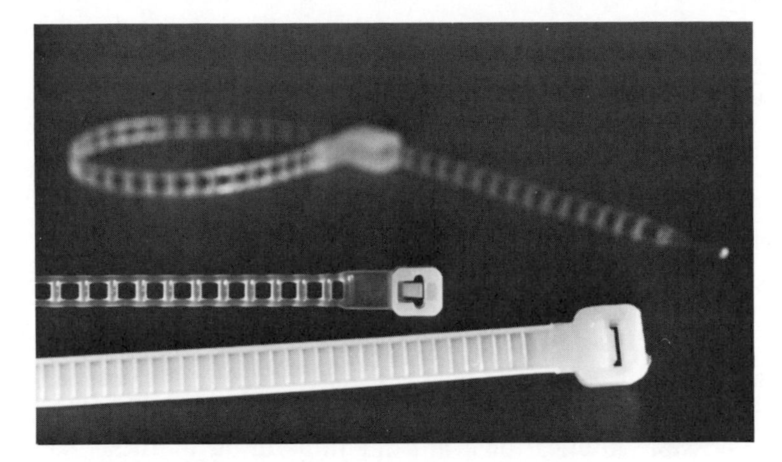

Figure 7.4. When a molded nylon filament is stretched it will neck-down, lengthen, and strengthen. The nylon ties shown on the top and in the middle have many holes and are mechanically deformed after molding. The nylon tie in the front requires less processing but weighs twice as much. (Dennison Manufacturing Co.)

Figure 7.5. For signal transmission, an optical fiber can replace a large copper cable. The optical fiber is made from sand, which is an abundant material. (Bell System)

using less material in making it. Galvanized steel and stainless steel are two processed materials with greater corrosion resistance and increased service-life.

For a design engineer, the choice of raw materials may depend on faith. A designer who is familiar with a particular material may use it very efficiently and rate it at a higher than actual value. If a designer is convinced that wood will be more available in the years to come, he might specify features or tools with wood in mind. Strong salesmanship by plastics suppliers may persuade people to believe plastics can outperform metals and natural materials. Assurance of a continued supply and technical assistance also influences the choice of materials greatly. When aluminum was first introduced, its properties, possible applications, and manufacturing methods had to be promoted to designers. It took years for people to get used to aluminum power cables, aluminum cans, and aluminum construction materials.

In this chapter, we will study specific topics in economical production and in design efforts to ease production. Some common ways a designer can aid manufacturing are by using simple and standard features, decreasing processing steps and cycle times, and minimizing the waste of material. The designer is undoubtedly in a position to made a part very costly unless he is careful.

7.2. CASTING AND DIE CASTING

Casting can produce strong, rigid components with complex curvatures and intricate internal details. A cast piece is usually relatively heavy for its size. The precision of casting varies over a wide range depending on the particular process. Casting is a basic manufacturing process as opposed to metal-forming processes or metal-removal processes. Machine tools are all built with a cast-iron base to get the weight, rigidity, and vibration-damping required for precision cutting.

In a casting process, the metal is melted and poured into a mold. After the metal has solidified, the mold is opened or crushed to remove the piece. The piece is then cleaned, trimmed, and machined to the final form. There are three areas that require the attention of the foundryman and the designer: (1) contamination by gases and dirt, (2) flow problem of the melt, and (3) shrinkage during solidification.

Contamination creates weak spots in the cast. The sources of contamination are generally within the control of the foundryman. A pouring basin is sometimes included in the mold to decrease the speed of the metal flow on entering the mold cavity. A hot metal can erode the mold material (sand). Screens are sometimes used to remove foreign matter from the melt. Passages and traps are sometimes built into a mold to allow the lower density dirt to float to the top. Gases dissolved in the molten metal may be removed by applying a vacuum to the melt. The piece can be designed with smooth contours to minimize turbulence and erosion. The critical parts of the cast should be placed near the bottom of the mold because impurities usually float.

For the melt entering a mold, the mold cavity is like a maze with intricate details. A good flow can be obtained by a smooth, streamlined cavity that produces minimal resistance and turbulence. Right-angle turns, sharp corners, and abrupt changes in sectional areas disturb the flow of the melt and can cause filling problems. The filling of a mold can be improved by using pressure from one of the following sources: (1) a higher rise (head), (2) a centrifugal force, or (3) a piston ram. Note that as the pressure on the melt increases, the shutting force required has to be increased so that it is greater than the product of the melt pressure and the projected area of the cavity. Vents are sometimes needed to allow trapped gases to escape. If a section is thin and large, it may impede the flow and produce an incomplete cast (mis-run). A rib can be added as an internal runner to channel the melt quickly to the far end. The front of the flow should progress smoothly. If the front splits into two flows, which rejoin at a different place, there is a chance that the two joining streams may not fuse properly (cold shut).

As the material shrinks, more and more melt can be fed into the mold. A riser and a sprue can supply an additional amount of melt after the pouring is stopped. The metal farthest from the gate should freeze first. The solidification boundary should propagate backwards to the gate. Once the gate is frozen no additional material can enter the mold cavity. If a pool of material is isolated by solidified material, the shrinkage in this pool cannot be replenished, and a void may form. The gate should be located close to the portion of the cast that is the last to cool. The designer should make all the wall thicknesses the same so that the shrinkage will be uniform. If a thick wall meets a thin wall, the latter will solidify first and will draw material from the thick section. To check the wall thickness, a cast is sawn apart and the sections inspected. The cast will continue to shrink after it has solidified. The shrinkage stress can be high enough to crack the cast. A hollow cylinder will shrink and wrap onto the core, which results in hoop stresses. The classical example is the Liberty Bell, which has a crack on its circumference caused by large hoop stresses. When the melt is in the process of solidifying it is very fragile. It will crack if high shrinkage stresses are set up against the mold. This is called *hot tear*. A way to avoid high shrinkage stresses is to decrease the stiffness of the locking and wrapping features. Figure 7.6 shows several ways to decrease the stiffness. Locking features should be avoided, and smooth curves should be used.

Normally, good design of parts and molds can decrease the number of rejects and lower the cost. There are also several methods to lower the material cost, labor cost, and machining cost. The material cost can be lowered by using techniques of minimum weight designs discussed in Chapter 4. Compact design and wide tolerances can also lower material consumption.

The zinc die casting companies are increasingly going into thin-wall casting, reducing 0.060 in. walls to 0.030 in. The filling and flowing problems previously encountered with thin walls are compensated for by improved design and better temperature control of molds. If serious difficulty does occur, the walls can always be increased by grinding on the die. Labor is decreased by different ways of automation and by using multiple-cavity molds. Machining is a major area of cost reduction in casting design. If the dimensions can be closely controlled and the surface finish can be improved, much of the machining can be eliminated. Considering the cost of machining, the extra effort in mold-making can be very rewarding. Precision investment casting is already used on hard-to-machine metals, such as titanium and stainless steel, to eliminate machining completely. Casting offers design flexibility not available in welding or forging. Side-cores can be used to save hole-drilling. Many separate parts can be combined into an integral design to eliminate assembly and machining of the fittings.

Die casting is a high-speed, high-volume production process for small to medium components. The cycle time of the process is short because both the mold material and the cast material are good thermal conductors. The process is less expensive than other casting

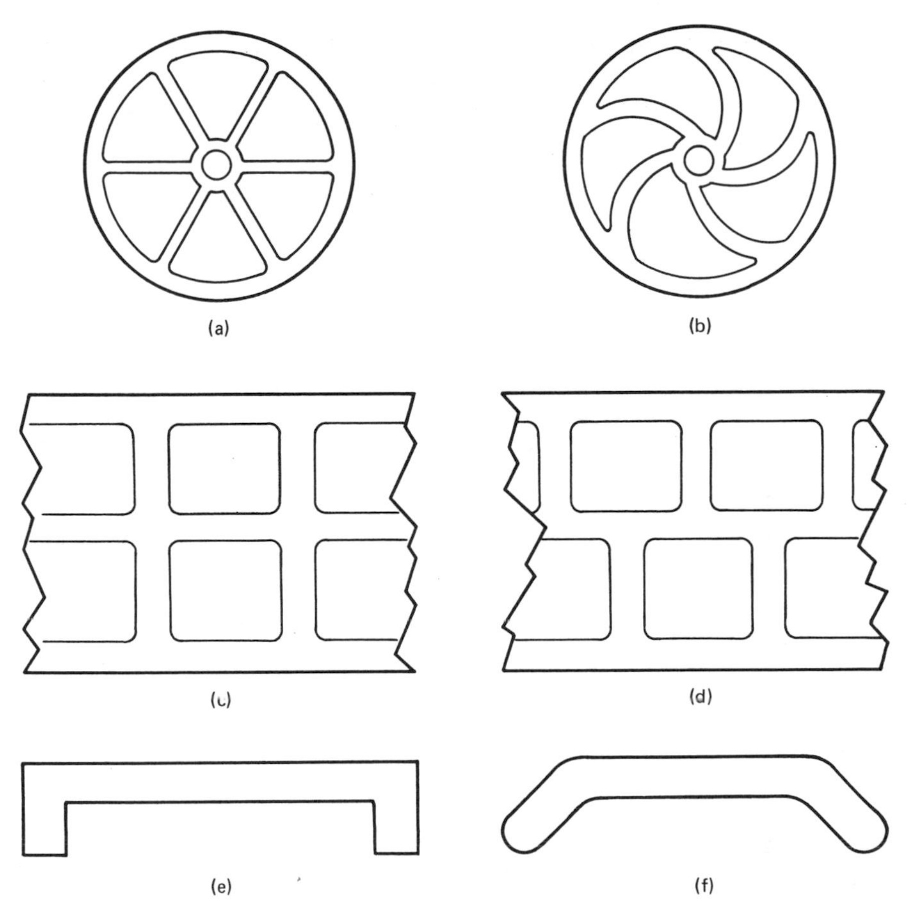

Figure 7.6. Ways to prevent hot tear by decreasing stiffness. In (a), an even number of radial spokes makes the member stiff and causes high radial shrinkage stresses, but an odd number of curved spokes (b) allows radial shrinkage. Four webs meeting at a point (c) make a stiff intersection, while staggered webs (d) decrease stiffness. The locking features and sharp corners of (e) cause high stresses. In (f) shrinkage stresses can be relieved by wedging through the curve.

methods when the production volume is large enough. Because the metal melt is forced into the die under pressure, very fine details can be duplicated. The physical properties of a die-cast piece are superior to a sand-cast piece of the same material. The precise dimensions and fine surface finish of a die-cast piece often eliminate machining. Die-casting is used to produce engine components and housings for power tools and business machines. It is also used to produce small components such as gears, flywheels, and decorative pieces. To have a long mold life, the cast material has to be a metal with a low melting point. Alloys of zinc, aluminum, lead, tin, magnesium, and copper are used. The die-cast metals are weaker than sand-cast steels. Also, low-melting metals are becoming expensive. Therefore, die casting is being squeezed out of the low-price market by plastic substitutes.

7.3. POWDER METALLURGY

Since the late 1960s, the powder metallurgy (P/M) process has been established as an economical method of mass production. Two of the outstanding features of this process are the small amount of scrap, and the ability to produce exotic combinations of metals or metallic compounds. Refractory metals such as tungsten, tantalum, and molybdenum can be fabricated only by this process. The production rate of the P/M process can be as high as several thousand pieces per hour.

P/M parts can have very close tolerances. They require little or no machining, and do not need a draft. They are usually small or medium in size with light or medium loads. Even though the P/M process is not expensive, it is not cheap enough to compete with screw machines. Furthermore, this process is not economical for products that require little machining in their sand-cast form.

The first steps in the P/M process are mixing different metal powders with die lubricants, and feeding the mixture into a press. The loose powders have densities that are about 40% of the theoretical maximum densities of the solid metals. The press cavity is more than twice the size of the final piece. The die is closed and the mixture of powders is compressed. The compressed piece can have a density as high as 90% of the solid density. The compressed piece, called the *green compact*, has enough strength to be handled and is ejected from the die. The green compact is then sintered in an oven with an inert atmosphere. The heating lasts for about 20 min at a temperature below the melting point of the metals. The process is completed after the sintering, though the sintered piece sometimes gets further processing to get additional properties. The piece may be repressed, hot-forged, or heat-treated (quenched in oil). It may also be impregnated with oils or plastics, or infiltrated with copper.

The porosity of a P/M piece is the complement of its density: % porosity = 1 - % density. The density of a P/M piece greatly influences its properties. With a 50% density, a P/M piece is permeable enough to be a filter. With a 75% density, it is porous and becomes a self-lubricating bearing when impregnated with oil. The yield strength and the ultimate strength of a powder metal increase in proportion to its density. Repressing and hot-forging can increase the density of a piece to more than 90% of the density of solid metal. The higher the density is, the more expensive is the piece. Plastic impregnating can seal the pores in the powder and prevent corrosion. Copper slugs can be placed on the green compact before sintering. During sintering, the copper melts and diffuses into the piece to increase its strength, hardness, and toughness.

The P/M technology is constantly changing and adapting to new applications. The processing of different suppliers may vary due to differences in their equipment. The design engineer should specify the performance and functions of a component rather than the chemical composition and processing details. P/M suppliers can often suggest changes and modifications to give comparable products at lower costs. Overdesign is wasteful. The materials, the production process, the strength, and the weight of a product should all be carefully matched to the function of a product.

If a product is to be manufactured by the P/M process, design efforts should be geared toward getting a proper fill, a uniform density, and a longer tool life. Wall thicknesses (at a fin or a web, or between a core pin and the edge) should be more than 0.060 in. to avoid filling problems (Figure 7.7(b)). Unevenly filled parts of a component have lower densities and lower strengths than the rest. Round holes can be used at noncritical parts of gears or pulleys to reduce weight. Sharp contours or internal key holes will result in a weak die, since the die must have the same shape as the pressed piece (Fig. 7.7(c)). Such a die is also expensive to build and requires more maintenance.

The female die is a hollow prism with the contour of the P/M component. The male die slides along this prism of constant area. Therefore, the P/M component should have a uniform thickness and straight sides. No draft is needed on the side walls and core holes. It is not possible to make a perfect sphere by P/M because the male die will have a feather edge. However, a sphere can be approximated by a shape with a cylindrical section between two spherical domes.

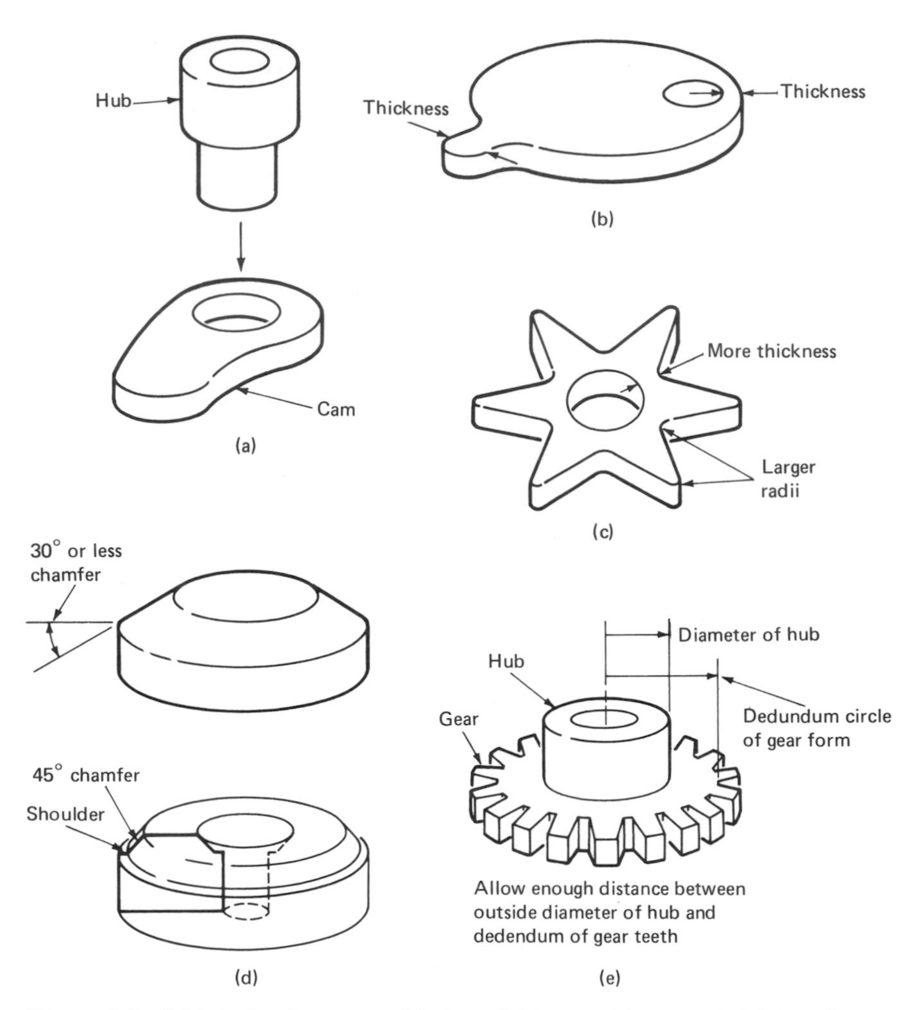

Figure 7.7 P/M design features. (a) A multistep part is separated into subcomponents to simplify the die. (b) The web should be thicker than 0.06 in. to avoid filling problems. (c), (d), and (e) Several features that strengthen the die.

Small indentations or protrusions on the top or bottom will present no difficulty if these features are thinner than 15% of the thickness of the piece. A P/M piece should have as few steps as possible, otherwise the nonflow property of the powder material will make the density nonuniform. Usually, one or two thicknesses are used. Changing a part with steps into two straight pieces can simplify the die and reduce the press capacity (Figure 7.7(a)). The two pieces can be assembled before sintering and they will bond together during sintering. Sometimes a small shoulder (0.010 in. wide) is added to a piece with 45° chamfers so that the die will have a stronger edge. A shoulder is not needed for chamfers less than 30° (Figure 7.7(d)). Gear teeth should be at a larger diameter than the hub so that the die will not be weakened (Figure 7.7 (e)).

P/M parts can also be made with controlled nonuniform densities. For many machine components, such as gears and engine connecting rods, it is desirable to have bearing properties at some locations and structural strength at others.

7.4. INJECTION MOLDING

Besides the cost of material, the cost of a molded piece includes the cost of the mold, the cost of the molding process, the cost of rejects, and the cost of secondary operations, such as trimming, degating, and drilling. A good mold is expensive, but it can reduce the molding cost. Careful and deliberate molding can eliminate rejects. A well designed mold can automatically degate and produce no flash, so that trimming becomes unnecessary. In this section, we will see how the decisions of a product designer can influence the cost of the molded pieces. We will mainly discuss injection molding because it is the least expensive method to manufacture thermoplastic products. In some cases, the product features for injection molding may be applicable to other molding processes.

The material cost may be reduced by adding regrind material, fillers (e.g., glass fibers), or foamed plastics. Every type of plastic has its own flow properties, shrinkage characteristics, and molecular characteristics. The mold may have to be designed to suit the properties of the material. For example, acetal polymers flow much easier and can be shot into the mold cavity at a higher speed than most of the common plastics. With a higher strength and rigidity, acetal plastics can have sufficient strength even in a thin section. The reduction in cycle time and thickness may compensate for the higher cost of the material. Figure 7.8 is a plot of cycle time versus wall thickness [126]. Crystalline materials solidify rapidly and gain enough strength to withstand the ejection forces. An engineer should always use the minimum amount of material that is functionally sufficient. Minimum weight means savings in material, cycle time, and processing energy. The engineering principles for minimum-weight designs are discussed in Chapters 1 and 4. Since it is easy to add webs or increase the thickness, initial design should be made thinner than normal. The sample pieces from the initial mold are tested and "beefed up" if necessary. A piece that is too thick may require welding and other costly mold changes. To check the wall thicknesses, sample moldings can be made with colored transparent plastic. The shade of color indicates the wall thicknesses.

A well made mold runs smoothly, reduces molding costs, eliminates secondary operations, and reduces the probability of rejects. Some

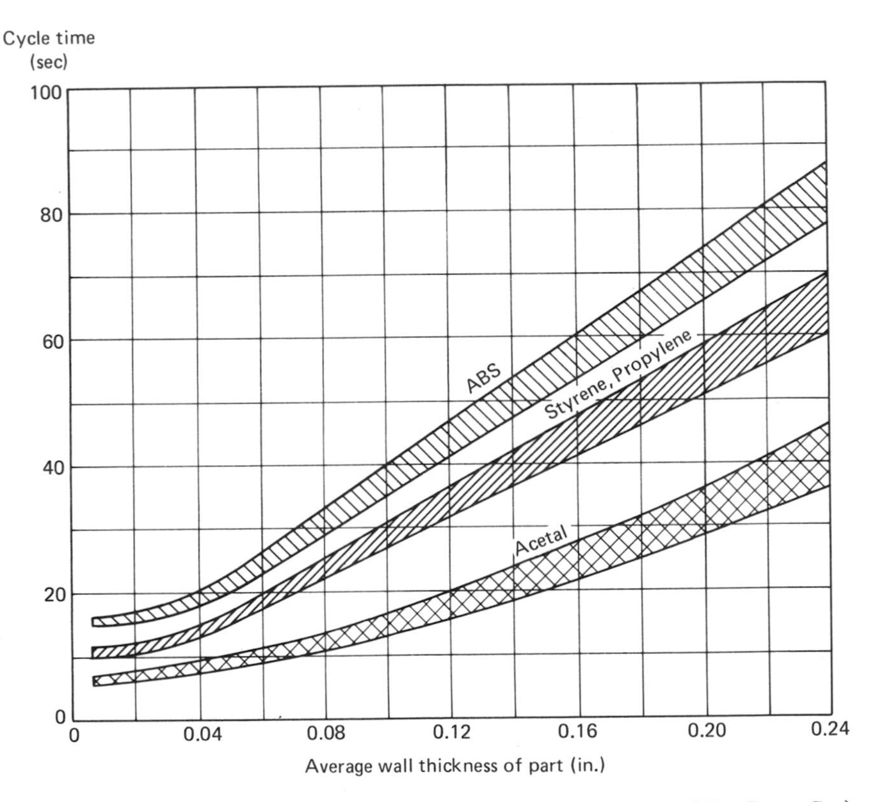

Figure 7.8. Thickness of a part versus cycle time. (Courtesy of Du Pont Co.). Some plastics have shorter cycle times for the same wall thickness because they have a low melt viscosity and they can quickly gain strength upon cooling. Both of these characteristics can be traced back to their crystalline molecular structure. Examples of acetals are Delrin by E. I. du Pont de Nemours & Co. and Celcon by Celanese Polymer Co.

design details are costly in mold construction. The cost of the mold can be reduced by using modular and symmetric designs, as shown in Figure 3.18. Small variations in a group of similar products can be handled by inserts in the mold. Section 7.10 discusses ways to simplify machining. Some expensive features are added to molds to reduce processing. A runnerless mold, also called *hot runner* or *insulated runner* mold, uses less material and processing energy. A mold with a good cooling system can be run faster because the plastic melt can freeze quickly. A well constructed system of ejectors can eject the part even if it is not chilled completely. Adhering to good practices in mold design is essential for trouble-free operation. A deep, thin web may be difficult to eject and may require longer chilling. Hard to mold features can also be costly in mold maintenance. If a web is torn off in a mold, the maintenance person will have to pry it out with a drill, screwdriver, or chisel. A weak core in the mold may break. The mold may slam shut on an unejected plastic part and cause serious damage.

Injection molding usually involves less trimming than blow molding and compression molding. A mold with a submarine gate (Figure 7.9) or a 3-plate mold (Figure 7.10) will degate automatically. Drilling in the molded piece is unnecessary if the mold has side-slides. Design features can be incorporated such that side-slides and drilling are both eliminated (see Section 6.1.3). Sometimes a limited amount of secondary operations does not add cost to the piece if the machine operator can do it within the cycle time of the molding operation.

A molded part may be rejected because of flash, warpage, or because it exceeds the design tolerance. A short shot (incomplete piece) occurs when there is filling difficulty due to improper gating, poor mold-balancing, or insufficient injection pressures. Molding defects can sometimes be traced back to the mold design.

The molding cost consists of the machine cost and the labor cost for one cycle of the process. Machines are chosen according to their capacities, e.g., 28 oz, or their clamping forces, e.g., 375 tons. The clamping force has to be larger than the projected area of the mold cavity times the injection pressure—about 10,000 psi. Figure 7.11 shows the various functions of a molding press on a time scale. Designers tend to neglect the importance of the cycle time. Molding vendors treat the cycle time as their bread and butter. They will do everything to speed up the cycle. Some ways to speed up the cycle

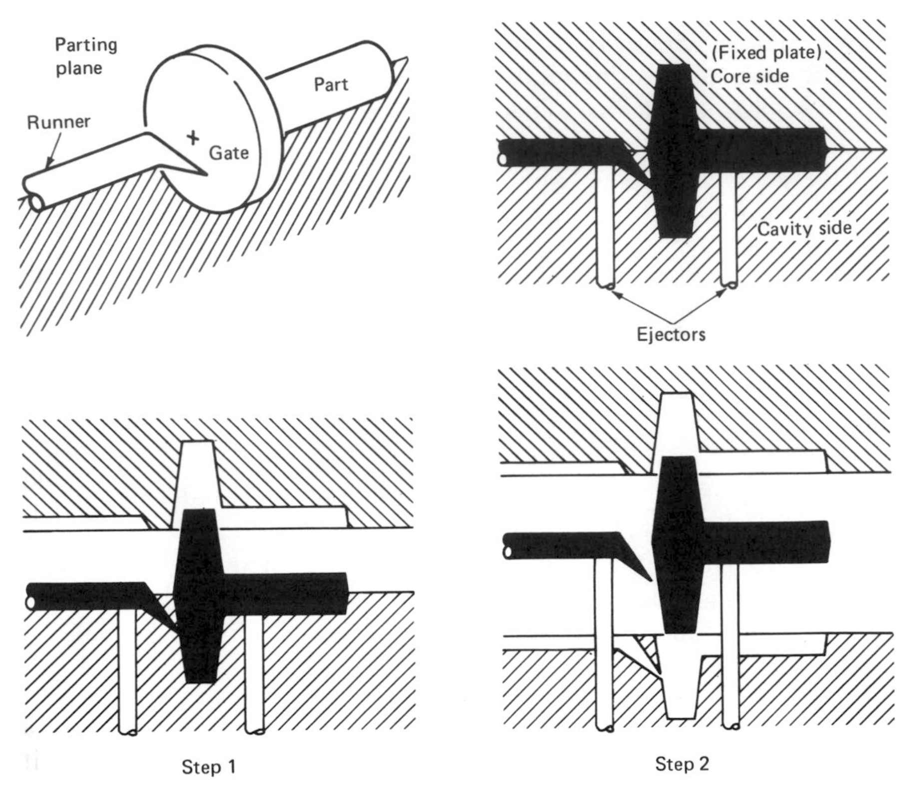

Figure 7.9. A mold with a submarine gate. In Step 1 the part separates from the fixed plate. In Step 2 the runner and the part eject from the moveable plate. The gate is broken.

are increasing the melt temperature, chilling the mold ice cold, and widening the gate. The maximum production volume in a given period of time is, of course, limited by the cycle time, which may depend on the capacity of the press and the curing time of the material. Sometimes, more mold cavities are needed to meet the production volumes required. Since rubber cures very slowly, a mold may be made with 500 identical cavities to reduce the operation cost. Design changes in such a mold will be extremely costly. If design features can be changed to speed up the cycle time, fewer molds will be needed to meet the production schedule.

Family molds are used for many products to lower costs. All the components of a product are put into one huge mold, as it is more economical to run one huge mold than to run several smaller ones.

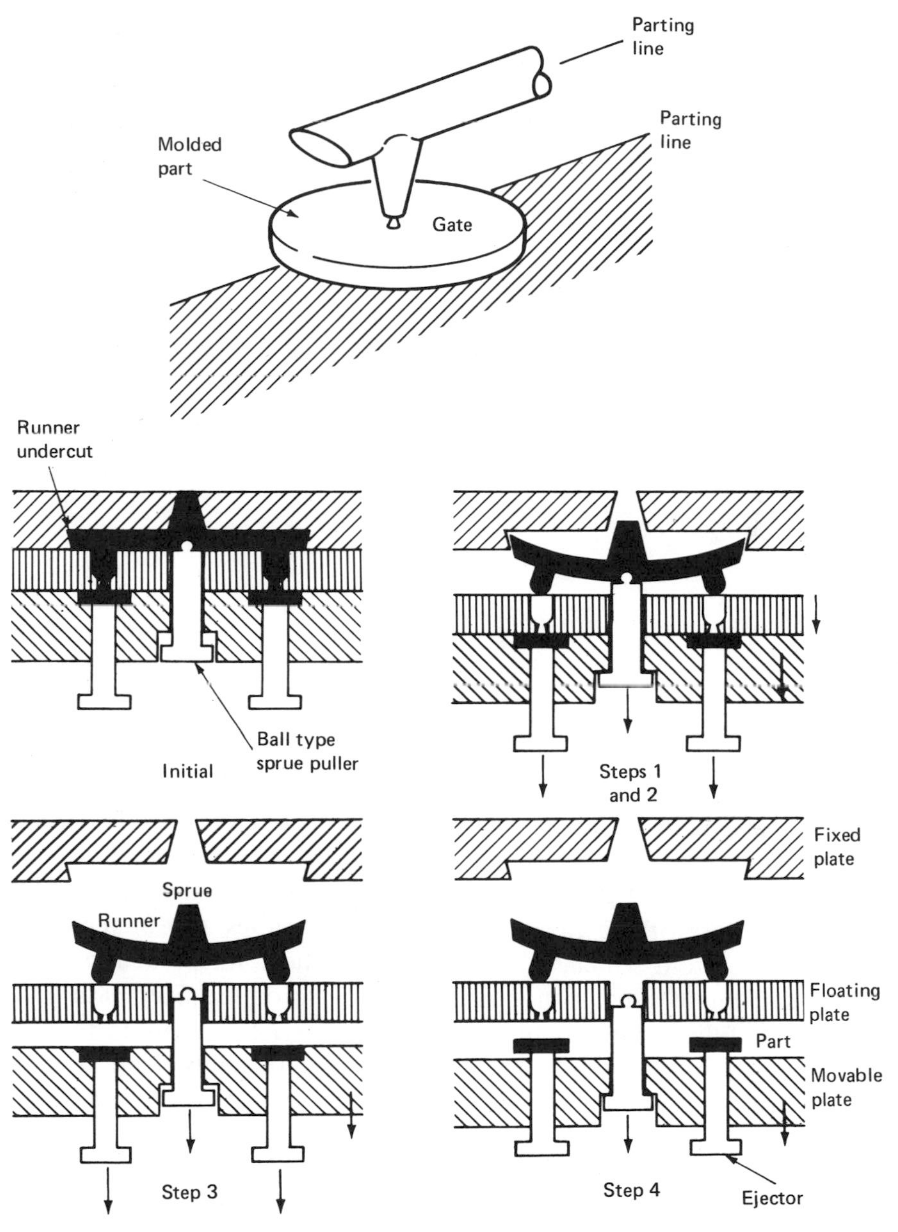

Figure 7-10. A 3-plate mold. Step 1, the runner separates from the floating plate. Step 2, the runner separates from the fixed plate. Step 3, the part separates from the floating plate. Step 4, the part separates from the moveable plate. Degating is achieved in Step 1. The runner falls out after Step 3. The part is ejected in Step 4.

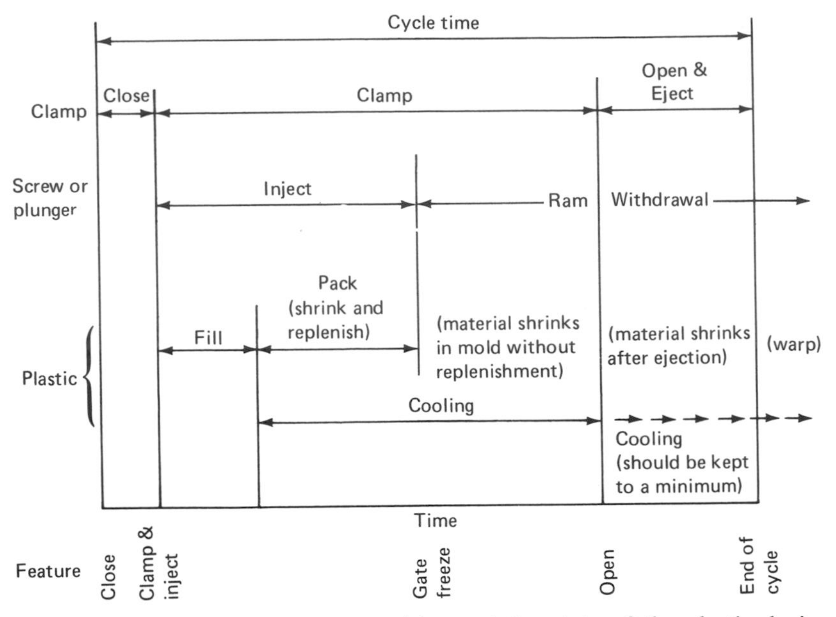

Figure 7.11. The functions of a machine and the state of the plastic during a molding cycle.

However, a family mold limits all the components to the same color. It is also difficult to balance a mold that has large and small components placed side by side. The design engineer can help to lower the molding cost by incorporating simple features and generous radii to improve the flow of plastics. A mold requiring side-slides takes longer to run because of the cam movements. Side-slides also increase the bulk of the mold which, in turn, increases the tonnage of the press. The location of a gate can change the length of the flow path and, thus, the filling time of the mold (Figure 7.12). A uniform wall thickness helps cooling, and the proper strength helps ejection. The quality of the molded piece depends strongly on the processing. Overheating the plastic melt degrades or burns the plastic material. Rapid injection may cause jetting and cold-melt. Early ejection usually results in shrinkage, distortion, and sink marks (Figure 7.13). Conversely, if the part is left in the mold longer, the mold will act as a shrink fixture to hold the dimensions to tolerance. Shrinkage distortion of a flat surface can be controlled and concealed by crowning, i.e., using a shallow spherical dome.

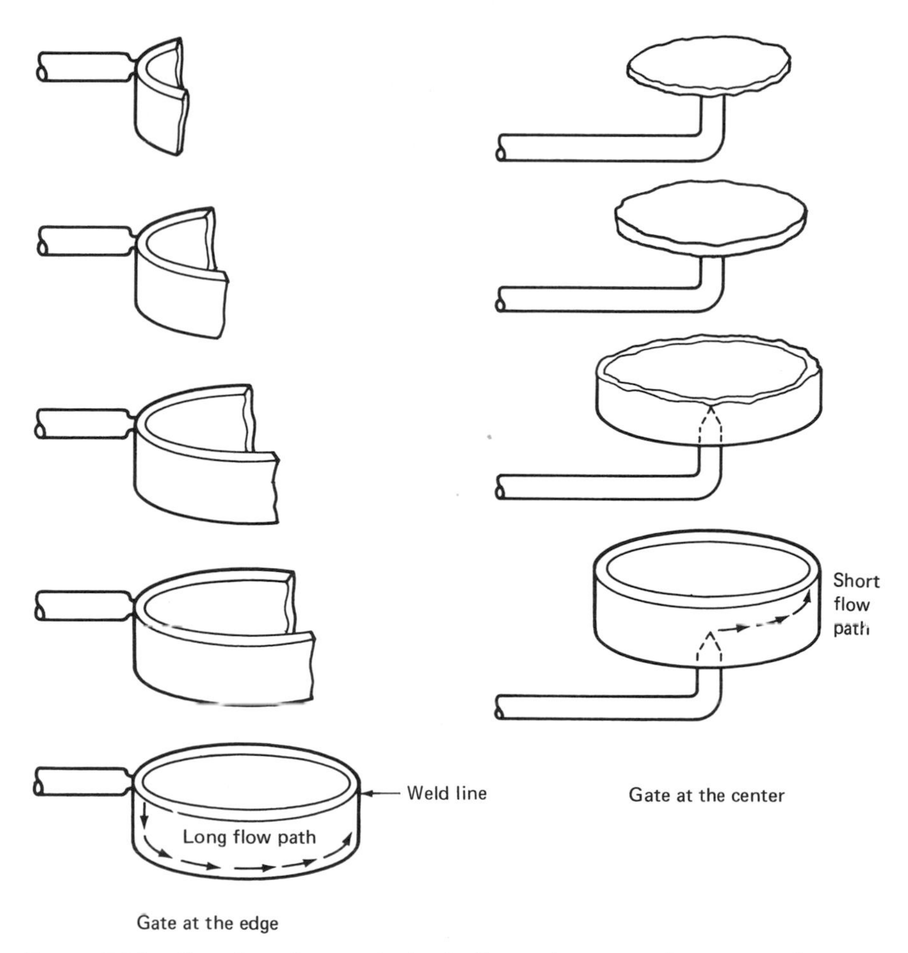

Figure 7.12. Gate locations and plastic flow. A gate at the center of a round piece shortens the flow path and produces a better flow pattern.

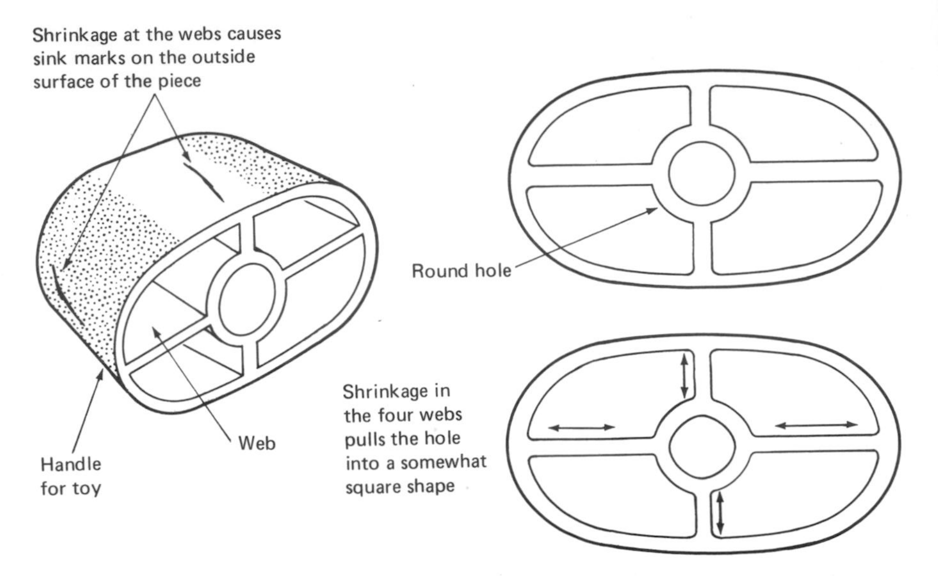

Figure 7.13. Postejection shrinkage causes distortion and warpage.

7.5. STRUCTURAL PLASTIC FOAMS

An integral plastic sandwich with a foam core between two solid faces is very strong and stiff in bending (Figure 4.58). Plastic structural foams offer the combination of lightness in weight and largeness in size. Large sizes (100 lbs or 9 ft X 9 ft area) can be made with structural foam and are not available from injection molding. Structural foams are used in the furniture industry. The foams can be molded into structures with smooth contours, round corners, and continuous forms, which are all very difficult to fabricate with wood. Structural foams are rigid enough to be used in housings for business machines (Figure 7.14), which used to be made of die-cast materials and sheet metal. In contrast to die-cast metals, foams are warm to the touch and light in weight. Molded-foam housings can provide internal ribs and bosses that simplify assembly. In comparable sheet-metal cabinets there are usually more parts requiring more assembly efforts. Structural foams also resist corrosion and damp-out noise. They are also used for outdoor construction and automobile applications.

There are many different methods for molding structural foams. The low-pressure molding of thermoplastics (by Union Carbide) has the widest application. This method can be used for almost any thermoplastic. The plastic material is mixed with compressed nitrogen or a chemical blowing agent. The mixture is then injected into the mold cavity. The injected volume is less than the cavity volume so that there is room for the material to expand. When bubbles are formed in the plastic material, the foam expands to fill the whole cavity. The bubbles collapse on coming into contact with the mold, which creates a dense, solid skin. Because the plastic material has gas dispersed throughout it, it has a milky, swirling look. This poor surface finish and a long molding cycle time are the greatest drawbacks of the process. When such a surface is placed on the outside of a product, it often has to be painted. The foaming produces numerous inherent voids such that overall shrinkage is eliminated. The molded part has excellent dimensional stability, no sink marks, and no internal shrinkage stresses. As an example, polypropylene is one of the materials that are very unstable dimensionally. For polypropylene objects with an area larger than 1 ft^2, warpage is so severe that even ribs and corrugations cannot help. However, polypropylene structural foam is very stable dimensionally. The large sizes of structural

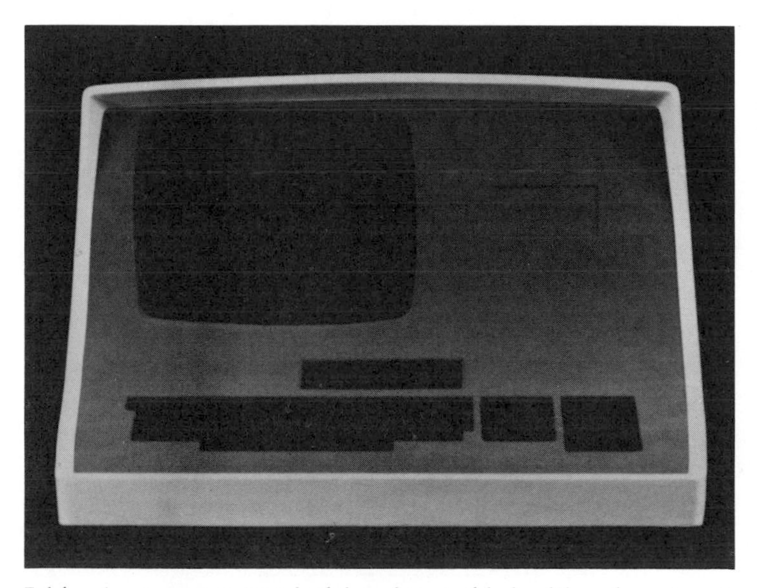

Figure 7.14. A computer terminal housing molded with polycarbonate structural foam. (Courtesy Mobay Chemical Corp.)

foams are possible because of the low molding pressures (200–400 psi), which permit low clamping forces and inexpensive tools to be used in the process.

There are several additives and variations used to improve structural foams. Fiber-glass reinforcement is used to further increase their strength. Chemical additives are available to make the foams fire-retardant. High-melting-point engineering plastics are used to get a higher heat deflection temperature. Increasing the mold temperature and the cycle time results in a better surface finish. The impact properties of foams may vary depending on the plastic resin and the method of testing.

The best wall thickness for low-pressure foams is about 1/4 in. If the wall thickness is further decreased, there may be filling problems. The density of the molded material increases because the foam core will be thinner. When the wall thickness is less than 0.18 in., the foamed material approaches the solid material. The fillets should have radii larger than 0.125 in. Bosses and ribs can be used extensively without producing shrink marks or internal stresses. Ribs can be used to strengthen a section, and bosses can help alignment and assembly. Standard assembly tools such as screws, snap-fittings, spin-welding, and solvents can be used for structural foams. The foamed core makes the material weaker in compression normal to the surface, tension normal to the surface, and shear parallel to the surface. A bolt-joint that compresses the surface may cause the core to collapse. A lap-joint using adhesives may cause the core to fail in shear. Sometimes the wall thickness is decreased in the vicinity of a fastener to increase the material density and eliminate the foam core.

7.6. WOOD

Wood is a versatile material that is used in structural products and fiber products; it is also a source of oils, chemicals, and energy. The discussion here is focused on the use of wood in structural products.

Since wood is a renewable material, under proper management wood resources can be inexhaustible. However, the sizes and grades of wood are constantly changing, as the emphasis shifts from old growth to managed forests. As the sizes of logs decrease, plywood manufacturers lower their minimum accepted diameter. This creates competition for raw material that was previously used only for making paper. Forest management develops fast growth species of "super" trees. Wood processors are concentrating on using the whole tree and reconstituted material. Wood users, such as the housing industry, are finding new ways to save lumber and plywood. The paper industry has introduced thinner and lighter papers to substitute for the heavier gages of paper with the same strength, stiffness, and opaqueness. Besides the changing raw material, there is a growing trend toward using foamed plastics and thermoformed plastics in furniture. The forest products industry has to improve production efficiency and maintain a high quality in its products to remain competitive. One outcome of boosting productivity is the integrated company that does everything: forest management, cutting, processing, paper, lumber, and reconstituted material. Improved coordination and better usage of waste may result from integrating these operations.

We will first examine the properties of wood. Then, ways to increase utilization and structural efficiency of wood will be discussed.

7.6.1. Properties of Wood

The properties of wood depend stongly on its species, grain direction, and moisture content. Wood can exchange water with the environment. In a humid environment, wood absorbs water and swells differently in the circumferential and radial directions. Anisotropic swelling causes warpage and cracks.

The highly directional properties of wood are due to its structure: wood behaves like a bunch of straws bonded together (Figure 7.15). Typical values of strength for pine wood are: axial tension 2000 psi, axial compression 1750 psi, transverse compression 450 psi, and shear parallel to grain 120 psi. The elastic modulus, E, of wood in the axial direction is about 1760 psi. Wood members are never intentionally placed under transverse tension because the transverse tensile strength of wood is as low as 3% of its axial tensile strength. The transverse tensile properties, such as strength and elastic modulus are rarely measured. Since the shear strength perpendicular to the grain is higher than that parallel to the grain, shear failure always occurs in planes parallel to the grain. In the case of a sheet of plywood under bending, shear failure appears as a delamination between layers. Designers never have to worry about shear failure perpendicular to the grain.

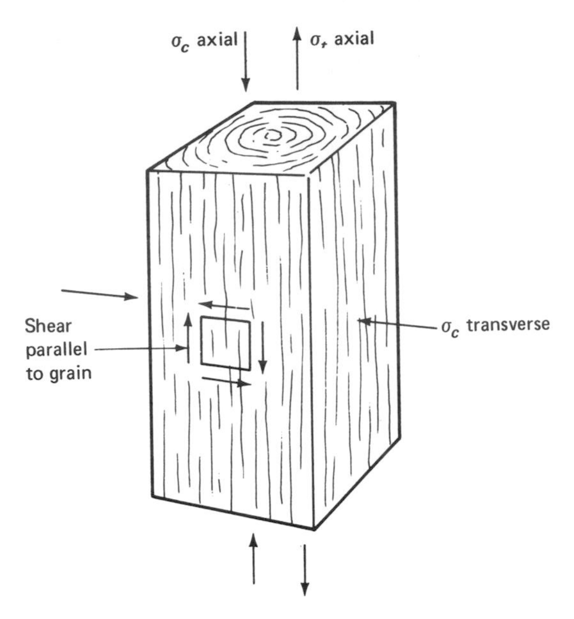

Figure 7.15. Properties of pine. $\sigma_{t\text{-axial}} = 2000$ psi; $\sigma_{c\text{-axial}} = 1750$ psi; $\sigma_{c\text{-transverse}} = 450$ psi; and $\tau_{\text{parallel to grain}} = 120$ psi.

The density of wood is in the range of 0.2 to 0.9 gm/cc. The strength of wood is directly proportional to its density. Because a heavy wood has more fibers, it is stronger.

7.6.2. Maximizing the Utilization of Wood

Amazing results are obtained by modern management of forests and development of "super" trees. On an equal time basis (27 yrs), the Georgia Pacific Corp. made trees grow 2.13 times larger in diameter and 1.23 times taller: a 560% increase in weight. This is achieved by removing underbushes, thinning, and applying fertilizers. The harvest time was cut down to a half of the previous time. On an equal land basis, the output of wood fibers has tripled.

The complete-tree method, which chips up the entire tree with leaves, branches, and bark, was first introduced in 1964 by H. E. Young. A newer machine under development can even pull the stump and roots out of the ground to increase production by 20%. Although the chips contain contaminants, such as bark and dirt, the removal of which costs more than debarking logs, the complete-tree method is still more economical than the conventional method. Compared to logs, chips require less labor and less space to transport and to store. The chipping machinery are selfcontained units that can operate even in bad weather (Figure 7.16). Since the complete-tree method processes small trees and branches, which were left untouched by conventional methods, more wood fibers can be harvested from the same forest. The finished site is also cleaner. At present, most of the chips are used for making paper. Because of the growing development of reconstituted materials, Young [131] projected that as much as 60% of the wood fibers might be used for reconstituted materials by 1984. On the product side, manufacturers and users of reconstituted material would pay less if the tolerance of bark and leaves is increased.

Since logs have a natural taper, a circular cross-section, and in-

Figure 7.16. The machinery for the complete-tree method. The equipment is self-contained, and requires very little labor. (Courtesy Morbark Industries, Inc.)

herent defects such as knots and checks, converting them to regular shapes (e.g., sheets, square blocks, and rectangular blocks) necessitates much loss. Computer simulation of feasible cutting schemes and other optimization techniques are used to minimize such loss. An example of using linear programming to maximize wood utilization is included in Appendix 6. Figure 7.17 shows the yield of different cutting methods. Among other things, the yield depends on the overall geometry of the tree. Most of the yield figures fall between 50 to 60%. Richards [133] summarized the factors that influence the yield in the following order of importance: (1) kerf, (2) board thickness, (3) edging, (4) diameter of the log, (5) taper in the log, (6) sawing method, and (7) length of the log. The kerf and the board thickness are particularly influential for thin boards. Changing the kerf from $\frac{3}{16}$ to $\frac{7}{16}$ in. decreases the yield from 66.9% to 53.6%. Increasing the board thickness from $\frac{1}{2}$ to $1\frac{1}{2}$ in. increases the lumber yield from 53.9% to 63.9%. If the output boards are 4×4 in.^{2} or larger in cross-section, the effect of board thickness is greatly decreased. The around-cutting method (also called 4-sides method, illustrated in Figure 7.17 (c)) can outperform the live-cutting method if the log has a larger taper. The around-cutting method first converts the tapered cylinder to a uniform rectangular cant, thus obtaining short side boards from the four sides. For most logs with a 0.3° taper, the live-cutting method yields 3% more lumber.

The thickness of the sawcut (the kerf) must be wider than the thickness of the saw blade to provide clearance. Without clearance, high friction will develop to impede the free movement of the blade. The amount of clearance depends on the springback of the wood after cutting, the flatness of the blade, and the regularity of the teeth. High-precision saws can reduce the clearance and decrease the amount of sawdust by 20% [134]. A slower feed and a smaller amount of blade protrusion (for circular saws) can improve the surface finish. When a circular saw barely protrudes the surface of a slab of wood, the cutting action at the top teeth along the grain of wood is somewhat like planing. Another important advantage of the above cutting method is the formation of long ribbon like chips that can be used for reconstituted material and paper. A single-tooth saw that makes coarse chips by a shaving action is also developed for that purpose [135].

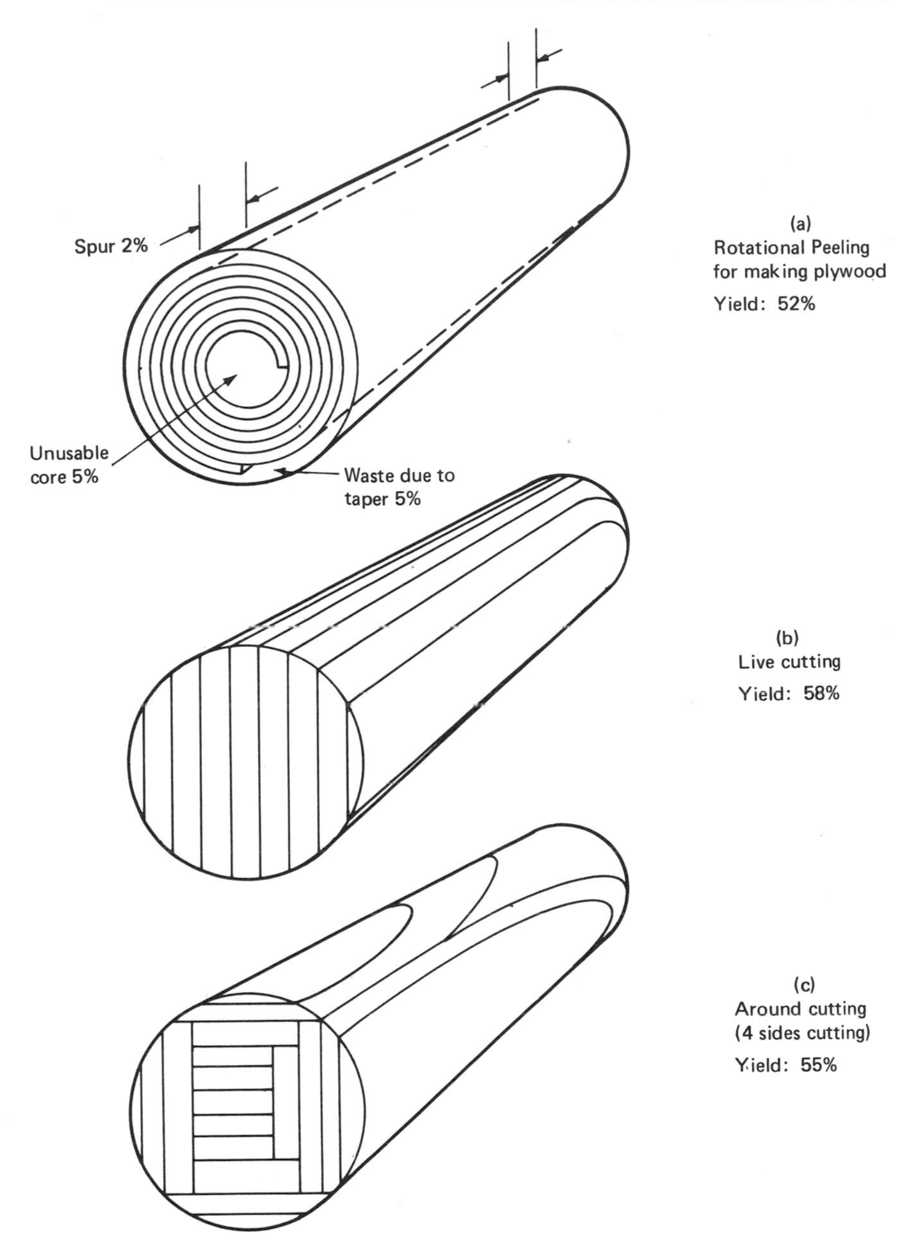

Figure 7.17. The yield of different cutting methods. Yield values can vary from log to log depending on the geometry. Live cutting requires less labor and often results in a higher yield. Around cutting is suitable for logs with large taper. The log is first converted into a cant with uniform rectangular section. Then the cant is sawn into planks.

7.6.3. Promoting the Strength of Reconstituted Materials

Reconstituted materials are adhesive-bonded wood materials such as plywood, flake board, particle board, and hard board. The advanced processing of such products eliminates local defects, averages out the directional properties, and improves stability, durability, and strength. Reconstituting is also a strategy to increase the utilization of scrap. In terms of marketing and consumer acceptance, reconstituted materials are made comparable in function, thickness, and price to commonly used materials. Interchangeability with well established materials is a common way to introduce and test-market a new material.

The means to promote strength are discussed in detail in Chapters 1 and 4. Figure 7.18 shows an I-beam and a reinforced beam. The I-beam has aligned flakes at the flange and random orientation flakes at the web. The fiber-glass-reinforced particle-board beam is based on the same concept as the steel-reinforced concrete. Figure 7.19 shows a composite plywood with flaky material laminated between two veneer faces. The veneers give extra strength and make the board look better. Figure 7.20 shows plastic-sheathed structural moldings with particle-board and plywood core.

Besides the static strength, other imnortant factors in the design of

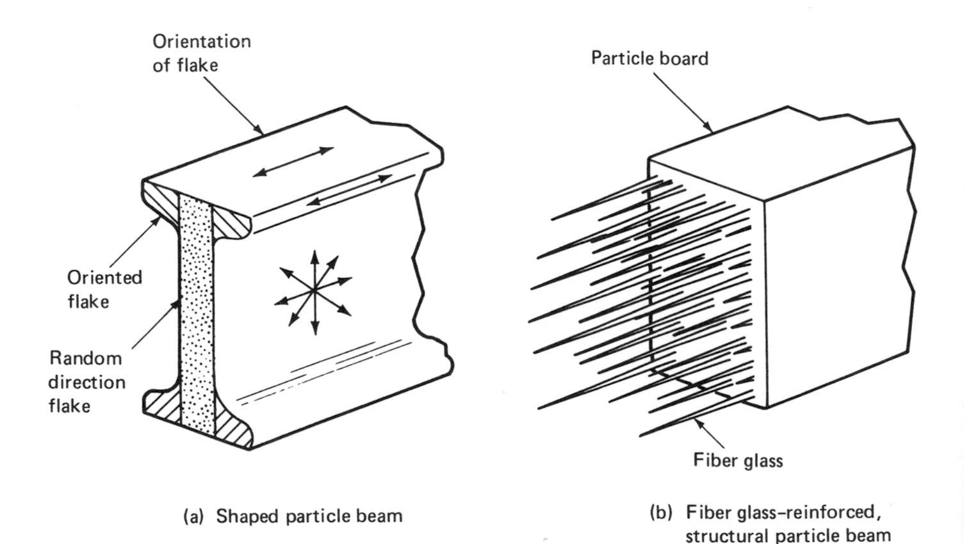

(a) Shaped particle beam

(b) Fiber glass-reinforced, structural particle beam

Figure 7.18. Strengthening reconstituted materials.

Figure 7.19. A composite plywood with flaky material laminated between two veneer faces. (Courtesy Potlatch Corp.)

composite materials are processibility, surface appearance, corrosion resistance, rigidity, and impact strength. Particle board is more rigid than lumber, plywood or flake board. However, particle board tends to be brittle and is seldom manufactured in a thickness of less than $\frac{1}{2}$ in. Hardboard is good in surface wear and corrosion resistance. Chemical treatments are available to increase the surface hardness and fire-retarding properties of wood.

Reconstituted materials can also be made with special curvatures, or made into tubular forms. Curved plywood is widely used in furni-

Figure 7.20. Examples of plastic-sheathed structural moldings with particle board and plywood core sections. (Courtesy Schock & Co., West Germany, and *Materials Engineering* [138].)

ture. The oval frames in tennis rackets are made of bent wood veneers.

7.6.4. Woodwork and Joint Techniques

Woodwork has become more and more expensive because of its high labor content. Joints are particularly important in wood constructions. Much cost reduction effort has been made in the area of mechanization. For example, the repair of defects (bore holes, apply glue, press-in plugs) is accomplished within a few seconds with automatic machines. High speed assembly methods such as RF glue (radio frequency) and staples are widely used. The strengths and weaknesses of various kinds of wood have to be closely matched with their application. To prevent mismatch and distortion, the product should be designed such that the tolerance is large enough to accommodate shrinkage variations. Figure 7.21 shows the modern way to install a cabinet door. The door will fit well even if it is off by 1 in. in dimensions and location.

Wood pieces can be joined together by glue, press-fit, and fasteners. Inserts and braces are also used. On a production scale, the complicated joint configuration is not a problem because it can be made by automatic machines. The consistency of joint strength and the reduction of manual work are major considerations in design (Figures 7.22 and 7.23).

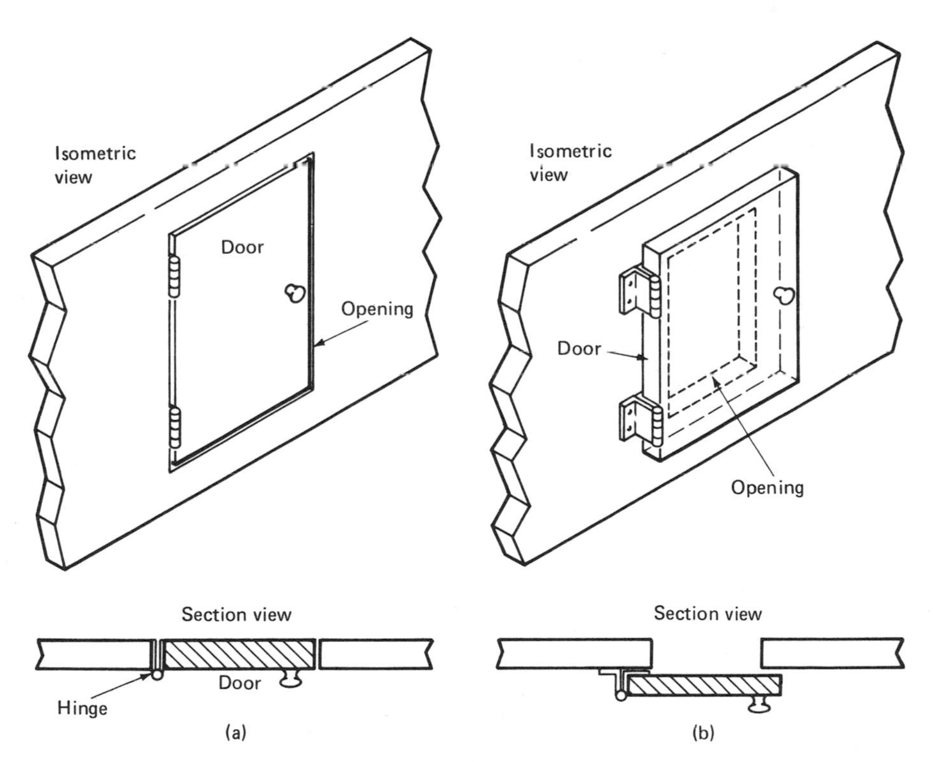

Figure 7.21. Design ideas to increase tolerance. (a) shows how a door that closes flush requires tight tolerances and careful fitting. An overlapping door (b) can vary in size and shift in position without its being noticed.

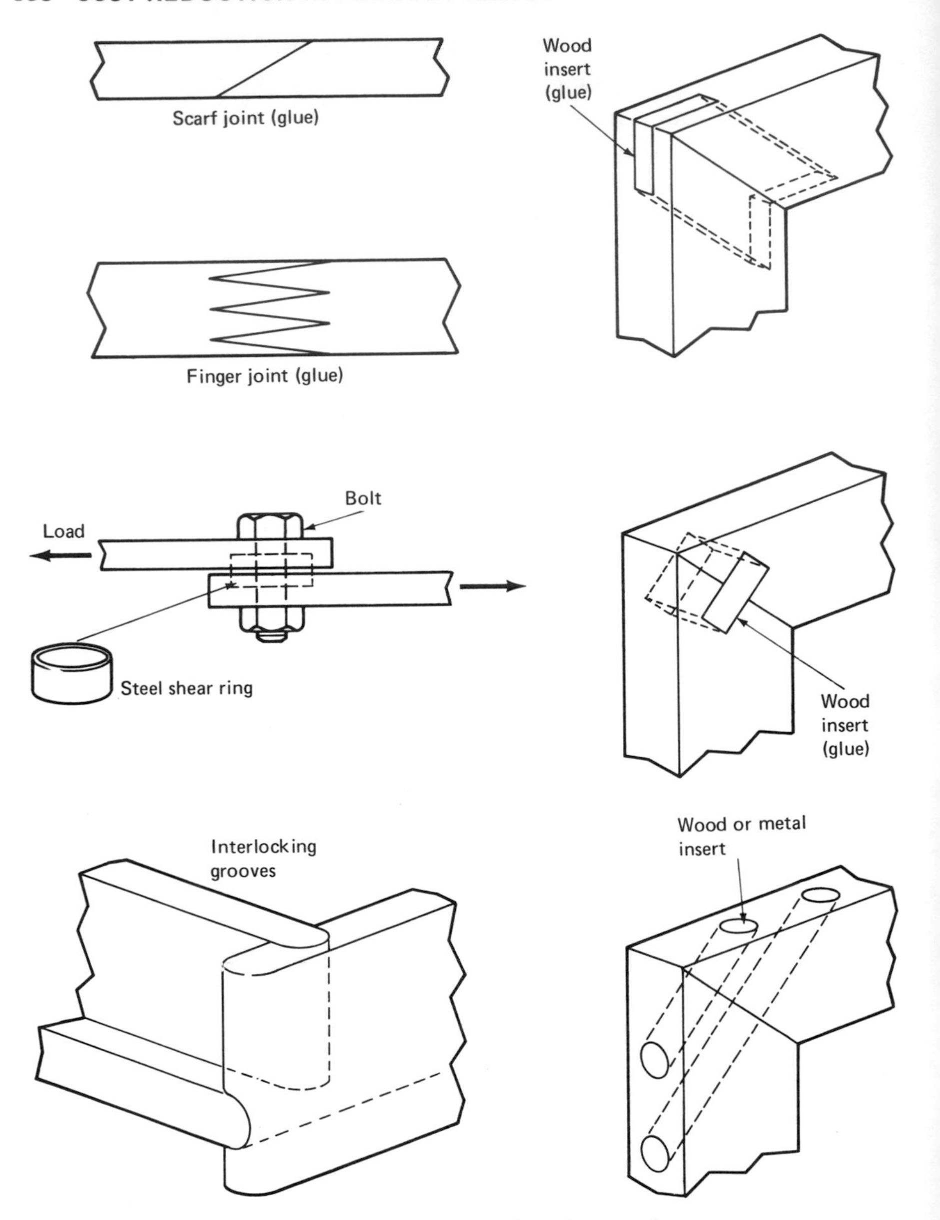

Figure 7.22. Joint designs for wood.

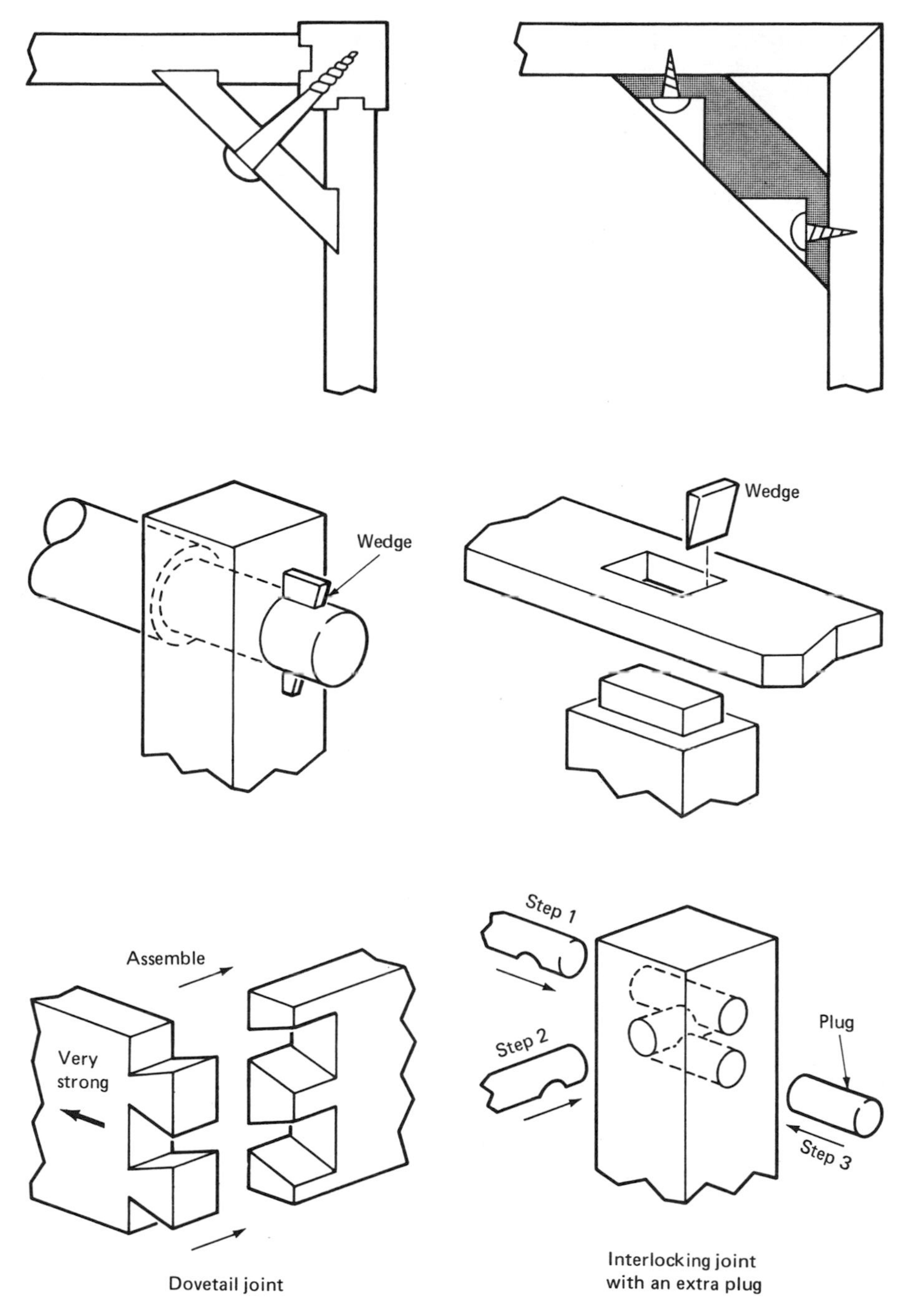

Figure 7.23. Several high-strength joints for wood.

7.7. METAL FORMING, WORKING, AND FORGING

If the stress in a material exceeds its yield strength, the material will deform permanently. If this deformation is carefully controlled, it can transform a workpiece into the desired shape. Provided there is a high enough force, something has to give. Since the tool is purposely made much stronger than the workpiece, the workpiece is forced to conform to the tool. The workpiece experiences tensile and compressive stresses at different locations. To form a piece successfully, one must avoid fracturing due to tensile stresses, and buckling (wrinkling) due to compressive stresses. The extent of forming is limited by the material's ductility. Residual stress and elastic springback are two other obstacles to the forming processes. The tool also introduces some problems, among which are burrs and die marks. There are often conflicting objectives between design and manufacturing. A high-strength material is desirable for product performance, whereas a low-strength material forms easier. The designer often has to compromise.

There are many forming and working methods, each of which is usually suitable only for a particular geometry and weight category. Figure 7.24 shows a bending process in which the workpiece is subjected to 3-point bending exceeding the elastic moment of the "beam." The bending force can be lowered if the distance between the supports is increased. Figure 7.25 shows a rolling process for thinning down a plate. Friction is very important to assure proper traction of the rollers. The amount of deformation is characterized by the percentage reduction in thickness. Figure 7.26 shows a spinning process. The workpiece is spun while a roller forces the workpiece to the

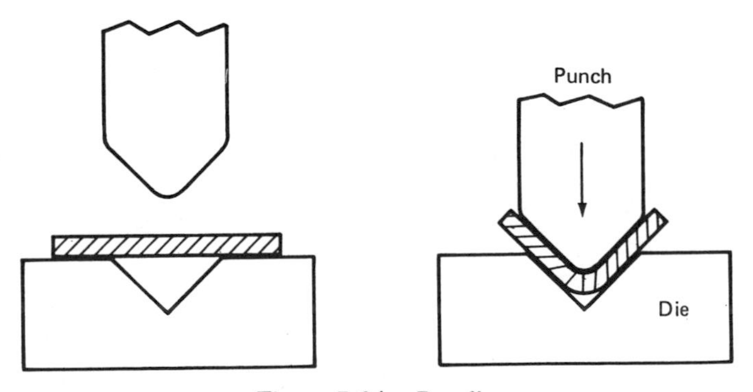

Figure 7.24. Bending.

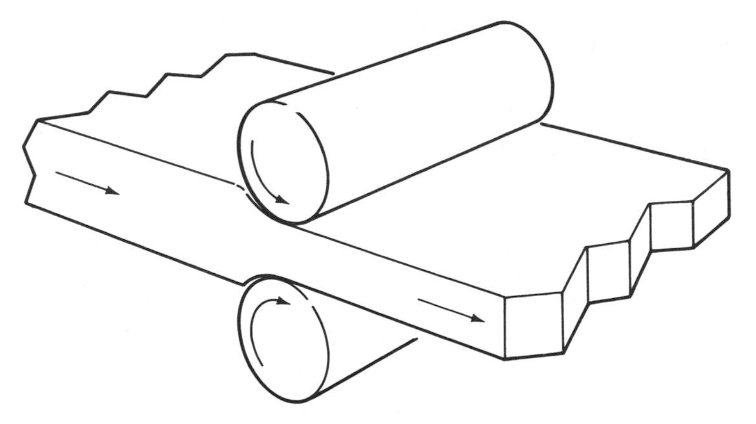

Figure 7.25. Sheet rolling.

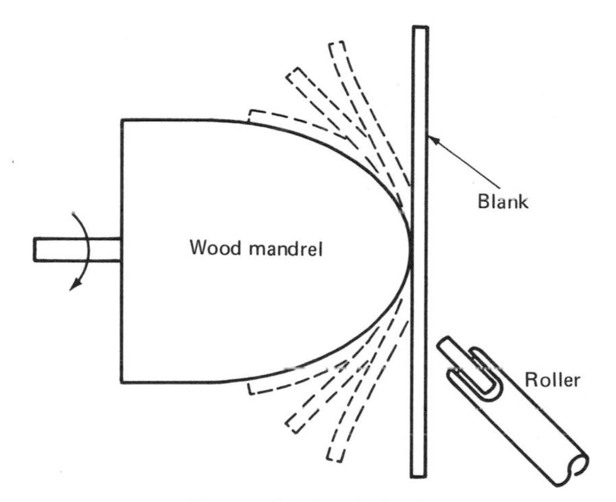

Figure 7.26. Spinning.

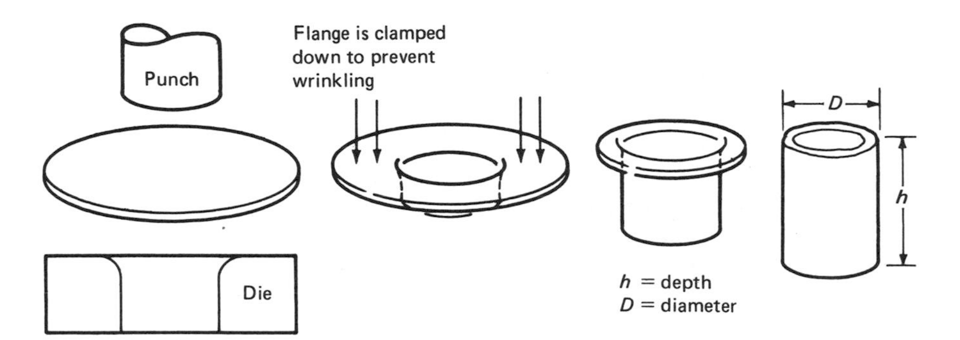

Figure 7.27. Deep drawing.

shape of the mandrel. Spinning is suitable for low-volume productions, since the mandrel can be made of wood. Figure 7.27 shows a deep-drawing process for forming cups, cones, and shell-type vessels. The blank experiences large compressive hoop stresses as its diameter is reduced. The bottom of the cup still has the thickness of the original blank, but the wall has become thinner. A depth-to-diameter ratio of 4 is usually the limit of the process. The flange may wrinkle due to the compressive hoop stresses. The tube may crack at the transition between the flange and the cylinder if the flange is too hard to pull through. Figure 7.28 shows a stretch-forming process. The workpiece is stretched to the yielding point and pulled against the die. The overall tension can eliminate compressive stresses. Hence, the workpiece will not buckle or wrinkle. Ironing is used to increase the depth and decrease the wall thickness of a deep-drawn piece (Figure 7.29). Extrusion produces prism shapes by forcing a billet through a die orifice. The material flows under hydrostatic pressure in the cylinder (Figure 7.30). The extruded materials are generally soft metals, which are suitable for cabinets, windows, and door moldings. Figure 7.31 shows thread rolling by squeezing a rod back and forth between two dies. Cold heading is very economical because it is a simple operation, has a short cycle time, and produces no scrap (Figure 7.32). Forging is a compression deformation process for high-strength materials (Figure 7.33).

The forming processes generally improve the material properties of a workpiece through work-hardening and an improved grain pattern

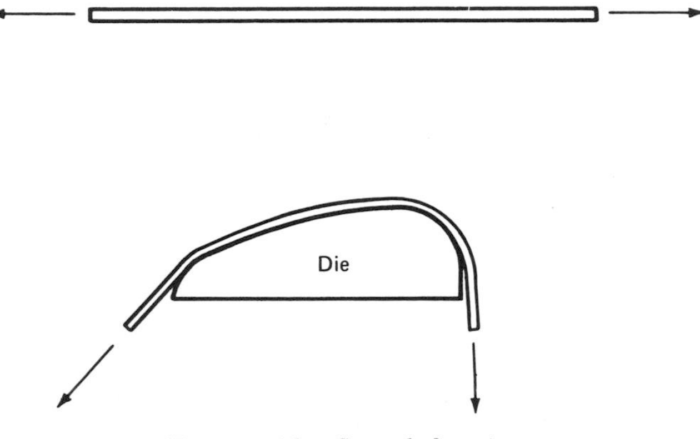

Figure 7.28. Stretch forming.

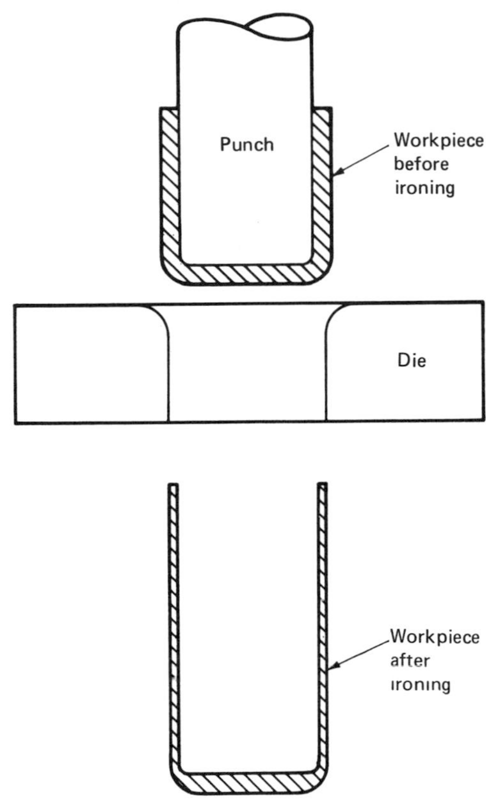

Figure 7.29. Ironing.

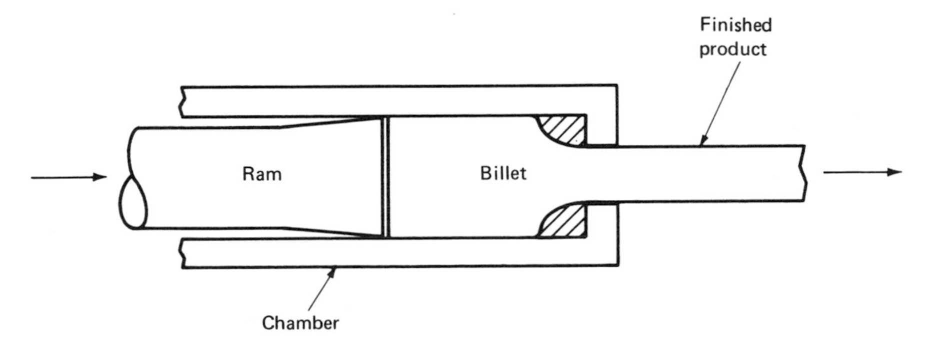

Figure 7.30. Extrusion.

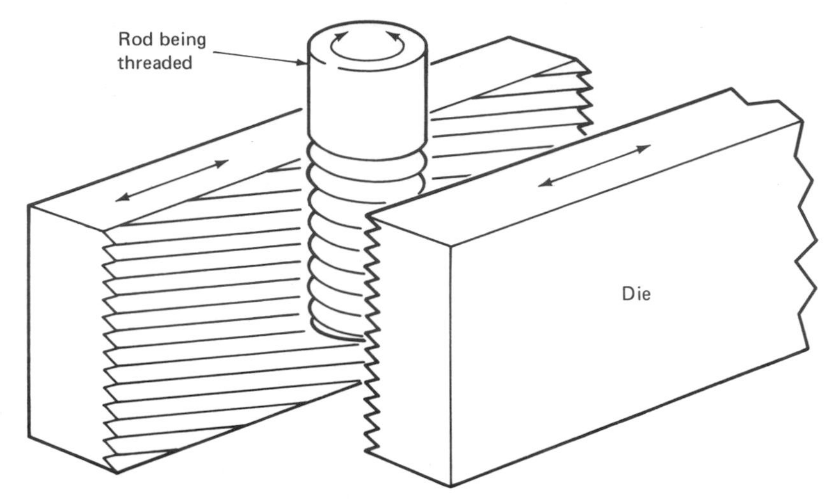

Figure 7.31. Thread rolling.

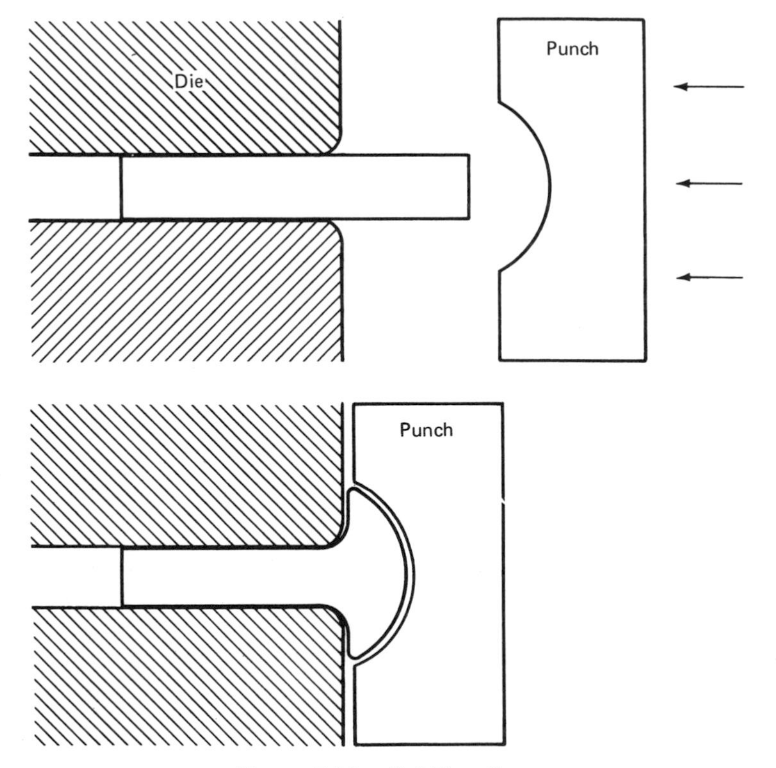

Figure 7.32. Cold heading.

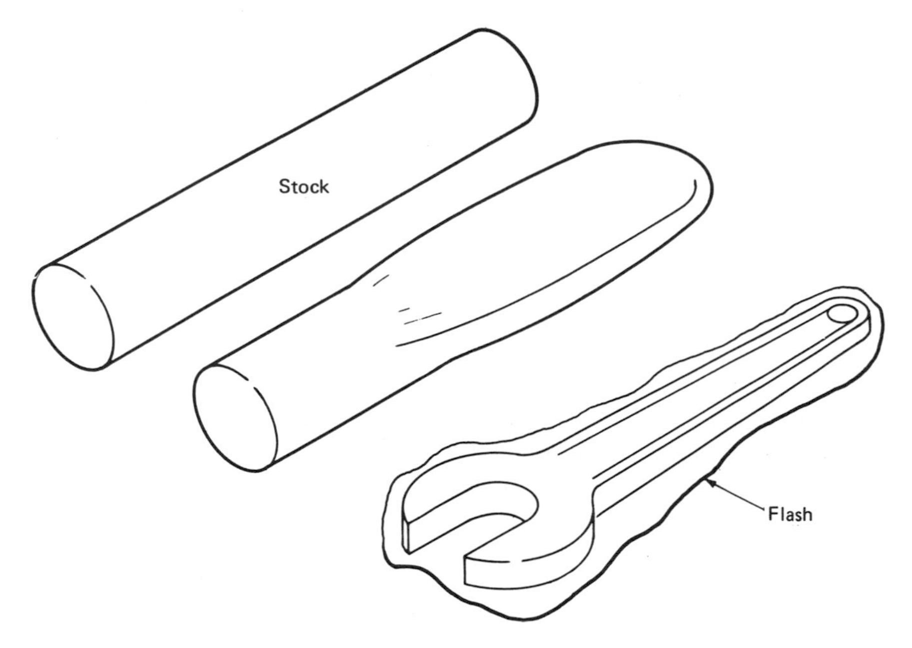

Figure 7.33. Forging.

(Figure 7.34). Forming is limited in that the strain in the workpiece must be less than that in a comparable tensile bar. Datsko's maximum-deformation theory [116] states "the maximum deformation that a material in a given condition can be subjected to in forming is that deformation that results in a natural tensile strain being induced in some direction in the part that is equal to the natural strain at fracture of a tensile specimen of that particular condition of the material." When the strain approaches the forming limit, the material needs to be annealed before further forming. Time and temperature influence

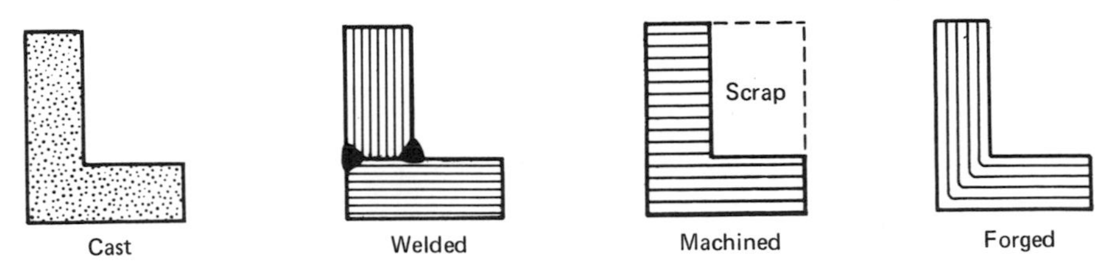

Figure 7.34. Grain pattern from several manufacturing processes.

the material's grain structure and, thus, the deformation process. High-energy-rate forming utilizes the impact of an explosive to get special fill and flow properties. Superplastic metals can deform up to 500% at an elevated temperature and a slow strain rate. The slow processing cycle of 2 min to 50 min makes the superplastic metals unattractive for mass production. Hot-forming and hot-forging are done at a temperature above the recrystallization temperature of the material. The main advantage is a low yield strength. The material will oxidize somewhat and may become brittle at higher temperatures. Lubricants are applied between the workpiece and the die if they slide against each other. Without lubricants, tool wear will be intolerable and sometimes drawing may be impossible. Some lubricants effect very low coefficients of friction; others can maintain a film even under an extremely high pressure.

Figure 7.35 shows a plate with a fracture strain (ϵ_f) of 0.5 in the grain direction and a fracture strain of 0.35 in the transverse direction. Because of this difference in fracture strain, the maximum bending curvature is $1/t$ in the grain direction and $0.7/t$ transverse to the grain, where t is the thickness of the workpiece. If the bending curvature exceeds these curvatures, multiple-step processing with annealing in between is needed. Anisotropy of sheet materials causes drawn cups to have ears (Figure 7.36). Spring-back can be a serious

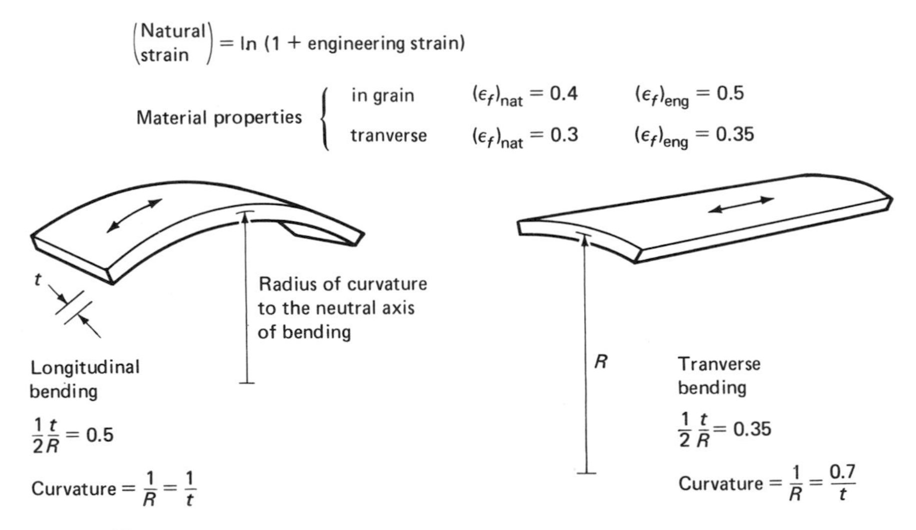

Figure 7.35. Bending of a plate with anisotropic properties.

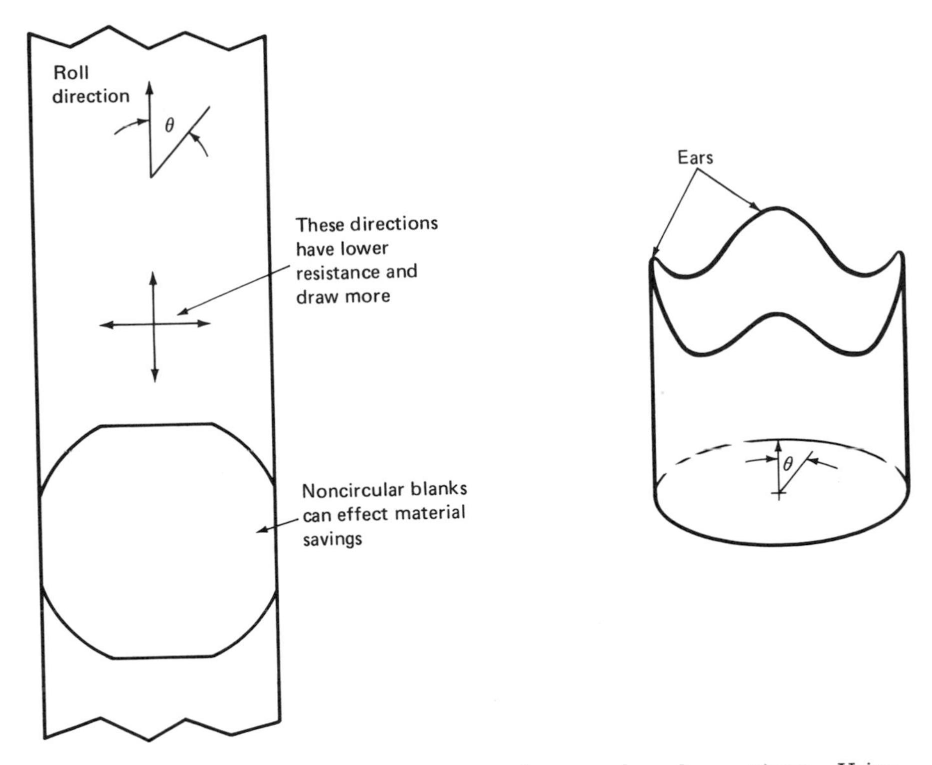

Figure 7.36. Anisotropy may cause ears to form on deep-drawn pieces. Using noncircular blanks can reduce scrap loss [114 and 139].

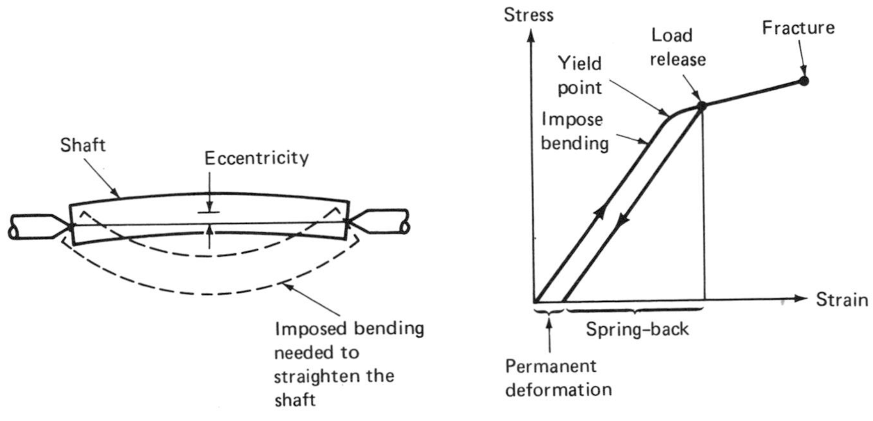

Figure 7.37. Straightening a shaft by bending.

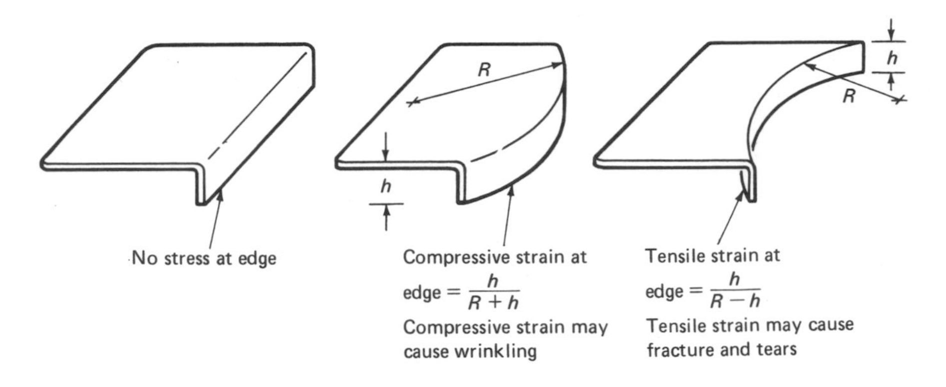

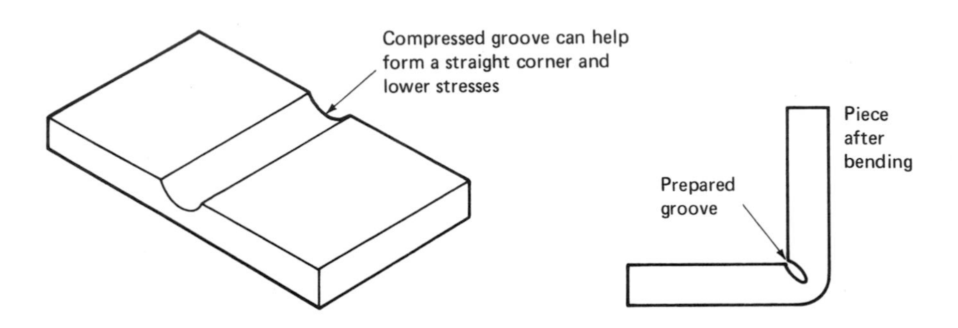

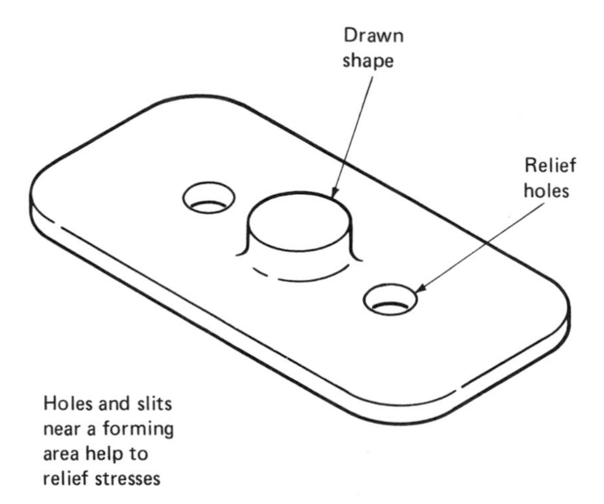

Figure 7.38. Bending and forming of plates.

problem in forming high-strength materials, brittle materials, or precision products. To get a 5-turn helical spring with a 1-in. ID, the steel wire may have to be wrapped 7 turns on a 0.7-in. dia pattern. Figure 7.37 shows a heat-treated shaft that is warped 0.010 in. The shaft may have to be bent 0.1 in. or more, and 90% of the imposed bending will be relieved by the elastic spring-back. Because of the large percentage of spring-back, precise bending is very difficult. The more precision we get from bending, the less grinding will be required. Cast-iron cam shafts are extremely difficult to bend because the fracture strain is very close to the yield strain. A slight overbending can cause the shaft to fracture. The designer can alleviate some of the problems by carefully observing the material deformation and the residual stress patterns. Figure 7.38 shows the strain resulting from bending an edge on a plate. Reliefs and cut-outs can often be used at noncritical areas to help forming.

Metal-forming processes are mostly high-volume processes that produce strong, accurate pieces that require little machining. Forging can be simplified if the workpiece is cast or sintered to a shape close to the final one. Precise forming can be economical if machining is eliminated. Material anisotropy can be used to get a larger depth-to-diameter ratio if the forming process starts with a noncircular blank (Figure 7.36). Compacted scrap metal can be hot-forged into useful products. Lubricant can be saved by carefully spraying the workpiece instead of wetting it excessively.

7.8. THERMOFORMING AND BLOW-MOLDING

Thermoforming and blow-molding are solid-state manufacturing processes that use air pressure. Thermoforming is also called *vacuum-forming* if negative pressures are used. In general, these processes are more expensive than injection molding because of their cycle time, the raw material, and the trimming required. To effect cost reduction with these processes, we must understand their characteristics and advantages. In specific products, such as bottles, blow-molding is the only practical manufacturing method.

The major steps of thermoforming and blow-molding are shown in Figure 7.39: (1) A sheet, tube, or bubble of warm plastic is produced by extrusion or injection, or a plastic sheet is simply warmed. (2) The warm material is pushed against the mold to produce an airtight

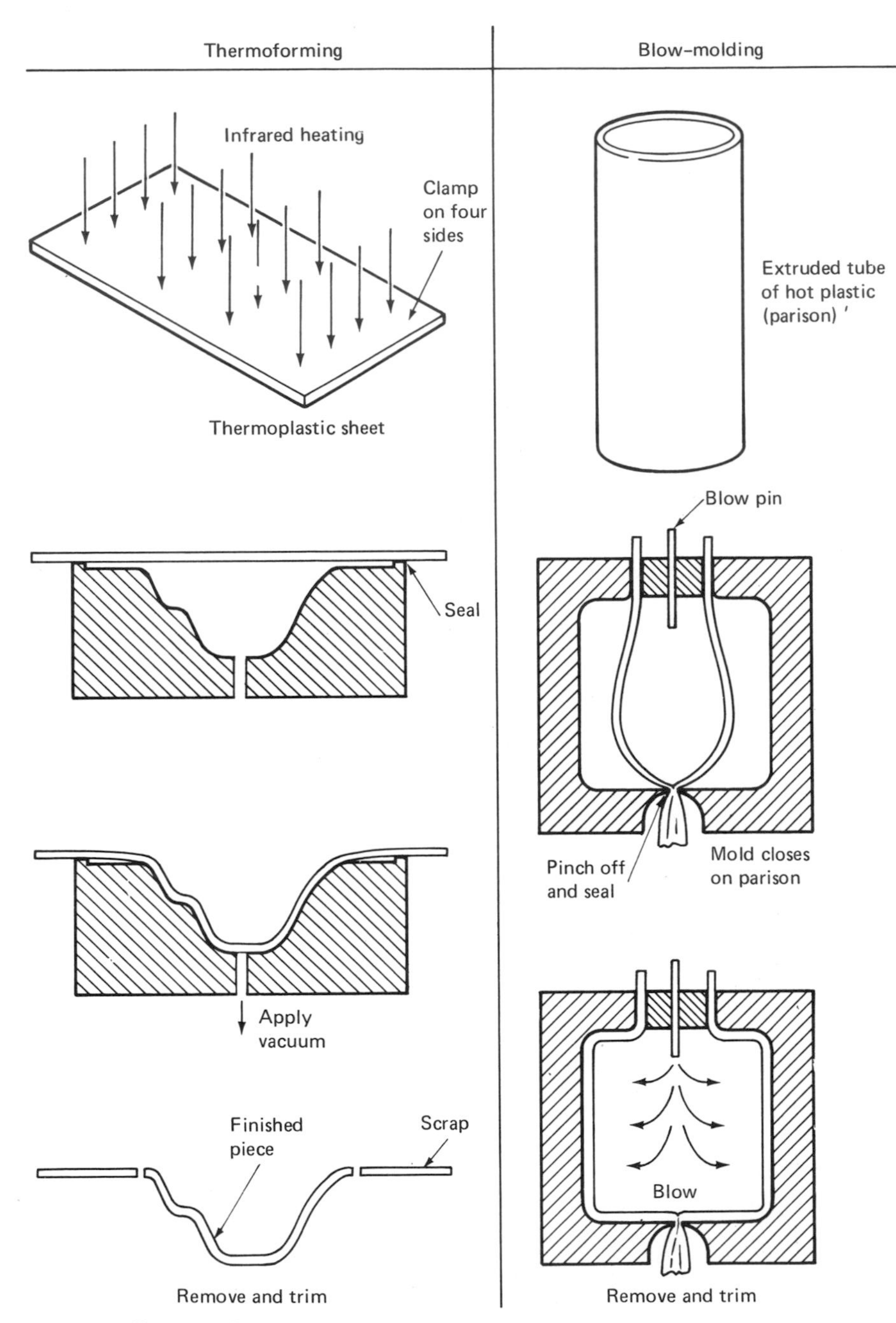

Figure 7.39. Major steps in thermoforming and blow-molding.

seal. (3) The material is blown or sucked into the mold by air pressure (or vacuum). (4) The part becomes cool and is ejected from the mold. (5) The finished part is trimmed to get rid of the excess material from the initial tube or sheet.

7.8.1. Product Design

Thermoforming is suitable for flat objects with a large area and a small thickness. Some products in this category are packaging blisters, commercial signs, briefcases, domes, windshields, canopies, and disposable containers. Figure 7.40 shows a foldable boat made from

Figure 7.40. A foldable boat made by thermoforming. (Courtesy Phillips Petroleum Co.)

thermoformed shells. Objects with large surface area and small thickness are difficult to mold by injection. For injection molding, a thin section is difficult to fill, and a large area requires a tremendous clamping force. Thermoforming also requires very little tooling and less equipment cost. The pattern for thermoforming may be made of wood. For plastic domes, no pattern or mold is needed.

Blow-molding can produce objects with closed, internal cavities: e.g., bottles, balls, tanks, hollow hands of dolls, buoys, curved, flexible tubes, bellows (Figure 3.16), and T-connections. If such objects are made by injection molding, two halves would have to be made separately and assembled. The blow-molded product would be stronger because of structural continuity. A hollow box is a strong configuration because it has the properties of a stiffened skin structure (see Section 4.5). Blow-molding does not permit ribs and webs. However, flat surfaces can be crowned (like a spherical shell), textured, or corrugated. If the product is an open half-shell, the mold may consist of two pieces matched together in a close shape (Figure 7.41). The finished product is obtained by cutting the blown piece in two. A certain blow-molding machine by Kautex can extrude two sheets of different colors. When the sheets are pinched together and blown, the finished product would have different colors on two sides of the parting line.

Blow-molding molds are usually made of aluminum, zinc alloys (kirksite), beryllium-copper, or cast iron. They are usually cheaper than injection molds. A mold may be made with inserts to allow easy variations in the features of the product. The tolerance at the pinch-off is less than that at a shut-off of an injection mold because the blow-mold does not have to contain a liquid. Since the soft plastic tube or bubble is a gummy solid, only limited details can be impressed by the mold. The surface finish of a product is affected by the texture of the extruded tube (parison) and the surface of the mold. Usually, a glossy finish cannot be obtained from blow-molding. Deliberate surface ripples or matte patterns are sometimes put in to hide the surface finish of a blow-molded piece. The mold cost can be reduced with a matte surface design in the product. The final operation for blow-molded parts is trimming. If the part is a bottle, the opening may have to be reamed to a closer tolerance. Uniformity of wall thickness is very difficult to achieve in blow-molding. The trimming and reaming often make blow-molding more expensive than injection molding.

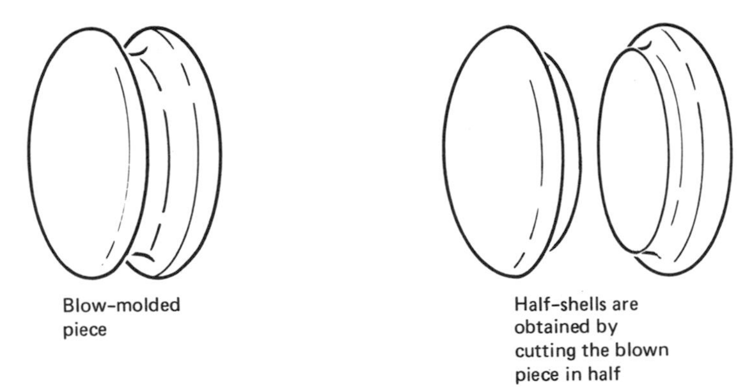

Figure 7.41. Half-shells can be blow-molded by matching two pieces to form a closed pocket.

7.8.2. Processing Control in Thermoforming

Thermoforming can be performed with a male or a female mold. A male mold (Figure 7.42(a)) impresses the details on the inside of the formed piece. A female mold impresses the details on the outside of the piece (Figure 7.42(b)). Figure 7.42 also shows the movement of the material under the forming pressures. For a deep piece, male molds will result in a thicker wall because a larger area of the initial blank will be formed onto the final shape. The bottom of the cup will have the full initial thickness. The flange of the cup may collect wrinkles because material from a larger radius is drawn inward to a smaller radius. Wrinkles are caused by compressive hoop stresses. A female mold uses a smaller blank area. Sections that are very thin may occur at sharp internal corners. If the material at the flange is allowed to flow into the cavity, the thickness of the piece will increase. Since material flowing from the flange into the cavity will decrease in radius, wrinkling may occur at the flange. Usually, there will be enough friction that the material ceases to form or move once it touches the cold surface of the mold. The thickness at a point depends on how soon that particular contact occurs. In Figure 7.42(a), the minimum thickness from the male mold is at the flange. The minimum thickness from the female mold is at the bottom circumference, which is the last place to come into contact with the mold. Generally a 1° to 3° draft (tapering angle to facilitate removal of the part from the mold) should be allowed. The ratio of final area to ini-

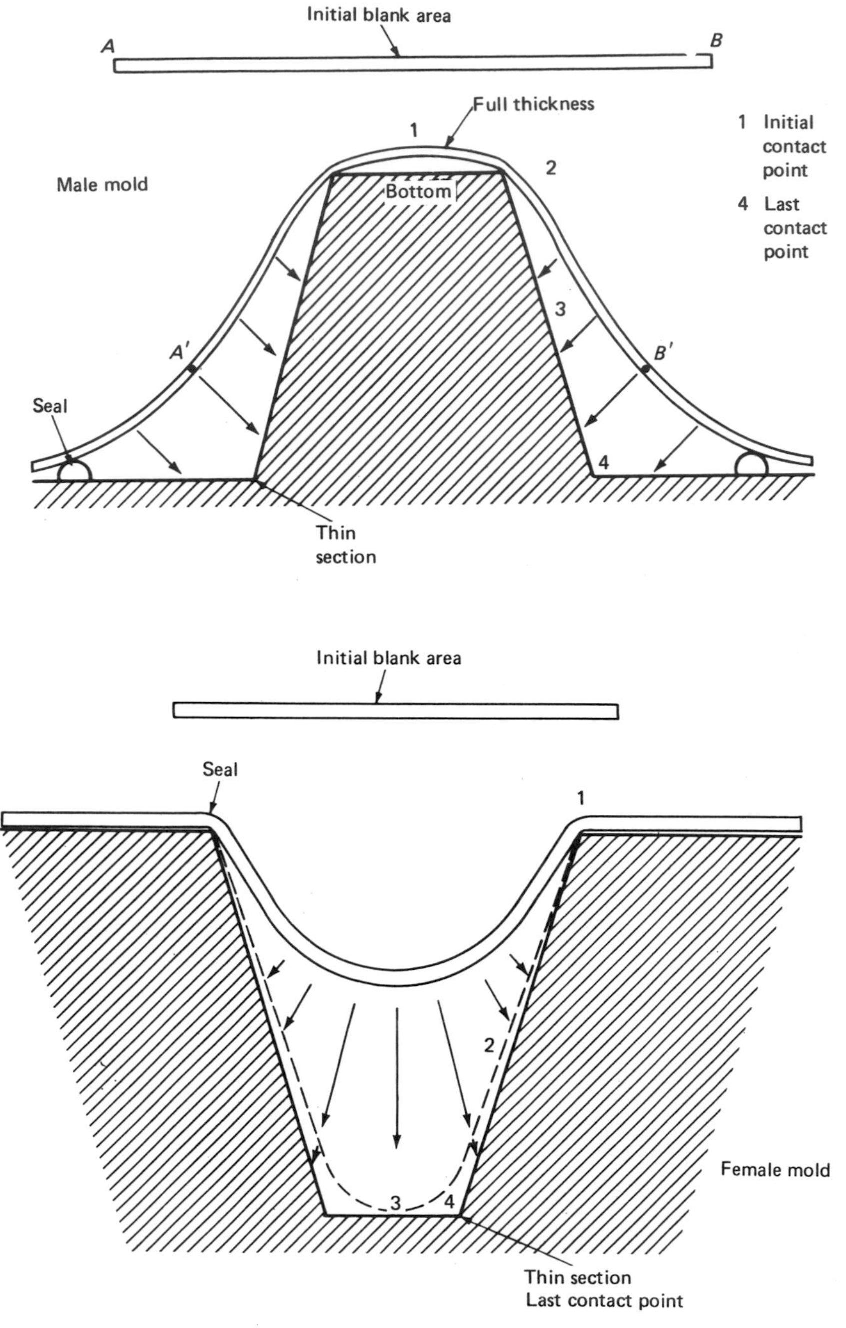

Figure 7.42. Thermoforming with male and female molds. The last point to touch the mold is the thinnest spot in the finished piece.

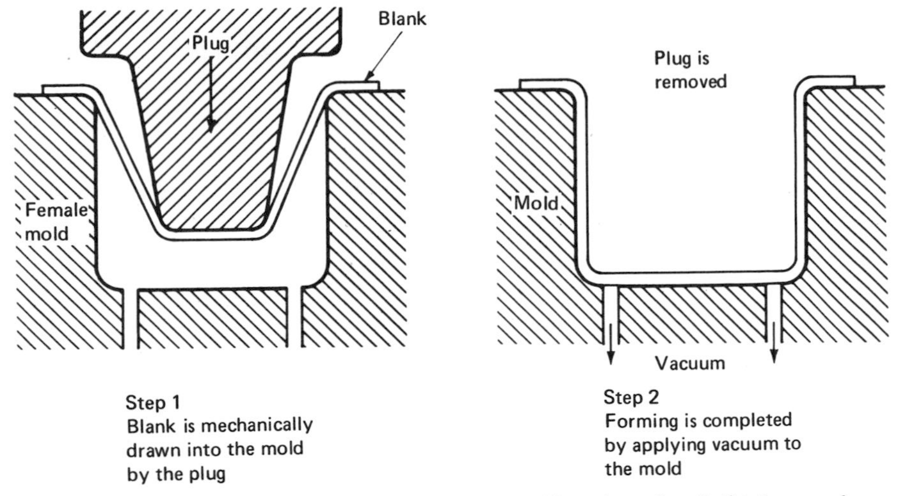

Figure 7.43. Plug-assist forming can improve uniformity of wall thickness when forming on a female mold.

tial area is generally less than 3. The ratio of depth to diameter (or narrowest width for rectangular parts) is usually less than 4.

One advantage of thermoforming is that decorative patterns can be printed on the blank before forming. Thickness control and pattern alignment can be achieved by shadow screens, which produce differential heating on the blanks. Hotter areas on the blanks can stretch more. Further control of thickness can be obtained by plug-assist

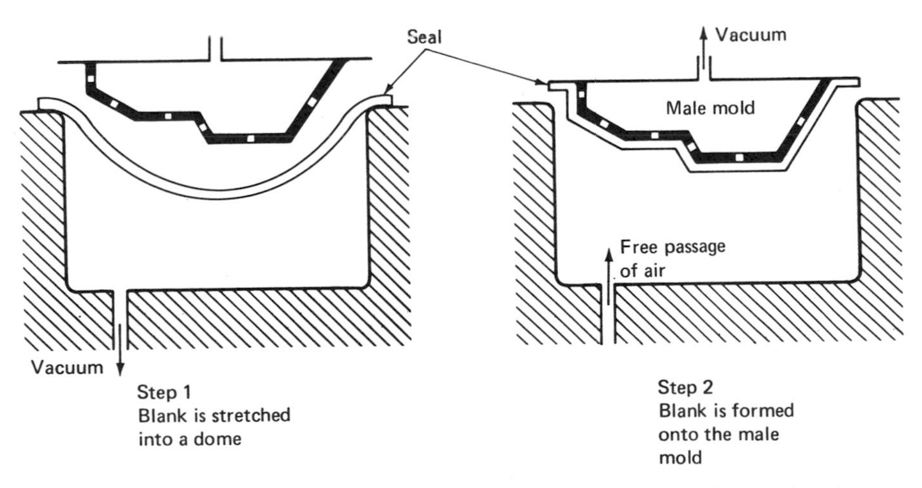

Figure 7.44. Snap-back forming makes the thickness more uniform when forming on a male mold.

forming (Figure 7.43) and snap-back forming (Figure 7.44). For plug-assist forming, the blank is first pushed onto a male mold, i.e., the plug, and then sucked into a female mold. The net effect is an averaging out of the thickness distributions of Figures 7.42 (a) and (b). Snap-back forming first transforms a flat sheet into a spherical shell and then forms the shell with a male mold. The intermediate step helps to redistribute the material and produce a more uniform thickness.

Thermoforming can also be performed without air pressure. Mechanical forces are used in some special cases: bending, drape forming, mechanical forming, and match die forming.

7.8.3. Processing Control in Blow-Molding

Blow-molding is very much like thermoforming using female molds. The material thickness distribution will be the same as shown in Figure 7.42(b). The pinch-off area has a small amount of residual after trimming. If the mold is closed first and a bubble of soft plastic is dropped in and blown, the wall thickness will be more uniform because the parison expands radially (Figure 7.45). Besides, there is no pinch-off, no trimming, and no residual. Very often the parison is a tube extruded from a piece of machinery called the head tool. The

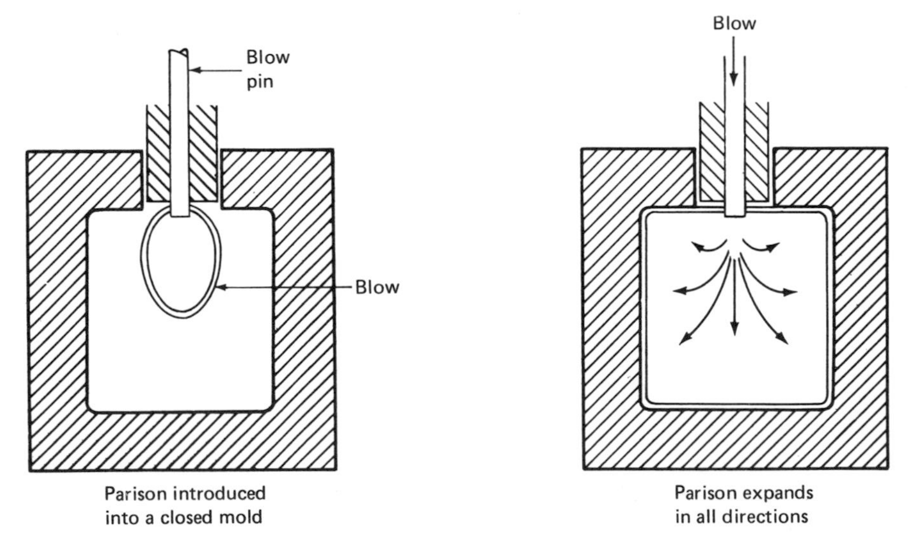

Figure 7.45. A better piece can be obtained if the parison can be dropped into a closed mold.

tube shrinks after extrusion, and depending on the material, the diameter of the parison may be much smaller than the diameter of the extruding head. Since the parison is under the pull of gravity, it will stretch longitudinally. As the parison gets long, the portion up at the top (immediately outside the extruder) will be supporting more weight and will stretch more. The material at the bottom of the parison will support less weight and will also be cooler. This nonuniformity in wall thickness can be utilized advantageously in making products that have a smaller radius at one end (Figure 7.46). In blowing a bottle with a rectangular cross-section, the parison can be shaped elliptical and nonuniform in thickness, so that it is thicker in areas that will be highly stretched (Figure 7.47). The parison is tailored so that the wall thickness is more uniform after the piece is blown. Using this technique, a tool box can be blown with the outside wall much thicker than the inside lining. The head tool can also be programmed to change thickness with time (to pulsate), so that it can blow objects with different diameters along its length. Another method is the differential cooling of the parison with jets of air. Areas that are cooled will be thick spots on the final object. A technique, called injection-blow molding, gives the best control over the wall thickness because it uses an injection molded parison (Figure 7.48). Injection molding can produce a surface with higher gloss, and can eliminate all the secondary operations at the neck and the bottom of a bottle.

Usually, the material used in blow-molding has a higher molecular weight and is less crystalline than that used in injection molding. The material has to stay in a gummy form for a range of temperature. A material with a low molecular weight tends to be watery when hot. Thus, the parison will not have enough strength to stay in shape after extrusion. Some materials are not "sticky" enough so that the finished piece splits along the parting line because of poor fusion. Medium molecular weight polymers may be used for injection-blow molding since in this process the parison strength is not as important.

7.9. EXPANDED METALS AND WIRE FORMS

Expanded metals and wire forms are lightweight and high-strength structures that are economical to manufacture. They allow free passage of air, light, sound, and heat. With wire forms, the designer can control the distribution of material very closely. The following discussion will further clarify the advantages and design details of these structures.

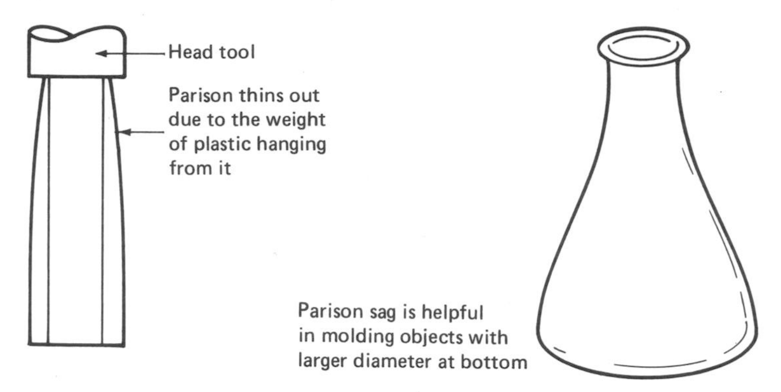

Figure 7.46. Parison sag is usually undesirable, although it can sometimes be utilized.

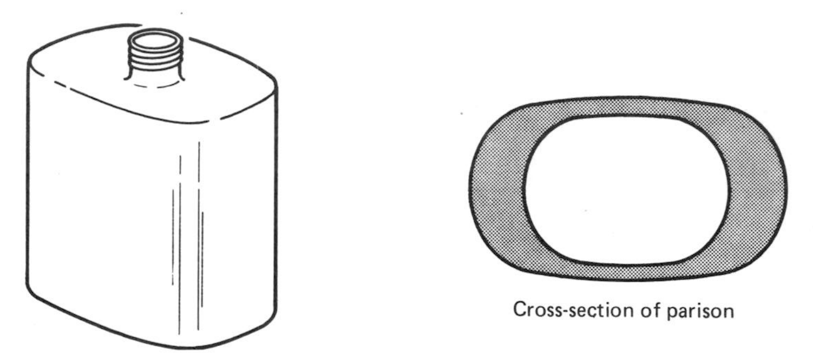

Figure 7.47. A bottle with noncircular cross-section will get a more uniform wall thickness with a noncircular parison.

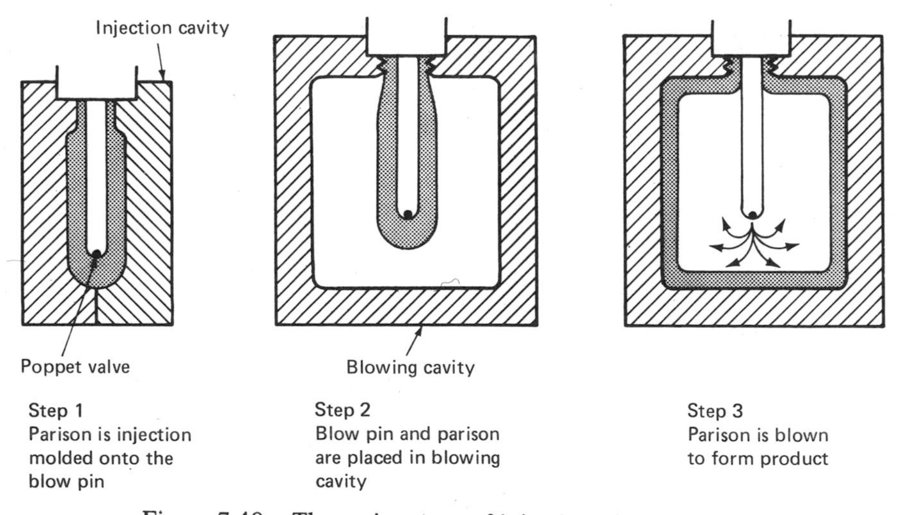

Figure 7.48. The major steps of injection-blow molding.

7.9.1. Expanded Metals

A sheet of metal (up to 5/16 in. thick) is cut with slits and expanded 50% to 400% in area. Unlike punching and perforating, the expanding process does not create scrap. Regardless of whether the expanding is achieved by stretching or back-and-forth bending, the production process operates at a high speed and a low cost.

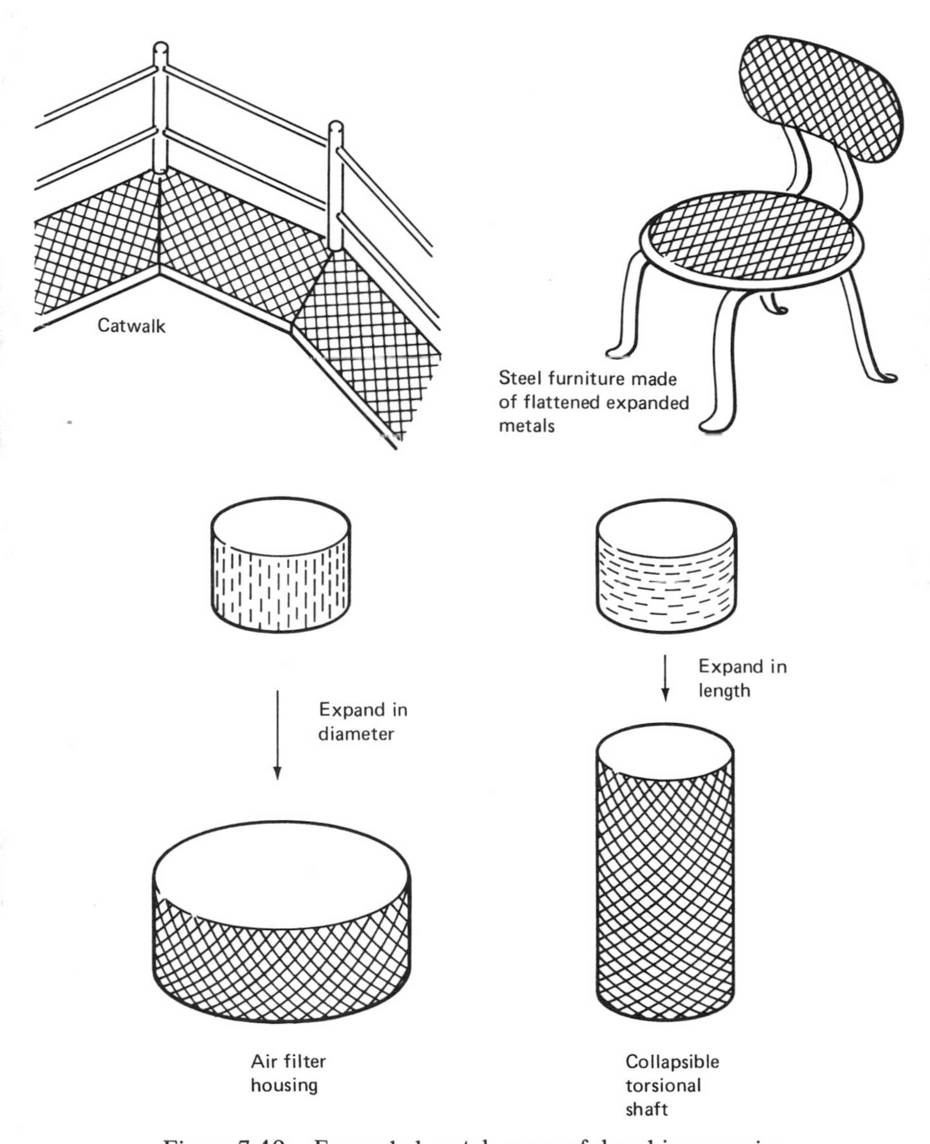

Figure 7.49. Expanded metals are useful and inexpensive.

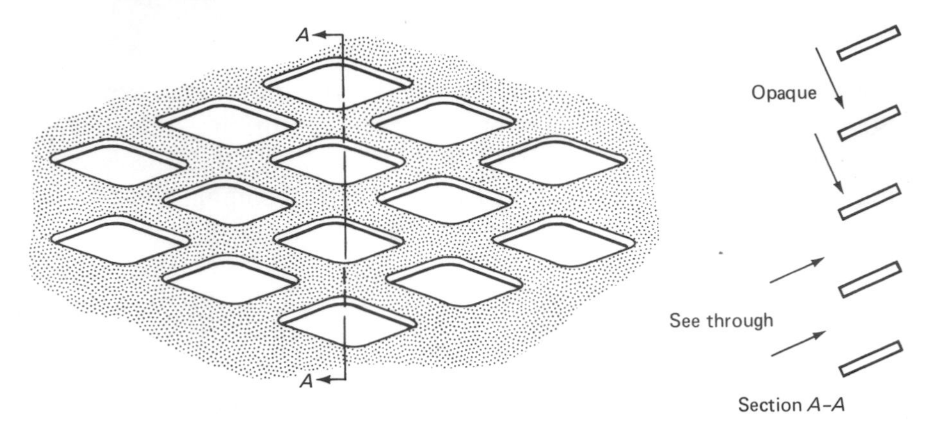

Figure 7.50. Expanded metals have directional visual effects.

Typical applications of expanded metals are: fences, catwalks, outdoor furniture, ventilator screens, and safety screens for fans, heaters, and the like. Figure 1.16 shows a collapsible steering column made by expanding a tube. Figure 7.49 shows several kinds of expanded metals for different applications.

Several inherent characteristics of the expanding process influence its properties and design. The standard expanded metals have diamond-shaped holes and 3-D textures. The 3-D textures make the materials more rigid in lateral bending. The diamond-shaped holes make the material extra strong in shear because forces are transmitted along 45° angled webs. In the expanded tube steering column, the pattern gives high torsional rigidity and strength. Some expanded metals are flattened to decrease thickness and to facilitate secondary operations such as bending and welding. The 3-D textures also provide good surface traction, which makes the materials prime candidates for catwalks. Figure 7.50 shows a section of one 3-D texture. Such a texture resembles the venetian blind in that at some angles it is transparent and at other angles it is opaque. Such a directional property can be used to shield light from an optical projector or provide decorative effects for a computer cabinet. Expanded structures with blank margins cannot be manufactured. Even though it is possible to leave a blank margin at the trailing edge of the expanded sheet, the transition between an expanded area and a unexpanded blank area (with 5 to 1 material density difference) is a potential weak spot. Perforation is usually used instead of expanding if blank margins are needed.

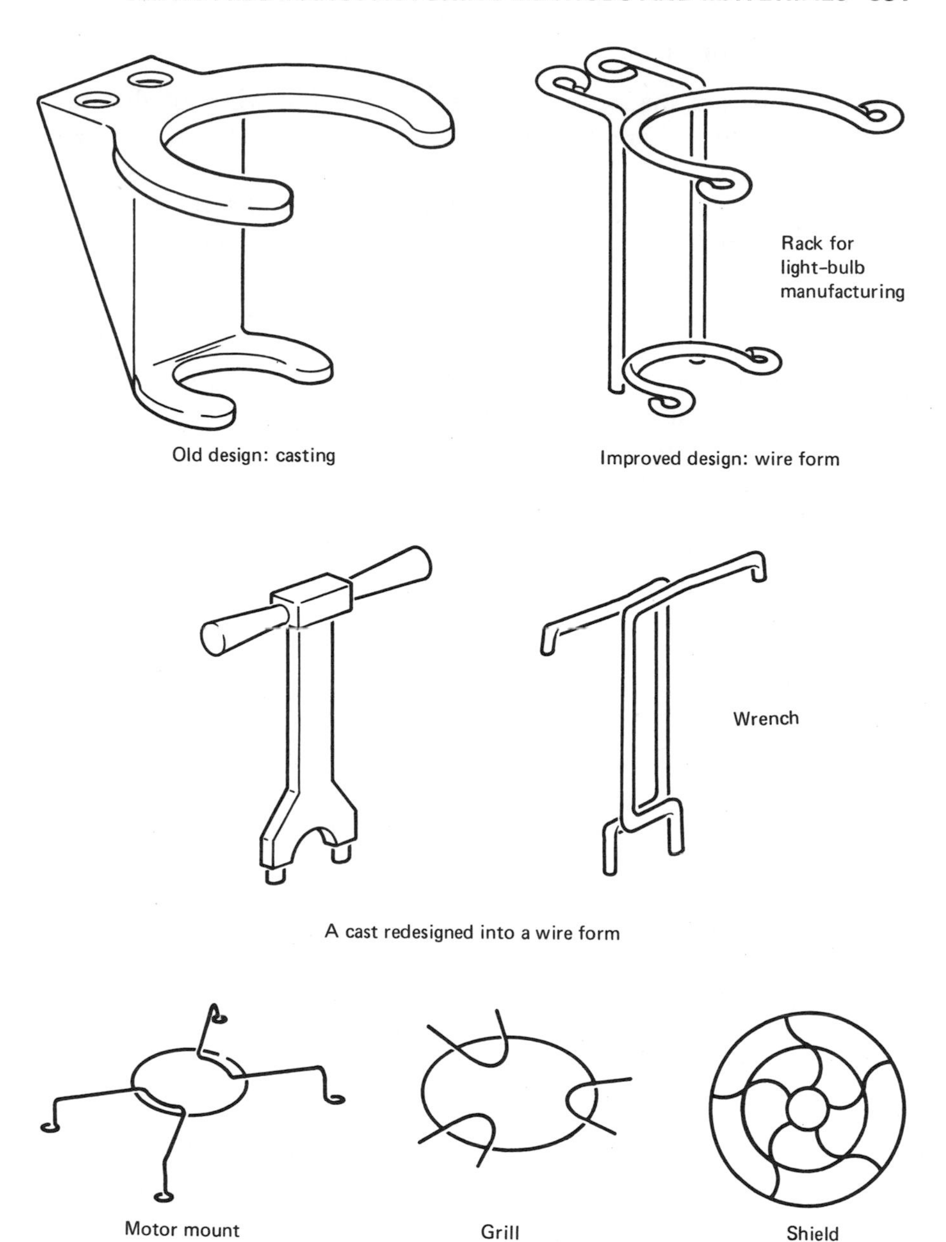

Figure 7.51. Examples of inexpensive wire-form products. ([143], [144] and technical literature from Titchener & Co.)

7.9.2. Wire Forms

Wire forms are made by bending and forming wires of different diameters into different geometries that are "resistance-welded" into structures. The designer has full control over the sizes or positions of wires for proper strength, rigidity, and stability. Wires are stronger than cast materials of equal rigidity. Therefore, rigidity may be the limiting design parameter. In many applications, a wire-form structure has the least weight and is inexpensive to construct. It is also rather easy to incorporate spring components, clips, bearings, hinges, cranks, and cams into a wire-form design, which means integral design is applicable to wire forms.

Examples of wire-form structures are display racks, cages, trays, grills, fan guards, refrigerator racks, supermarket carts, and baskets. Figure 7.51 shows several wire-form designs that can replace similar cast-metal or sheet-metal products. Repeated usage of wires of one specific geometry within a product can reduce tooling and fixture costs. In Figure 7.51(b), the wrench is made from two indentical pieces.

Wire form designs are limited by the bending, forming, and welding processes. Figure 7.52 shows some of the limitations and the proper design methods. Since different vendors may have different equipment, accepting design suggestions from the vendors can often reduce the piece cost. A wire form-design should be less restrictive to accommodate changes.

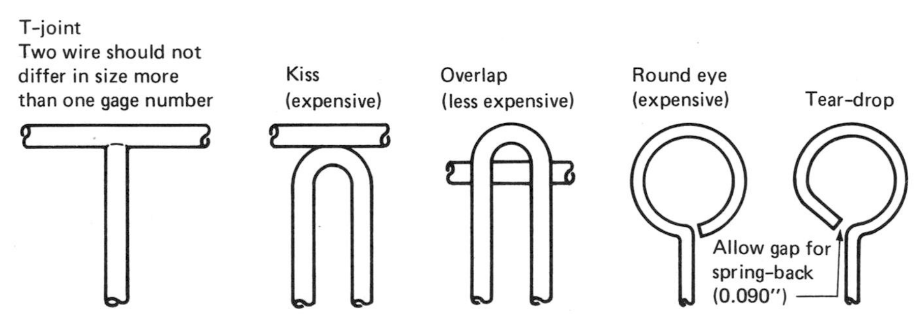

Figure 7.52. In specifying the shape of wire forms, an engineer should consider the ease of forming and welding. ([144] and technical literature from Titchener & Co.)

7.10. MACHINING

Machining is the removal of material from a workpiece to obtain the desired geometry and specifications. Machining processes can be classified into: (a) chip-forming processes such as drilling, reaming, turning, milling, and shaping; (b) chipless processes such as grinding, lapping, and electric-discharge machining. The machining cost is influenced by (1) the design characteristics such as material, geometry, tolerance, and surface finish; (2) the amount of operator control in loading, unloading, repositioning, inspecting, and tool-changing; and (3) the capability of machine tools such as the tool size, the cutting speed, and the required lubrication. The design characteristics are specified by the design engineer on a drawing. The other operating factors are controlled by industrial and manufacturing engineers.

$$\begin{aligned} \text{total cost} &= \text{cost of workpiece} + \text{cost of tool wear} \\ &\quad + \text{cost of machine and operator time} \\ &= C_1 + \frac{C_2}{N} + C_3\ (\text{total time}) \\ &= C_1 + \frac{C_2}{N} + C_3 \left(t_1 + t_2 + \frac{t_3}{N}\right) \end{aligned} \tag{7.1}$$

where

C_1 is the cost of the workpiece,
C_2 is the cost of the tool,
N is the number of workpieces completed with each tool,
C_3 is the cost of machine and labor per unit time,
t_1 is the loading, unloading, inspecting, and repositioning time,
t_2 is the actual cutting time, and
t_3 is the tool-changing time.

7.10.1. Simplifying Machining in the Design Stage

A machinist making a part from a blueprint cannot introduce any alterations because he does not know what effect a minor change (shortcut in machining) may have in the overall design or in the interaction between the components. The design engineer, who is completely responsible for the blueprint, should make sure that every little detail on the drawing is as easy to manufacture as possible. There are quite a few places where a designer can help in holding down the cost of machining, as discussed below.

The amount of metal removal. Machining is generally considered a precision process not suitable for removing large amounts of metal. The cost of machining is directly related to the cutting time, which is proportional to the amount of metal to be removed. A large amount of metal removed also means a large amount of valuable material converted into chips. A large weight difference before and after machining is a signal for a potential area of cost reduction. Machining can be reduced if precision die casting is used instead of sand casting. Similarly, fine blanking, precision stamping, forging, and deep drawing often reduce or eliminate the machining operations. A shaft that is warped due to heat-treatment should be bent back close to being straight before the final grinding. Since bending is much cheaper than grinding, some extra effort in bending is a good practice. In selecting sizes for bars and tubes, it is common to specify the next larger standard size to avoid machining. Where a shoulder is needed to act as a *stand-off* (also called a *spacer*), machining can be minimized by assembling pieces from the standard stock (Figure 7.53). Whenever possible, material should be left untouched to reduce the machining cost. Places that are to be hollowed out might be left solid (Figure 7.54). If a design is changed such that the machining

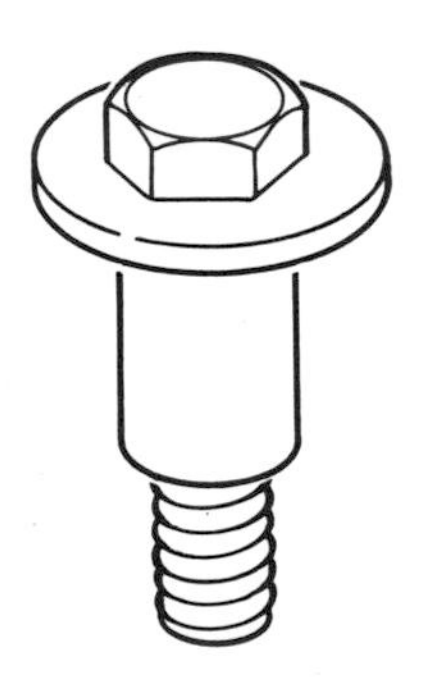

One-piece construction requires too much machining; this design is good for cold-heading

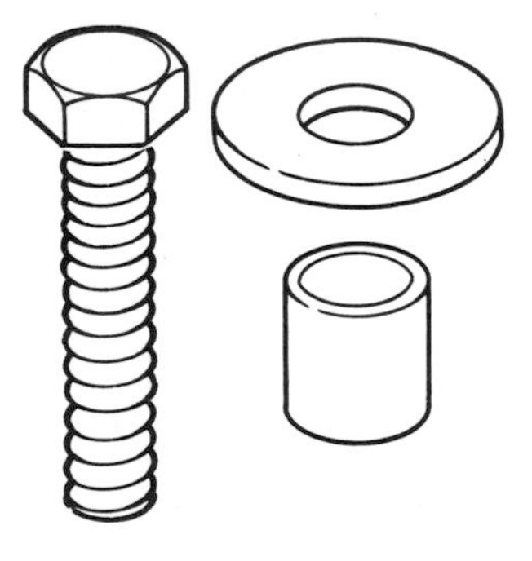

Assembly from several pieces of standard stock minimizes machining

Figure 7.53. The choice of design depends on the production method.

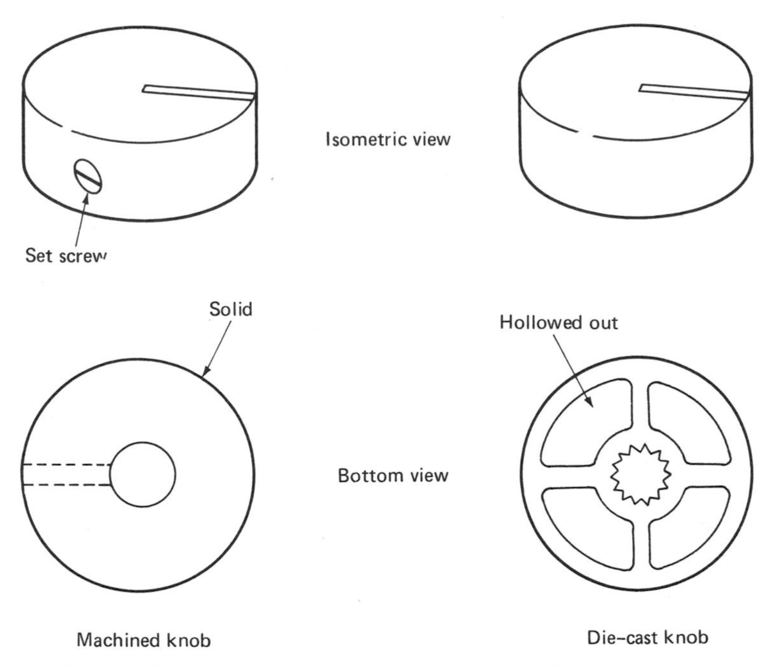

Figure 7.54. Another example of interrelation of design with the production method.

process is replaced by forging or die casting, the new design would definitely conserve more material.

Specifications of tolerance and finish. It is extremely important not to overspecify tolerance and finish. The easily obtainable accuracy is normally 0.01 in. (two places of decimal accuracy). Accuracies better than 0.01 in. start to get costly very quickly. For example, an accuracy of 0.005 in. is about twice as expensive. The cost of an accuracy of 0.001 in. is six times that of the 0.01 in. accuracy. The unit of surface finish is a root-mean-square microinch. A piece that is in as-cast condition or saw-cut finish would have a surface finish of 2000 μin. Finish-turned pieces from a lathe would have a surface finish of 63 μin. Again the cost of finishing skyrockets from here. A ground surface (32 μin.) costs about twice as much as the 63 μin. finish, but a honed surface (10 μin.) costs 4 times as much as the 63 μin. finish.

Selection of material. The strength, hardness, heat capacity, and grain-size of material contribute to the machinability of the material, which is also affected by the friction coefficient between the material and the cutting tool. Machinability can be quantified by the tool life for a given cutting speed and condition. Alternatively, machinability can be characterized by the cutting speed for a tool life of 60 min. An approximate rating based on fixed cutting speed is: magnesium 500, aluminum 300, zinc 200, brass 200, steel 100, stainless steel 80, bronze 80, nickel 55, beryllium 50, titanium 40, zirconium 35, molybdenum 8, and tungsten 8. The machining time can be considered inversely proportional to the machinability rating. As an example, machining an aluminum piece would take ⅓ the time of machining a steel piece of the same shape. A material with a good machinability rating may balance the higher cost of the material by decreasing the manufacturing time. The ease of machining has to be balanced against the ease of other portions of the manufacturing processes. For example, steel with a high sulfur content machines well but welds very poorly. Such steel cannot be used where welding is involved.

Easy clamping and lining up. These two factors contribute to the time of machining and the accuracy of the machined part. A large flat surface is not good for positive positioning because any roughness or warpage causes the piece to rock. A 3-point support is much better because it is less sensitive to casting defects. A piece with two feet or with a tilted bottom may be equipped with a temporary tab on the casting. After the piece is completely machined, the tab can be cut off. It is nice to stay in one position or one alignment as much as possible. Figures 7.55, 7.56, and 7.57 show three examples in which the work in positioning and aligning the workpiece is reduced by a careful design. The number of shoulders on shafts should be kept to a minimum. Curved surfaces should be excluded from machining.

Reducing tool changing and allowing for larger tools. Reducing the frequency of tool changing and allowing for larger tools shorten machining time. Larger tools are stronger and, therefore, they cut faster. They also wear longer because they have more cutting surfaces. To allow for larger tools means to leave enough clearance and overshoot space. It is also desirable to make the radius on the product the same as the radius of the tool so that no extra maneuvering is needed (Figure 7.58). If the holes have the same diameter, and all radii on the workpiece are the same (Figure 7.59), the need to change tools would be reduced.

Minimizing the machining area and the tool travel. Instead of ma-

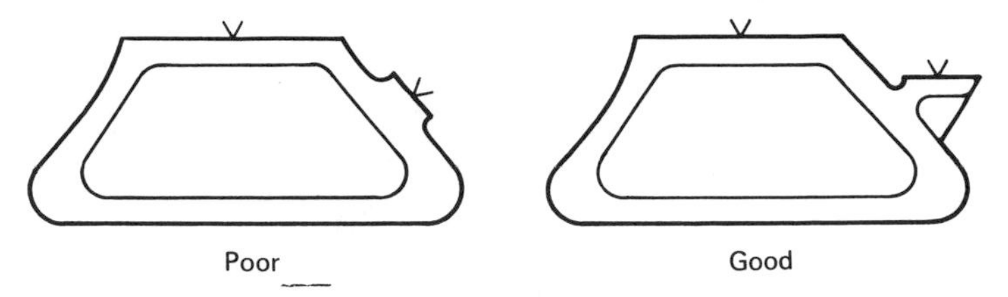

Figure 7.55. Eliminate a repositioning by making all machined surfaces in one direction.

Figure 7.56. Eliminate an alignment by making several machined surfaces in one plane.

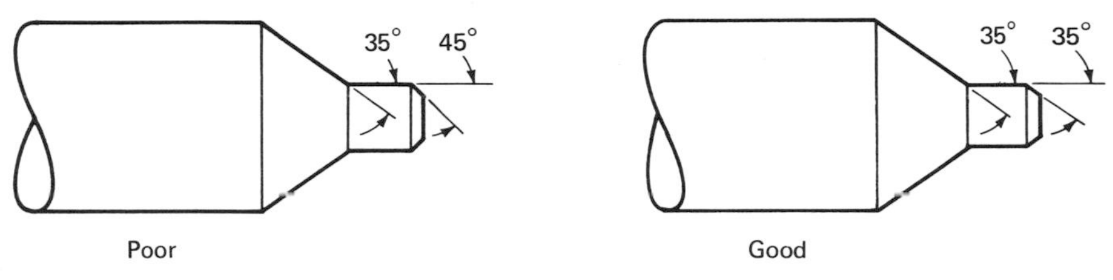

Figure 7.57. Eliminate an alignment by standardization of features and angles.

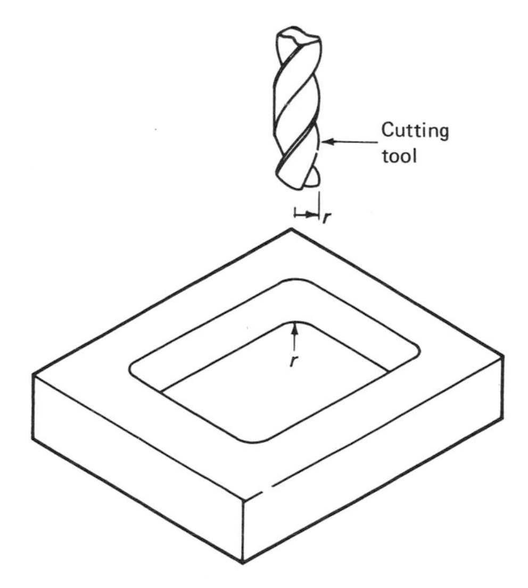

Figure 7.58. A cutting pattern is simplified by making a radius the same as that of the tool.

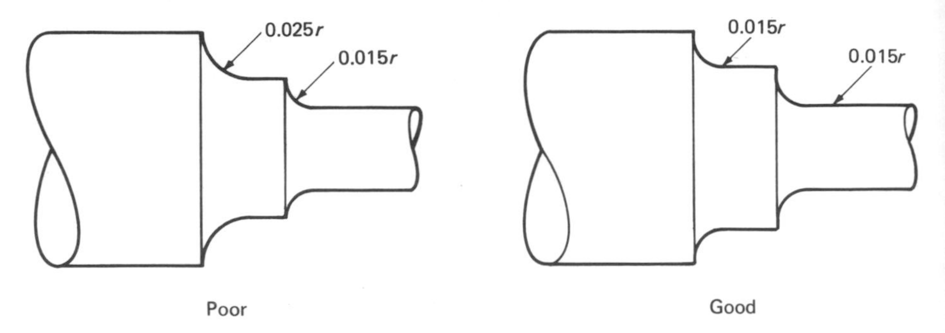

Figure 7.59. Eliminate a tool change by standardization of radii.

chining a whole flat surface, 3 pads are usually sufficient (Figure 7.60). Spot facing can be used around a bolt (Figure 7.61). Figure 7.62 shows how a small change in shape can reduce the amount of machine travel in cutting a slot for a pin linkage [146]. Figure 7.63 shows how a change in design eliminates most of the machining [146].

Preferred shapes and operations. Because of the operations of the machines and the shapes of the tools, some forms are easier to produce. Dimensions and tolerances that come from functional designs may not lead to straightforward manufacturing. A linear dimension may have to be converted into an angular measurement in actual production. Cumulative tolerances and reference lines often make a piece very difficult or impossible to machine. To simplify machining, manufacturing drawing is often needed.

Tapered alignment should be used carefully. Figure 7.64 shows how a taper and a shoulder may create redundant alignment and bad fit. If several countersunk bolts are used on a surface, even the slightest misalignment would be visible (Figure 7.65). With regular hexagonal bolts, as long as there is clearance between the shank of the bolt and the hole, misalignment would not be noticed. When there is a need for a shoulder to be on a shaft, a relief should be put near the shoulder to help grinding (Figure 7.66). Since external circular grooves are easier to cut, oil grooves should be placed on the shaft instead of on the bearing (Figure 7.67). Noncircular holes are

Figure 7.60. The use of pads reduces the machining area.

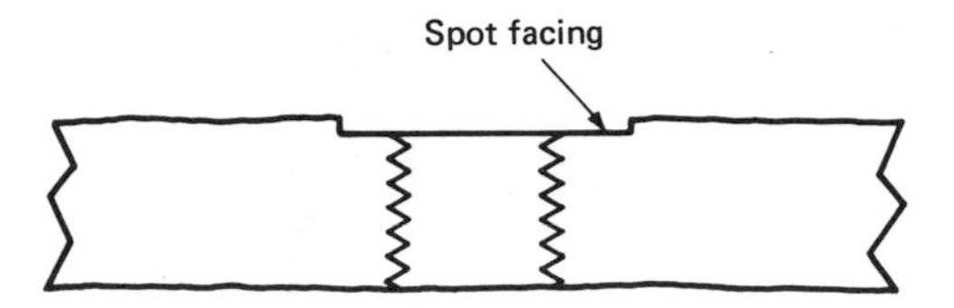

Figure 7.61. Machine the minimum area needed for proper seating of a bolt.

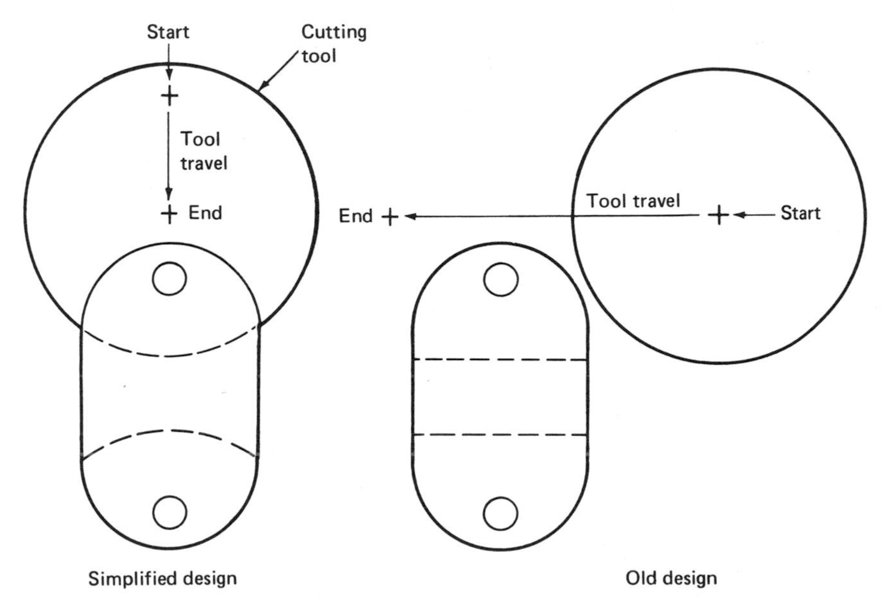

Figure 7.62. A design change that greatly reduces tool travel. (Strasser [146].)

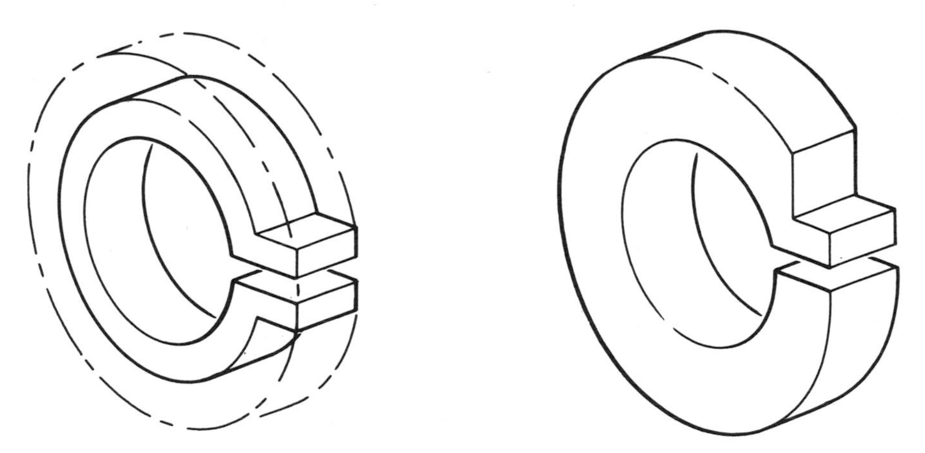

Figure 7.63. An alternative design to reduce tool travel. (Strasser [146].)

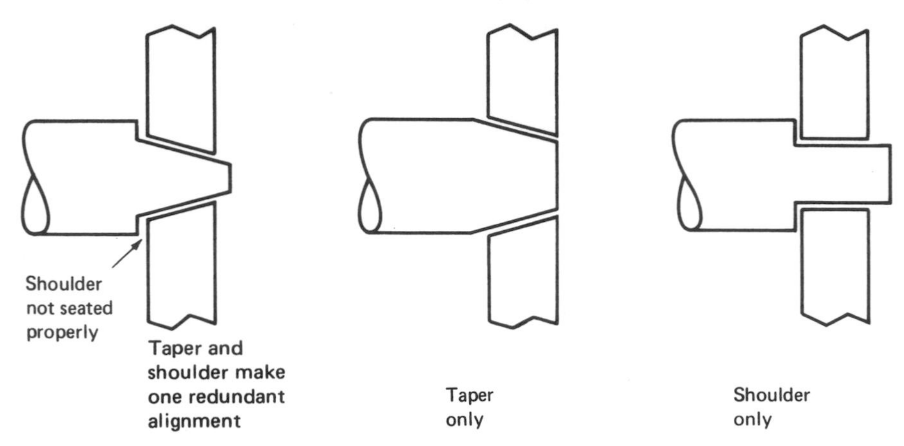

Figure 7.64. Redundant alignments create difficulty in machining and assembly.

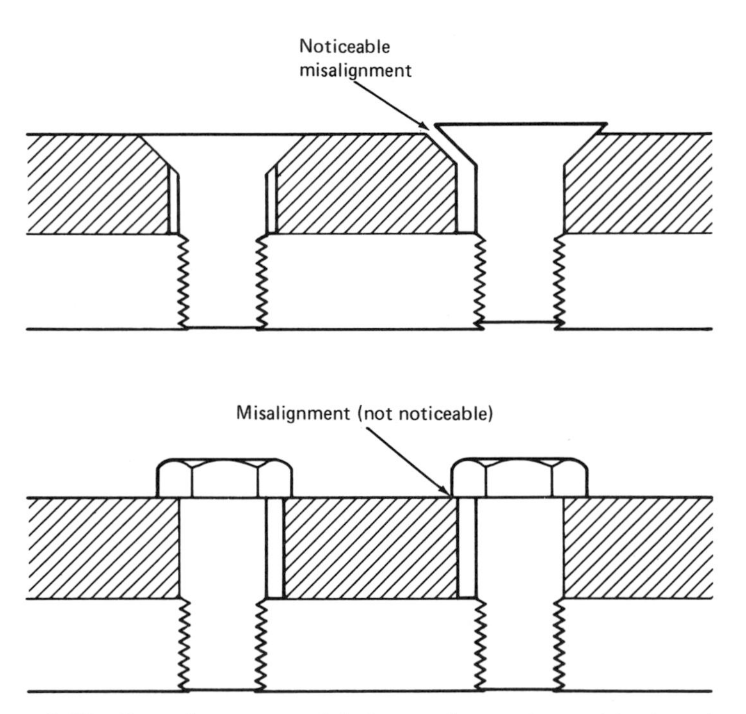

Figure 7.65. Several countersunk bolts require precise positioning of holes.

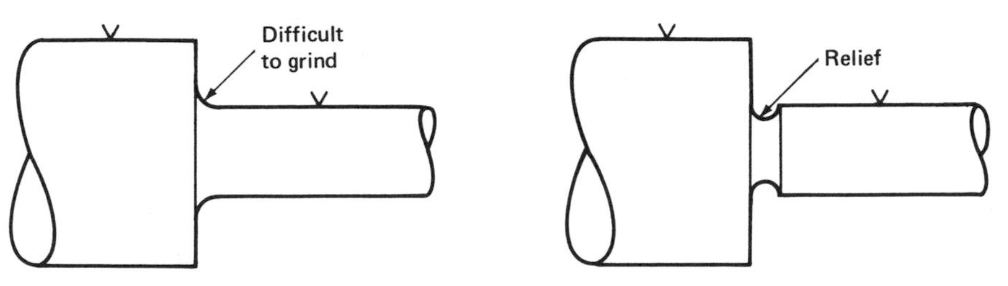

Figure 7.66. Relief at shoulder helps grinding of the small diameter.

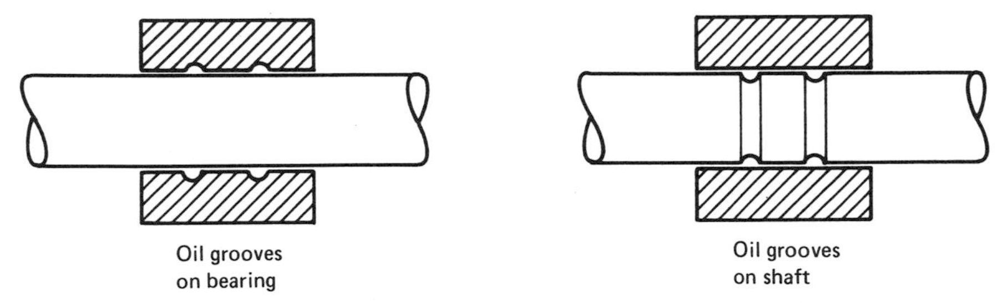

Figure 7.67. Oil grooves should be placed on shafts because external grooves are easier to machine.

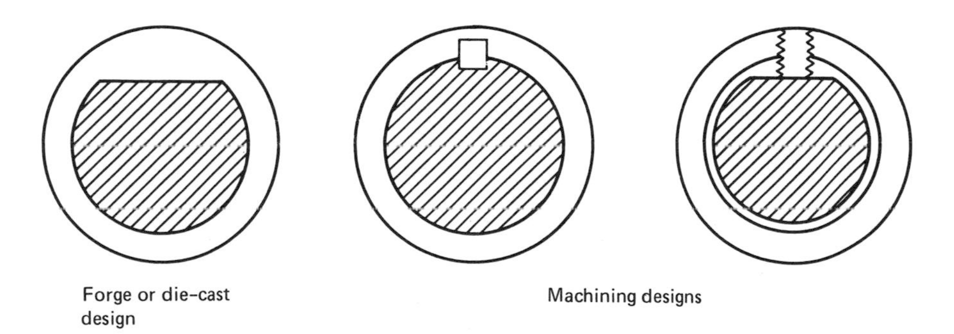

Figure 7.68. Noncircular holes are very difficult to cut.

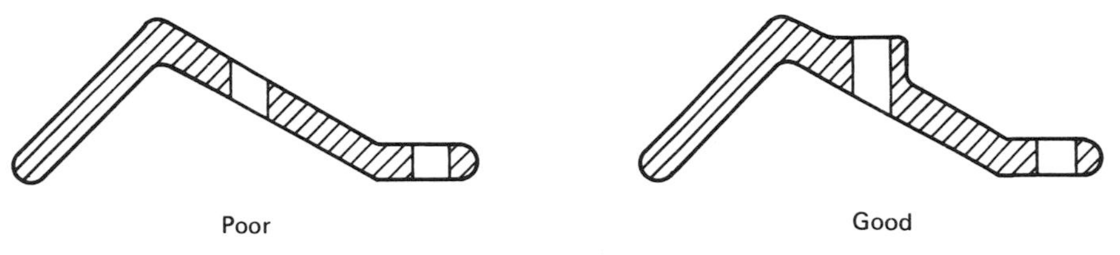

Figure 7.69. Drilling on a sloping surface is difficult.

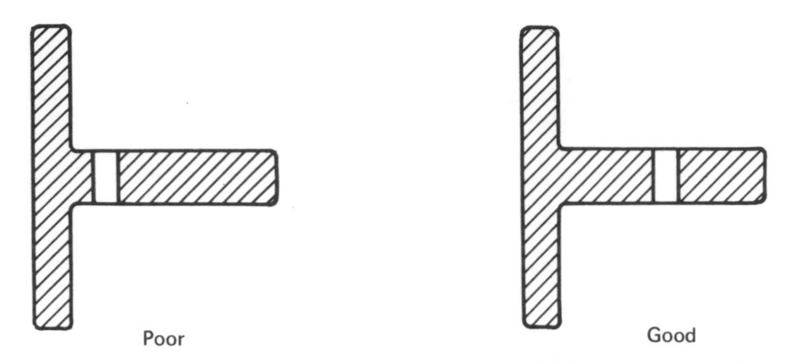

Figure 7.70. A hole placed too close to a web is difficult to reach for cutting and assembly.

more costly to get than two slots and a key (Figure 7.68). When lettering on plastic products, raised letters should be used. It is easier to inscribe letters in a mold than to emboss them.

A turning operation is preferred over a shaping operation. Centerless grinding is cheaper than grinding on centers. A sloped surface is difficult to drill because the cutting force tends to deflect the tool and cause the hole to be mislocated (Figure 7.69). A large hole on a thin sheet of material is also difficult to drill.

Easy chip removal and accessibility of the area to be machined. If the depth of a hole is more than 4 times its diameter, it becomes difficult to get the chips out and away from the cutting area. Blind

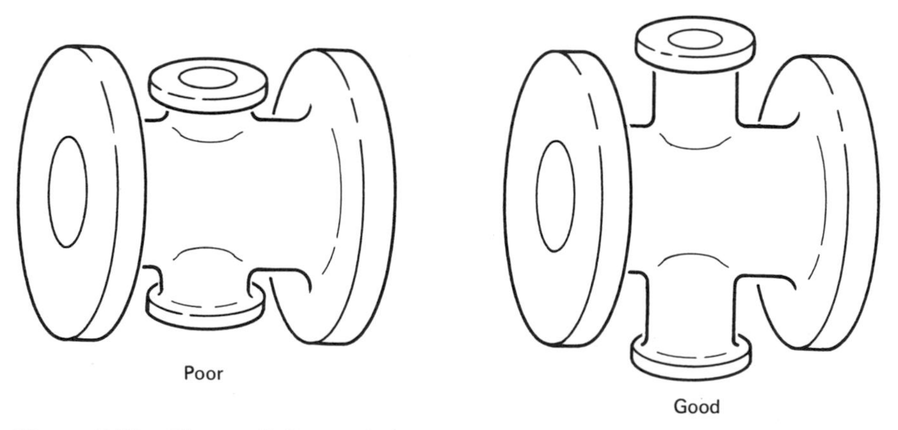

Figure 7.71. The small flange that extends farther is more accessible and easier to clamp.

holes are difficult to tap. If there is reason to thread a blind hole, extra depth is needed to provide room for the tap tool. Holes should not be placed too close to a web (Figure 7.70). Figure 7.71 shows a cross-connection pipe joint. Making the small-diameter flange extend higher would increase its accessibility and facilitate machining and clamping.

7.10.2. Selecting a Cutting Speed for the Minimum Cost and the Minimum Time

As early as 1907, Taylor [145] published studies on the optimization of the cutting speed in machining operations. First we need to express the cost as a function of the cutting speed. Referring to Equation 7.1, which appeared early in Section 7.10,

$$\text{total cost} = C_1 + \frac{C_2}{N} + C_3\left(t_1 + t_2 + \frac{t_3}{N}\right) \tag{7.1}$$

where the variables are as defined in the beginning of Section 7.10, the cutting time, t_2, the number of workpieces completed with each tool, N, and the tool cost, C_2 can be further described by the following three equations:

$$t_2 = \frac{L}{V} \tag{7.2}$$

where L is the length of tool travel (in feet); for a turning operation, it is equal to (axial length) $\times$ $\pi D/(\text{feed} \times 12)$, and V is the cutting speed in feet per minutes

$$N = \frac{T}{t_2} \tag{7.3}$$

where T is the tool life in minutes

$$C_2 = \frac{\text{the cost of the tool blank}}{\text{number of regrinds possible}} + \text{the cost of regrinding.}$$

According to Taylor's equation of tool life and cutting speed:

$$VT^n = C \tag{7.4}$$

where C and n are constants. The tool life decreases with increasing cutting speed if everything else is held constant. The constant n (typically 0.1 to 0.4) depends strongly on the material of the tool and the workpiece. The constant C (about 180 to 1000) depends on the material and the physical factors of the cut. Applying Equations 7.2 and 7.4 to 7.3 we get:

$$N = \frac{T}{t_2} = \frac{(C/V)^{1/n}}{(L/V)} = \frac{C^{1/n}}{L} V^{(1-1/n)} \tag{7.5}$$

Substituting 7.5 and 7.2 into 7.1:

$$\text{total cost} = (C_1 + C_3 t_1) + \frac{C_2 + t_3 C_3}{C^{1/n}} V^{(1/n-1)} + (LC_3)\frac{1}{V} \tag{7.6}$$

Equation 7.6 contains three groups of factors: those that are constant, those that increase with increasing cutting-speed, and those that decrease with increasing cutting speed. These three groups of factors are plotted against the cutting speed in Figure 7.72. A minimum-cost point exists at the condition $d(\text{total cost})/dV = 0$. The differentiation method of optimization was discussed in detail in Section 2.2. By differentiation, we get

$$\text{the cutting speed with the minimum cost} = \frac{C\,C_3^n}{\left[\left(\frac{1}{n} - 1\right)(C_2 + t_3 C_3)\right]^n} \tag{7.7}$$

Substitute 7.7 into 7.4,

$$\text{the optimum tool life for the minimum cost} = \left(\frac{1}{n} - 1\right)\left(\frac{C_2 + t_3 C_3}{C_3}\right) \tag{7.8}$$

Aside from the optimization method described above, we can also achieve a minimum-time operation, which is used to prevent a bottleneck situation or to meet a production deadline. A minimum-time production might involve a higher cost because the cost of the workpiece and the cost of the tools are assumed to be insignificant. In order to speed up an operation, more machinable material is chosen. Tools are run at higher speeds than the speed used in the minimum-cost operation. Better and more expensive tools are used whenever they can help reduce the frequency of tool change. A minimum-time

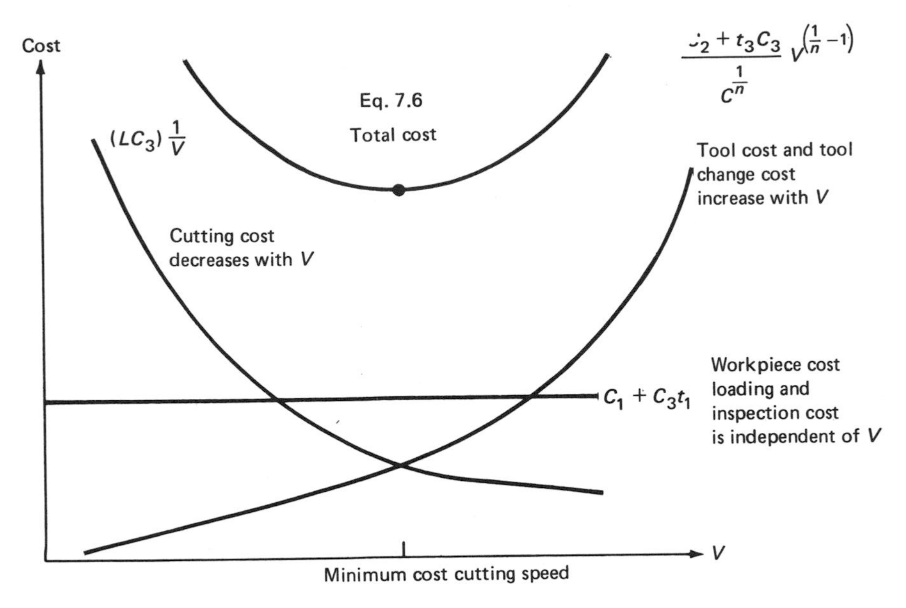

Figure 7.72. Plot of cost versus cutting speed.

operation may pay off in terms of preventing overtime pay or the need of adjusting personnel and machinery with the fluctuation in the production level. The minimum-time operation is formulated in terms of the time component of Equation 7.1:

$$\text{total time} = t_1 + t_2 + \frac{t_3}{N} \tag{7.9}$$

Put 7.2 and 7.5 in 7.9

$$\text{total time} = t_1 + \left(\frac{t_3 L}{C^{1/n}}\right) V^{(1/n-1)} + L \cdot \frac{1}{V} \tag{7.10}$$

Equation 7.10 is similar to Equation 7.6 in that there are constant factors and factors that increase or decrease with the cutting speed. The optimum value of cutting speed V is obtained by differentiating $d(\text{total time})/dV = 0$

$$\text{the cutting speed for minimum time operation} = \frac{C}{\left[\left(\frac{1}{n} - 1\right) t_3\right]^n} \tag{7.11}$$

The optimum tool life for minimum time operation $= \left(\frac{1}{n} - 1\right) t_3$

(7.12)

The minimum-time optimization is shown in Figure 7.73. The cutting speed of a minimum-time operation is higher than that of a minimum-cost operation for the same workpiece.

7.11. UTILIZATION OF SHEETS AND PLATES

Efficient utilization of sheets and plates can be considered at the early stages of design. A piece with an odd shape may be separated into two simpler pieces. Clever nesting can reduce the amount of scrap. The scrap itself may be used for other stampings or may be recycled. Usually, scrap materials are sold back to the suppliers at market prices. The content of the following discussion applies to sheet metal, sheet plastics, wood, rubber, cardboard, paper, cloth, and leather. The manufacturing process for such materials may be stamping, blanking, flame cutting, saw cutting, die cutting, or shearing.

7.11.1. Nesting

Depending on the labor content and the speed of the process, the material cost may be quite a large percentage of the total cost. With sheet-metal stamping, the material cost is more than 50% of the total cost. Therefore, a moderate percentage of saving in the material cost can reduce the total cost significantly. Nesting and staggering are ways to get more pieces out of a blank. The actual amount of waste depends on how well the contours mate together and how well they fit on the blank.

The shipbuilding industry uses many curved beams of different curvatures. Because of the differences in curvatures, the pieces will not mate at all. One way to increase utilization is to cut all the beams straight or with a slight curve. The final curvatures can be obtained by bending the pieces.

The nesting of *T*, *L*, and *U* shapes is shown in Figure 7.74. Note that staggering circles closely (Figure 7.3) is similar to making hexagons with round corners. Many practical shapes can be arranged with

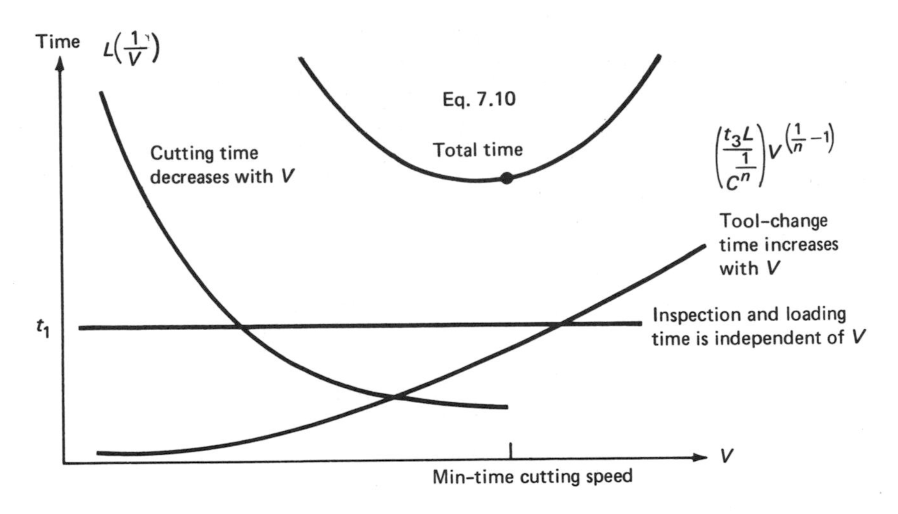

Figure 7.73. Plot of time versus cutting speed.

several cardboard templates put in different mating positions. When several shapes are to be arranged onto a sheet, the large pieces would be positioned first and the smaller pieces fit into the gaps. Scrap areas can be used for other components (Figure 7.74(d)). If the internal slot of the U shape (Figure 7.74(c)) is enlarged without changing the external dimensions, the number of pieces per roll of material will not increase, but the weight of scrap will increase. In this case, scrap is maximized while the other production parameters remain the same.

Some shapes can interlock perfectly on the blank. Examples are triangles, hexagons, and rectangles. The total angle at one intersection point is 360°. When this angle is divided into 3 equal portions, we get a hexagon. When the angle is divided into 4 and 6 portions, we get rectangles and triangles, respectively. Many variations can be derived from these basic shapes. A rectangle can be distorted into a parallelogram or sectioned into various irregular shapes. A triangle may be a rectangle cut into two or a hexagon cut into six sections.

Figure 7.75 shows an adaptation from one of M. C. Escher's periodic drawings [151]. In this shape, the left and right sides are identical and the top and bottom sides are identical. Such a shape meshes so perfectly that no scrap is produced. Figure 7.76 shows several examples of waste reduction by design modifications. These de-

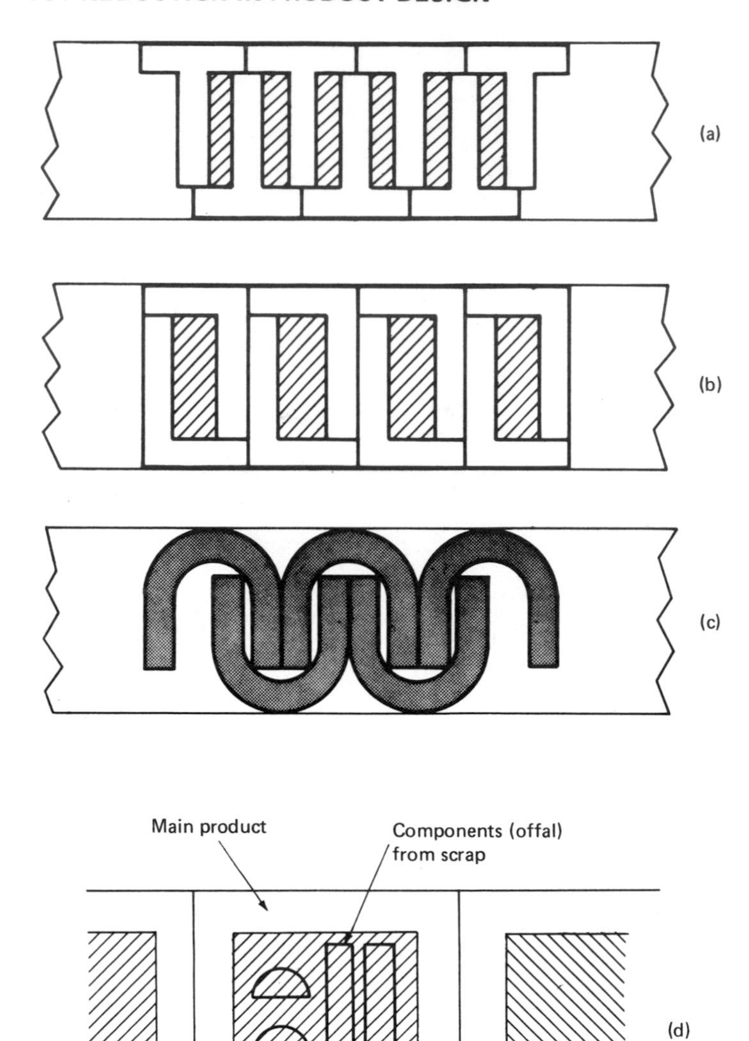

Figure 7.74. Nesting of T, L, U shapes, and utilizing scrap. If one component can be made from the scrap of another, as in (d), the two components are complementary in raw material.

signs also have identical left and right boundaries to facilitate tight packing. Both of Figures 7.75 and 7.76 belong to a type of mathematical shapes called TTTT, where T stands for translation of line segments. Each of these shapes is composed of 4 translation lines (two pairs) and hence it is called TTTT. The junction of these 4 line segments are located at the vertices of a parallelogram (or rectangle or square). The movements that can generate interlocking shapes are: translation T, 180° rotation C, 120° rotation C_3, 90° rotation C_4, 60° rotation C_6 and gliding mirror reflection G [155]. The secret of the repetitive figures is that about half of its boundary has the same shape as the other half. Figure 7.77 shows 7 more types of figures that are readily applicable to stamping on a strip. Figure 7.77(a) shows a CCCC shape. All four lines (C-lines) have

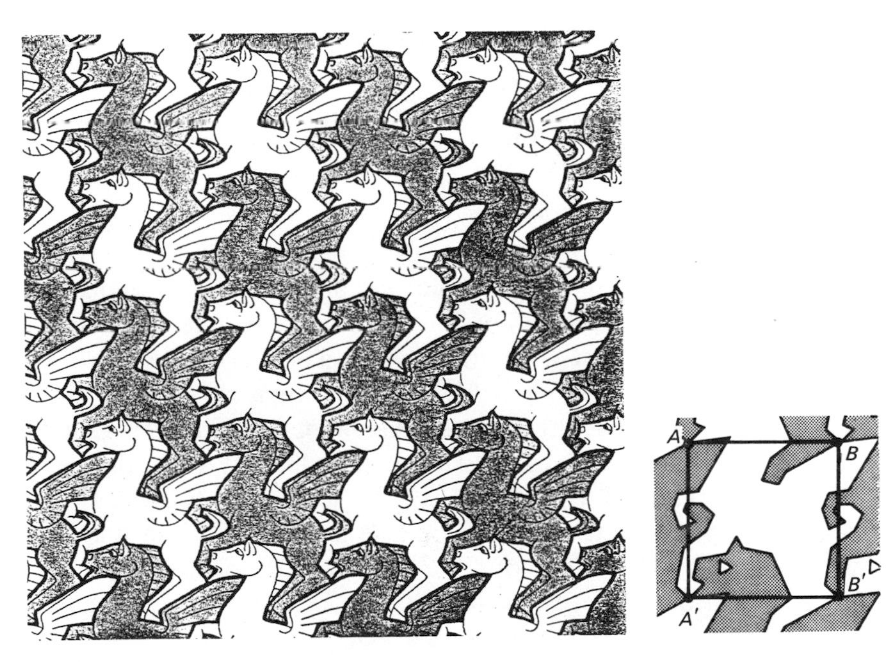

Figure 7.75. This elegant interlocking Pegasus (winged horse) may be considered as a rectangle in disguise. (From "Fantasy and Symmetry: The Periodic Drawings of M. C. Escher." The Escher Foundation—Haags Gemeentemuseum—The Hague) $AA' = BB'$ translated to the left one unit. $A'B' = AB$ translated down one unit.

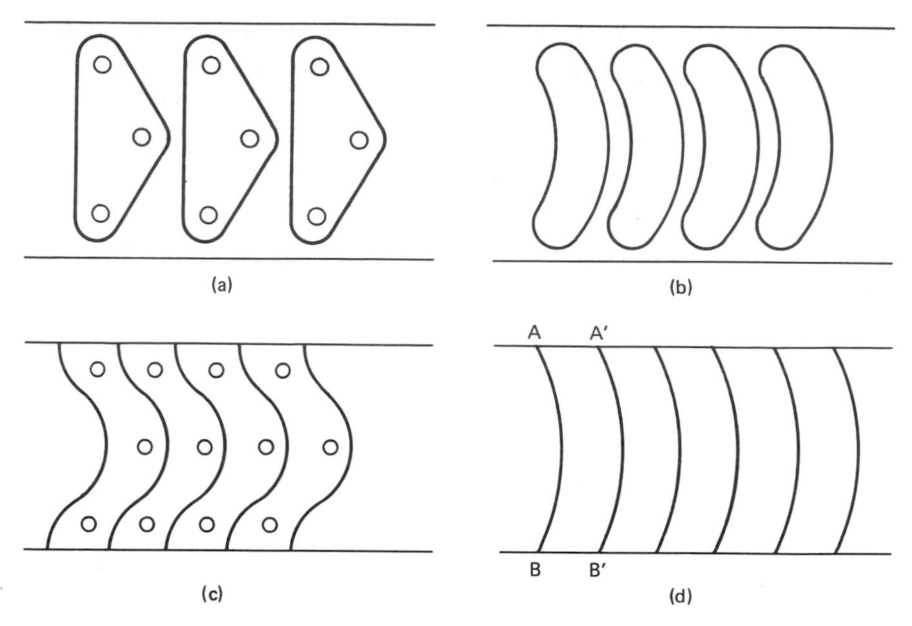

Figure 7.76. Design changes to reduce scrap. (b) is an improved version of (a), and (d) is an improved version of (c). (1) The protrusion on one side should fit into the concave part of another. (2) Noncritical areas may be hollowed out to facilitate tighter packing. (3) The intersection of a cutting line with the edge of the roll should be close to 90°.

half-turn symmetry—they will coincide perfectly with themselves when rotated 180° about their mid-points. Any straight line is a C-line. When the line CD in Figure 7.77(a) is degenerated to zero length (a point), the shape CCC is obtained (Figure 7.77(b)). In (c) and (d) of Figure 7.77, the line pair AB and CD is a translation pair. The figure in (d) is obtained by degenerating line DE from the figure in (c). In (e), (f), and (g) of Figure 7.77, the lines AB, BC and A′B′ are G lines (gliding mirror reflection line pair). They can map into one another by reflecting about the median of the strip and translating along it. For all the shapes in Figure 7.77, the whole strip can be made to overlap onto itself with a half turn or gliding mirror reflection about the median of the strip. Numerous practical examples of scrap-free shapes are given by Heesch and Kienzle in Reference [155].

Usually, areas that are not critical in functional design are altered to conserve materials. The use of geometric symmetry to reduce

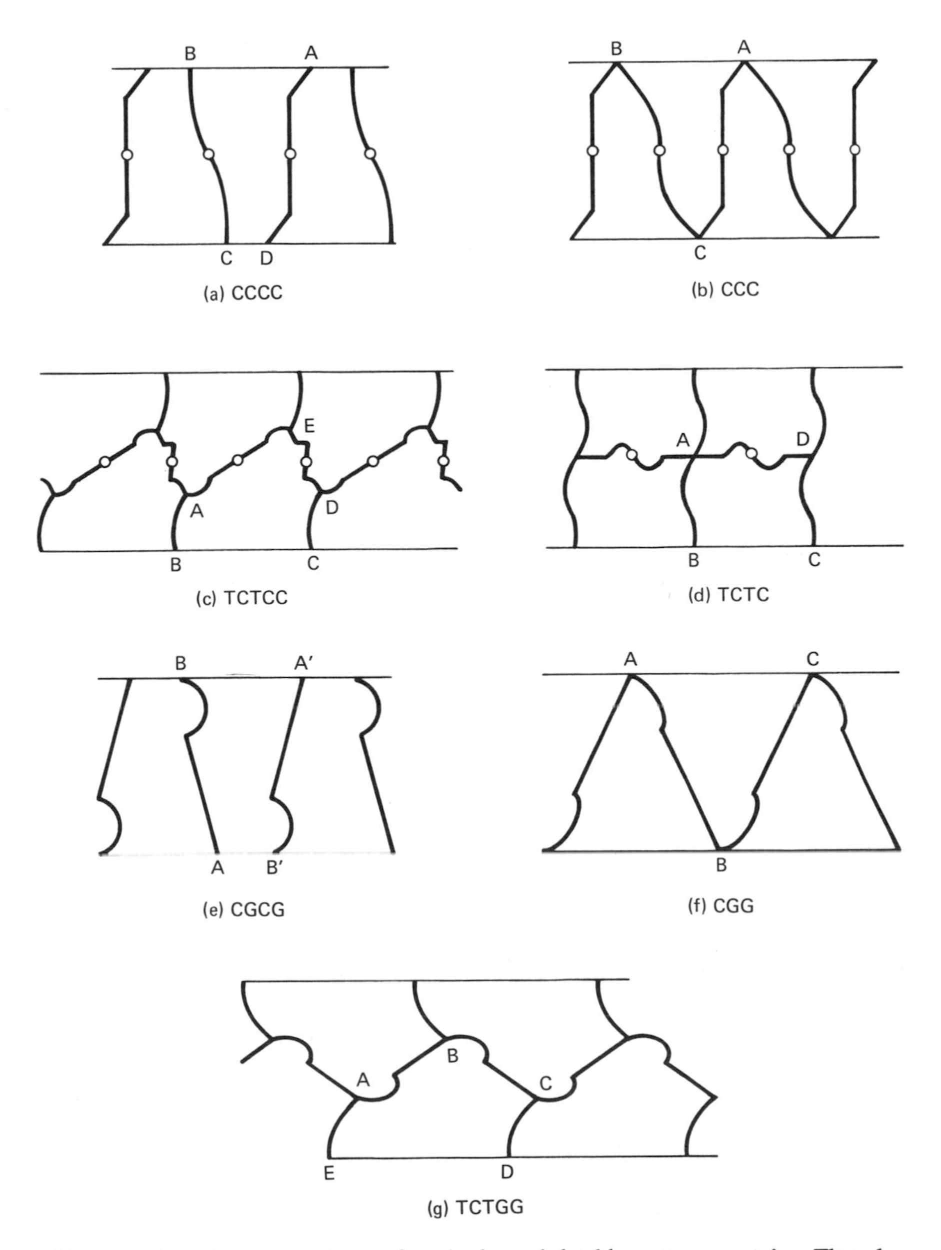

Figure 7.77. Scrap-free shapes for single and double rows on a strip. The edges of the strip contribute C lines to the shapes. The shapes in (e) to (g) contain G lines and they require the strip material to have identical faces. The T line, G line and half of the C line can have any arbitrary shape as long as they appear in pairs.

scrap can sometimes handicap design since half of the boundary is dictated by mathematical rules. It is easier if the design engineer seeks a scrap-free form that encloses the desired shape. A small trimming (or gap) would then furnish the desired shape.

An optimization approach to materials conservation, called a "knapsack" problem, was developed by Gilmore and Gomory [156, 157]. A large rectangular sheet of material can be converted to groups of smaller rectangles and squares with minimum waste. When many irregular shapes have to be nested, they can be grouped to form rectangular modules [160, 161]. These rectangular modules can then be arranged onto a large blank.

7.11.2. Processing Features and Speed

If material is conserved at the expense of increasing the processing time, the net effect may be an increase in cost. We will discuss some processing features that should be considered in the design stage.

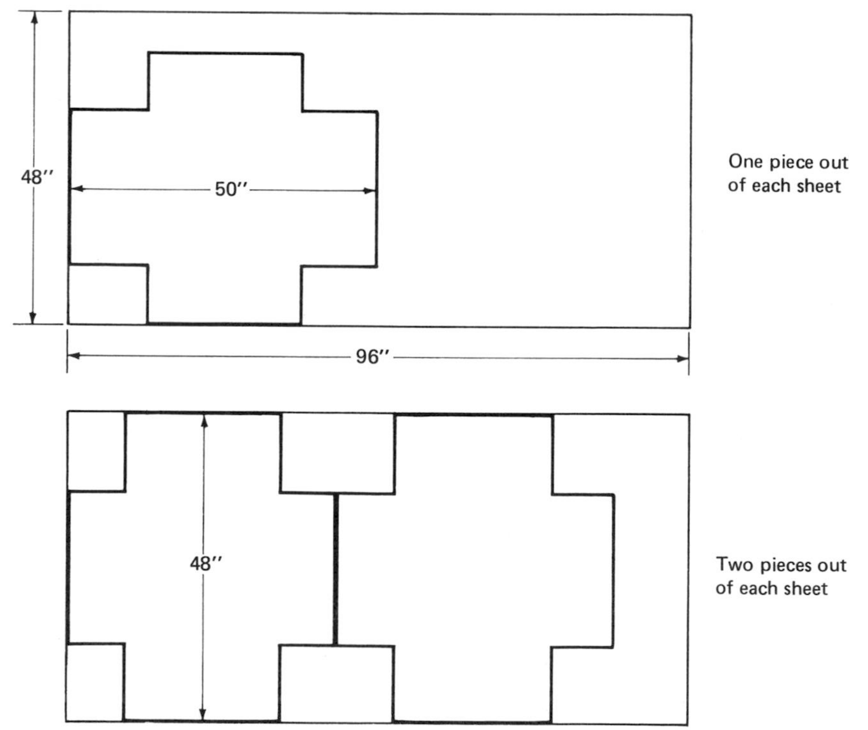

Figure 7.78. A reduction of 1 to 2 in. may reduce the material cost by half.

Most raw materials are made in sheets or rolls for easy handling and storage. The width of a sheet is a design limitation. Wider blanks permit more nesting. However, a wide blank requires more sets of dies and larger presses. Materials such as corrugated cardboard and plywood, which cannot be rolled up, are severaly limited in their overall size. In some cases, reducing the dimension of a piece by 1 or 2 in. can cut the cost by almost half (Figure 7.78).

In most stamping jobs, small strips of material are left at the edges and between the pieces so that the scrap can be easily handled and removed. The width of a gap is governed by the strength required. The width of a strip should be larger than 1/32 in. and between 0.7 to 1.5 times the thickness of the sheet. If the strip is not too long and is not located at a sharp corner, one thickness is enough. Long strips tend to be floppy and 1.25 to 1.5 times the thickness may be needed to get enough strength.

The loss from cutting wood can be reduced by using a thin saw blade. Similarly, the kerf has to be considered in flame-cutting. The high material utilization shown in Figure 7.76 requires special design features and may require a slower processing speed. The lines intersecting the edge of the stock should make an angle close to 90°. The stamping process for the shapes in Figures 7.76(b) and (d) is like a cut-off operation. Because of the smaller cutting circumference, less tooling and less cutting energy are required. Cutting these shapes is less accurate than cutting the shapes shown in Figure 7.76(a) and (c).

Multiple dies increase the initial cost but reduce the required amount of labor and processing time. In flame-cutting, multiple torches are often used because the cycle time is long (Figure 7.79). It is necessary for all pieces to have the same orientation, which may conflict with nesting. Triangular patterns tend to be nested with two opposite orientations. For stamping, two passes can be used to process one blank. On the second pass, the blank is inverted and fed in for the second time.

Pilots are needed for progressive dies. These are holes to register the material in order to get proper alignment. Pilots may be put in the workpieces if there are convenient holes at proper locations. Usually, pilots are just placed in the scrap. Progressive dies can do many jobs: piercing, forming, drawing, shaving, embossing, blanking, etc. Progressive dies are very inexpensive for high-volume productions because they perform many operations at a high speed.

Figure 7.79. Flame cutting. To compensate for the slow cutting speed, multiple torches are often used. Such operation may prevent effective nesting of components on the plate. (Courtesy Airco Welding Products.)

8. FROM THEORIES TO PRACTICE

Cost reduction is intrinsic to engineering and business. In the design of products, cost reduction is often implied. Because of this relationship, the techniques of cost reduction are diverse and scattered through all fields of engineering and business. This book has been written with a wide range of engineers and designers in mind. Readers with a particular background may find it difficult to accept an unfamiliar approach. Yet they may find, in their special fields, that the subject may not be fully developed. The aim of this book is to introduce all the cost reduction methods in product design. Readers who feel interested in a specific topic are encouraged to read the books suggested in the list of references.

There are certain similarities in all approaches. Design objectives and failure criteria are treated in Chapter 1 (Section 1.3). The objective function, discussed in Chapter 2, serves a similar purpose for optimization. Value-engineering advocates analyzing functions. The balance of stress and strength goes with the philosophy "put the material where it is needed." Balance means the right amount and distribution of resources in a system. Value-engineering has a goal: to eliminate unnecessary costs. Optimization is the technique that explicitly determines the minimum weight design and the best balance. The Law of Diminishing Returns is an economic aspect of system balance. So, we arrived at the same goal from different directions.

Which method should we use? When is the best time to use a particular technique? How can we get the most out of cost-reduction projects? We can generally say the design process goes in the following order: conceptual design, mathematical modeling, and production detailing. Value-engineering seems to lean toward concepts, optimization toward models, and the other methods more toward details. Value-engineering is often used to identify the problem and the other methods are used to solve it. In many consumer products, the ability of the product to satisfy the customer's needs is the responsibility of marketing research. The definition of objective would be a combined effort of engineering and marketing.

If cost reduction is applied to the redesign of a product, care must be taken in choosing the project. There is always room for improvement, but we should consider whether the improvement is worth the effort. That the cost of a product can be reduced does not mean that the original designer did an inadequate job. He might have been pressed for time. The circumstances may have changed. The cost of raw materials, certainly, is not constant. After a product is on the market, manufacturing costs and sales figures can be better estimated. Data on failures and excessive strengths are available from consumer experience and reaction. Carve-out design can be applied to eliminate such excess strength. To pick a cost-reduction project, first study several products with good and steady sales. The one with the largest potential savings would be taken first. Often a costly product or component is given higher priority. On an equal percentage basis, an expensive component would yield more dollars of savings. Savings from an incremental investment can be effectively studied by marginal analysis.

There are also management and operational aspects of cost reduction that are not covered in this book. Efficient execution of programs and good allocation of personnel certainly make a great difference. Planning, coordinating, and forecasting are all important. Obsolete raw materials have to be phased out while new raw materials are stocked. New labels, new packaging, and instructions have to be coordinated with other activities. The purchasing department also has to gear up for new bargains. Because of these hidden expenses, a cost-reduction project has to produce substantial savings within a short time in order to justify its execution. If a project cannot pay for itself in one or two years, it may not be worth pursuing. Though these operational factors are not within the scope of engineering design, an engineer should be aware that other people are also trying to do good jobs and they deserve credit for it.

Successful cost reduction of a product might hurt the feelings of the engineer who originally designed it. Elimination of labor may cause union problems. Laying off workers can arouse insecurity in the remaining employees. A shift from one component to another definitely influences the suppliers. Even a small reduction in cost decreases the vendor's profit, which is usually a percentage of the component's cost. There would be no initiative for vendors to participate in cost-reduction programs unless revenues are shared. Typically, part of the savings that result from a vendor's suggestions would go to reward the vendor. In a cost-reduction project, an engineer often needs cooperation and approval from other departments. It is wise to involve the appropriate departments from the very start so that they feel they are part of the team.

REFERENCES

Chapter 1

1. Juvinall, R. C.: "Engineering Considerations of Stress, Strain and Strength," McGraw-Hill Book Company, New York, 1967.
2. Shigley, J. E.: "Mechanical Engineering Design," McGraw-Hill Book Company, New York, 1977.
3. Matousek, R.: "Engineering Design," Blackie & Son Limited, London, 1974.

Chapter 2

4. Lyusternik, L. A.: "Shortest Paths, Variational Problems," translated and adapted from the Russian by P. Collins and R. B. Brown, Pergamon Press, the MacMillian Co., New York, 1964.
5. Almgren, F. J. Jr. and J. E. Taylor: "The Geometry of Soap Films and Soap Bubbles," *Scientific American*, July 1976, pp. 82–93.
6. Hiller, F. S. and G. J. Lieberman: "Operations Research," Holden Day Inc., San Francisco, 1974.
7. Siddall, J. N.: "Analytical Decision-Making in Engineering Design," Prentice Hall, New Jersey, 1972.
8. Fox, R. L.: "Optimization Methods for Engineering Design," Addison-Wesley, 1973.
9. Johnson, R. C.: "Optimum Design of Mechanical Elements," John Wiley & Sons, Inc., New York, 1961.
10. Johnson, R. C.: "Mechanical Design Synthesis with Optimization Applications," Van Nostrand Reinhold Co., New York, 1971.
11. Stark, R. M. and R. L. Nicholls: "Mathematical Foundations for Design: Civil Engineering Systems," McGraw-Hill Book Company, New York, 1972.
12. Nelder, J. A. and R. Mead: "A Simplex Method for Function Minimization," *Computer Journal*, 7, 1965, pp. 308–313.

13. Bellman, R. E. and S. E. Dreyfus: "Applied Dynamic Programming," Princeton University Press, 1962.
14. Duffin, R. J., E. L. Peterson and C. Zener: "Geometric Programming: Theory and Application," John Wiley & Sons, Inc., 1967.

Chapter 3

15. Gage, W. L.: "Value Analysis," McGraw-Hill Publishing Company, Ltd., London, 1967.
16. Mudge, A. E.: "Value Engineering, A Systematic Approach," McGraw-Hill Book Company, New York, 1971.
17. Hammer, W.: "Design for Safety—or else!" *Machine Design*, Mar. 23, 1972, pp. 108–115.
18. Klein, S. J.: "Negligent Design Can Land You in Court," *Machine Design*, Nov. 11, 1971, pp. 122–126.
19. Jacobson, R. A.: "Are You Personally Liable for a Defective Product," *Machine Design*, June 13, 1974, pp. 136–139.
20. Carliss, O. S. and W. W. Olsen: "OSHA Factors in Design," *Mechanical Engineering*, Dec. 1973, pp. 18–21.
21. Federal Register, Washington D. C., Government Printing Office, Dec. 8, 1972, Jan. 22, 1973.
22. Toy Manufacturers of America, Proposed Voluntary Product Standard TS 215, National Bureau of Standards, Washington D. C. 20234.
23. Regulations published under the Federal Hazardous Substance Act, Part 1500, Chapter 2, Title 16.
24. "Occupational Noise Exposure," Code of Federal Regulations, Subpart G, Section 1910.95, Occupational Safety and Health Act, 1970.
25. "Specifications for Paints and Coatings Accessible to Children, to Minimize Dry Film Toxicity," American National Standards Institute, 1430 Broadway, New York. z66.1-1964 (R1972).
26. Hammer, W.: "Product Safety versus Reliability: Know the Difference," *Machine Design*, July 8, 1971, pp. 26–28.
27. Norquist, W. E.: "How to Increase Reliability of Consumer Products," *Mechanical Engineering*, Jan. 1973, pp. 25–28.
28. Mauritzson, B. H.: "Cost Cutting with Statistical Tolerances," *Machine Design*, Nov. 25, 1971, pp. 78–81.
29. Morse, W. L.: "Recycling British Style," *Machine Design*, July 11, 1974, pp. 20–26.
30. Jacobson, R. A.: "Resource Conservation: The Coming Design Parameter," *Machine Design*, July 25, 1974, pp. 92–96.
31. Brown, H.: "Human Materials Production as a Process in the Biosphere," *Scientific American*, Sept. 1970, pp. 116–124.
32. Rondeau, H. F.: "The 1-3-9 Rule for Product Cost Estimation," *Machine Design*, Aug. 21, 1975, pp. 50–53.
33. Roberts, G. M.: "Forging Flexible Costumes of Steel," *Product Engineering*, Mar. 15, 1971, p. 66.

34. "Choosing Among Modules Builds Custom Microscope," *Product Engineering*, July 1973, p. 10.
35. Gallagher, C. C. and W. A. Knight: "Group Technology," Butterworth & Co. London, 1973.
36. Khol, R.: "Group Technology: A New Approach to Cost Control," *Machine Design*, Feb. 18, 1971, pp. 114–119.
37. Edwards, G. A. B.: "The Principle of Composite Components," *Machine Shop and Metalworking Economics*, Feb. 1970, pp. 2–8.
38. Grant, E. L. and W. G. Ireson: "Principles of Engineering Economy," the Ronald Press Company, New York, 1970.
39. Riggs, J. L.: "Economic Decision Models for Engineers and Managers," McGraw-Hill Book Company, New York, 1968.
40. Maffei, N.: "How to Slash In-Process Inventory Costs," *Assembly Engineering*, Nov. 1975, pp. 14–16.

Chapter 4

41. Owen, J. B. B.: "The Analysis and Design of Light Structures," American Elsevier Publishing Company, Inc., New York, 1965.
42. Hemp, W. S.: "Optimum Structures," Oxford University Press, printed in Great Britain, 1973.
43. Miura, K.: "Proposition of Pseudo-Cylindrical Concave Polyhedral Shells," Report No. 442, Institute of Space and Aeronautical Science, University of Tokyo, Nov. 1969.
44. Marshek, K. M. and M. A. Rosenberg: "Designing Lightweight Frames," *Machine Design*, May 15, 1975, p. 88.
45. Jones, J. W.: "Limited Analysis . . . An Inquiry into the Ultimate Performance of Structural Parts," *Machine Design*, Sept. 20, 1973, pp. 146–151.
46. Seshadri, T. V.: "Design Beyond the Yield Point," *Machine Design*, Sept. 19, 1974, pp. 134–139.
47. Calladine, C. R.: "Engineering Plasticity," Pergamon Press, printed in Great Britain, 1969.
48. Marshall, T.: "Ultra-light Column Takes Ton After Ton of Loading," *Product Engineering*, Nov. 17, 1969, pp. 72–74.
49. Blodgett, O. W.: "Paper and Scissors Show How Parts React to Stress," *Machine Design*, Mar. 6, 1975, pp. 87–91.
50. Agrawal, H. N.: "A Simple Way to Visualize Torsional Stress," *Machine Design*, Sept. 18, 1975, pp. 98–101.
51. Hartog, J. P. D.: "Advanced Strength of Materials," McGraw-Hill Book Company, New York, 1952.
52. Durelli, A. J., E. A. Phillips and C. H. Tsao: "Introduction to the Theoretical and Experimental Analysis of Stress and Strain," McGraw-Hill Book Company, New York, 1958.
53. Scipio, L. A.: "Structural Design Concepts," Technology Utilization Division, National Aeronautics and Space Administration, Washington D. C., 1967.

54. Lifshey, A. L.: "A Time-Saving Method for Designing Ribbed Plastic Parts," *Machine Design*, May 30, 1974, pp. 52–56.
55. Slysh, P.: "The Isogrid: King of Lightweight Design," *Machine Design*, Apr. 19, 1973, pp. 102–107.
56. Phillips, E. A.: "Quick Calculations for Corrugation Stiffness," *Product Engineering*, Sept. 12, 1960, pp. 75–77.
57. "Simple Plastic Process Yields High Performance Cellular Cores," *Product Engineering*, Aug. 1973, pp. 40–41.
58. Knapp, R. H.: "Pseudo-Cylindrical Shells: A New Concept for Undersea Structures," *Journal of Engineering for Industry, Transactions of the ASME*, May 1977, pp. 485–492.

Chapter 5

59. Beck, R. D.: "Plastic Product Design," Van Nostrand Reinhold Company, New York, 1970.
60. Wood, A. S.: "Resin Savings Through Product Redesign," *Modern Plastics*, July 1974, pp. 40–43.
61. Billmeyer, F. W. Jr.: "Textbook of Polymer Science," John Wiley & Sons, Inc., New York, 1971.
62. Faupel, J. H.: "Engineering Design," John Wiley & Sons, Inc., New York, 1964.
63. Baer, E.: "Engineering Design for Plastics," Robert E. Krieger Publishing Company, Huntington, New York, 1975.
64. Walton, N.: "Plastic Bearings," *Machine Design*, July 13, 1972, pp. 119–122.
65. "1972–1973 Bearings Reference Issue," *Machine Design*, June 22, 1972, pp. 41–42.
66. Andersen, D. A.: "Predicting Wear in Plastic Bearings," *Machine Design*, July 10, 1975, pp. 85–87.
67. Theberge, J. E., B. Arkles and P. J. Cloud: "How Plastics Wear Against Plastics," *Machine Design*, Oct. 31, 1974, pp. 60–61.
68. "Design and Production of Gears in Celcon," Celanese Plastics Company, Newark, New Jersey 07102.
69. Theberge, J. E., P. J. Cloud and B. Arkles: "How to Get High-Accuracy Plastic Gears," *Machine Design*, Sept. 6, 1973, pp. 140–145.
70. "1973-1974 Fastening and Joining Reference Issue," *Machine Design*, Nov. 22, 1973, pp. 132–134.
71. Richlin, S.: "Plastic Parts that Remember," *Machine Design*, Jan. 10, 1974, pp. 87–91.
72. "Disposable Forceps," *Machine Design*, May 4, 1972, p. 38.
73. "Shmoo-Inspired Closure Is Bistable," *Machine Design*, Nov. 11, 1971, p. 153.
74. "Minor Tooling Change Brings Major Product Improvement," *Modern Plastics*, Aug. 1971, p. 32.

75. "Spring Action Adds Snap to Living Hinges," *Plastics World*, Cahners Publishing Co., Jan. 20, 1975.

Chapter 6

76. Greenwood, D. C.: "Product Engineering Design Manual," McGraw-Hill Book Company, New York, 1959.
77. Greenwood, D. C.: "Engineering Data for Product Design," McGraw-Hill Book Company, New York, 1961.
78. Greenwood, D. C.: "Mechanical Details for Product Design," McGraw-Hill Book Company, New York, 1964.
79. Benes, J. J.: "The Cost of Fastening," *Machine Design*, June 10, 1971, pp. 115–119.
80. Semmerling, W.: "Upgrade Manual Assembly Operations," *Assembly Engineering*, Feb. 1974, pp. 38–41.
81. Barnes, R. M.: "Motion and Time Study, Design and Measurement of Work," John Wiley & Sons, Inc., New York, 1968.
82. Mallen, S. E.: "The Design/Manufacturing Interface," *Assembly Engineering*, Jan. 1974, pp. 22–27.
83. Boothroyd, G. and A. H. Redford: "Mechanized Assembly," McGraw-Hill Publishing Co., Berkshire, England, 1968.
84. Leaman, J. M.: "Piece-Part Designs for Reliable Automatic Assembly," *Assembly Engineering*, May 1975, pp. 24–27.
85. Riley, F. J.: "Design Parts for Automatic Assembly," *Assembly Engineering*, May 1973, pp. 40–43.
86. Strasser, F.: "3-D Stampings," *Machine Design*, Nov. 13, 1975, pp. 143–144.
87. Strasser, F.: "How to Stamp the 'Hole' Thing," *Machine Design*, Nov. 1, 1973, pp. 103–104.
88. "Guidelines for Using the Injected Metal Assembly Process," *Assembly Engineering*, April 1976, pp. 42–46.
89. Jay, F. H.: "Joining Parts by Die-Casting," *Machine Design*, Apr. 15, 1971, pp. 93–95.
90. "Molding Assembly Cost Savings into Plastic Parts," *Assembly Engineering*, Nov. 1973, pp. 32–35.
91. Chow, W. W.: "Snap Fit Design," *Mechanical Engineering*, July 1977, pp. 35–41.
92. Lee, M.: "Economical Snap-Fit Designs Simplify Plastic Assemblies," *Product Engineering*, Aug. 1973, pp. 31–32.
93. Graves, B. P.: "Taking the Guesswork out of Press and Snap Fits," *Assembly Engineering*, Feb. 1975, pp. 20–22, Mar. 1975, pp. 20–21.
94. "Delrin Acetal Resin Design Handbook," by E. I. duPont de Nemours & Co., Plastics Department, Wilmington, Delaware, copyright 1967, pp. 72–90.
95. Sprow, E. E.: "The Cost of Fastening Plastics," *Machine Design*, July 27, 1972, pp. 72–77.

96. "Controlling Fastening Reliability and Costs," *Assembly Engineering*, Jan. 1973, pp. 26–30.
97. White, E. M.: "Identifying and Solving Rivet Clinch Problems," *Assembly Engineering*, Dec. 1974, pp. 20–23.
98. Strasser, F.: "20 Common Riveting Errors and Their Causes," *Assembly Engineering*, June 1974, pp. 26–30.
99. Fryberger, C. T.: "The New Look in Wiring Hardware," *Machine Design*, Mar. 20, 1975, pp. 70–73.
100. "Improving Productivity with Terminals," *Assembly Engineering*, July 1975, pp. 26–28.
101. Dillan, G. O. and A. Seaman: "Component Outlines on Circuit Cards Cut Insertion Costs," *Assembly Engineering*, July 1973, pp. 42–45.
102. Ruhl, R. L., L. D. Metz and R. Heckman, "Reducing the Low Level Incidence of Solder Shorts in Wave Soldering," *Electronic Packaging and Production*, Feb. 1977, pp. 82–85.
103. Petrie, E. M.: "A Guide to Successful Adhesive Bonding," *Assembly Engineering*, June 1976, pp. 36–41, July 1976, pp. 18–22, Aug. 1976, pp. 26–29.
104. Schneberger, G. L.: "15 Ideas for Choosing and Using Adhesives to Cut Costs, Problems," *Materials Engineering*, Oct. 1974, pp. 28–31.
105. Kroehs, A. R.: "Flux Coatings Help Control Preform Soldering," *Assembly Engineering*, Mar. 1974, pp. 26–28.
106. Van Dyke, C.: "Brazing Preforms Speed Assembly and Reduce Costs," *Assembly Engineering*, Nov. 1975, pp. 18–21.
107. Morgan, G.: "Brazing and Soldering Update," *Appliance*, Nov. 1974, pp. 40–43.
108. Prudden, P. D.: "Joining Methods for Small Assemblies," *Machine Design*, June 14, 1973, pp. 127–131.
109. Strasser, F.: "Ten Tips for Better Weldments," *Machine Design*, April 22, 1976, pp. 72–75.
110. Wadleigh, A.: "Inertia Welding Saves," *Assembly Engineering*, Dec. 1974, pp. 34–36.
111. Chookazian, M. and W. Tyrrell: "Ultrasonic and Electromagnetic Bonding of Thermoplastics," *Manufacturing Engineering*, April 1976, pp. 46–47.
112. Dreger, D. R.: "Welding Plastics by Vibration," *Machine Design*, Dec. 26, 1974, pp. 42–44.
113. "Plastics Reference Issue," *Machine Design*, Dec. 12, 1968, pp. 23–27.

Chapter 7

114. Dallas, D. B.: "Tool and Manufacturing Engineers Handbook," Society of Manufacturing Engineers, McGraw-Hill Book Company, New York, 1976.
115. Grand, R. L.: "The New American Machinist's Handbook," McGraw-Hill Book Company, New York, 1973.
116. Datsko, J.: "Materials Properties and Manufacturing Processes," John Wiley & Sons, Inc. New York, 1967.

117. Bolz, R. W.: "Production Processes, the Productivity Handbook," Conquest Publications, Novelty, Ohio, 1974.
118. Trucks, H. E.: "Designing for Economical Production," Society of Manufacturing Engineers, Dearborn, Michigan, 1974.
119. "From Abundance to Scarcity, Is that the Materials Story?" *Product Engineering*, Oct. 1974, pp. 28–33.
120. Rusinoff, S. E.: "Foundry Practices," American Technical Society, Chicago, 1970.
121. Kearney, A.: "Castings Without Defects," *Machine Design*, June 26, 1975, pp. 42–45, July 10, 1975, pp. 82–84.
122. "Powder Metallurgy Design Guidebook," by the Metal Powder Industries Federation, P. O. Box 2054, Princeton, New Jersey, 1974.
123. Altemeyer, S.: "Multi-Part P/M," *Machine Design*, May 15, 1975, pp. 80–82.
124. Altemeyer, S.: "Density/Economy Relationships of P/M Parts," *Machine Design*, June 29, 1972, pp. 72–76.
125. Rubin, I. I.: "Injection Molding, Theory and Practice," John Wiley & Sons, 1973.
126. "Delrin Doodle Kit," by E. I. duPont de Nemours & Co., Plastics Department, Wilmington Delaware, 1975.
127. Williams, V. A.: "A New Shot from Structural Foam," *Production*, Aug. 1975, pp. 39–45.
128. Archer, E. W. and D. H. Bergstrom: "Structural Plastic Foams: Current Status," *Mechanical Engineering*, Nov. 1974, pp. 23–28.
129. Dreger, D. R.: "Structural Parts from Plastic Foams," *Machine Design*, May 17, 1973, pp. 136–142.
130. Abelson, P. H. and A. L. Hammond: "Materials: Renewable and Nonrenewable Resources," American Association for the Advancement of Science, Washington D. C., 1976.
131. Young, H. E.: "Complete Tree Concept 1964–1974," *Forest Products Journal*, Dec. 1974, pp. 13–16.
132. "Manual of Lumber and Plywood Saving Techniques," by NAHB Research Foundation, Inc., P. O. Box 1627, Rockville, Maryland, June 1971.
133. Richards, D. B.: "Hardwood Lumber Yield by Various Simulated Sawing Methods," *Forest Products Journal*, Oct. 1973, pp. 50–58.
134. Johnston, J. S. and A. St. Laurent: "Tooth Side Clearance Requirements for High Precision Saws," *Forest Products Journal*, November 1975, pp. 44–49.
135. St. Laurant, A.: "Improving the Surface Quality of Ripsawn Dry Lumber," *Forest Products Journal*, Dec. 1973, pp. 17–24.
136. Anderson, E. A.: "New Ways in Wood Products," *Forest Products Journal*, Sept. 1973, pp. 56–58.
137. Hanover, S. J., A. G. Mullin, W. L. Hafley and R. K. Perrin: "Linear Programming and Sensitivity Analysis for Hardwood Dimension Production," *Forest Products Journal*, Nov. 1973, pp. 47–50.

138. McNatt, J. D.: "Engineering Properties of Ten Particleboards," *Materials Engineering*, Oct. 1974, pp. 42–43.
139. Galinowski, J.: "Using the Planar Anisotropy in Sheet Metal to Facilitate the Production of Deep Drawn Cups," *Sheet Metal Industries*, Feb. 1975, pp. 79–87, Mar. 1975, pp. 132–137, Apr. 1975, pp. 185–189.
140. Bosche, E. R.: "Thermoformed Plastic Parts," *Machine Design*, Jan. 7, 1971, pp. 94–97.
141. Jones, D. A. and T. W. Mullen: "Blow Molding," Robert E. Krieger Publishing Co. Inc., Huntington, New York, 1971.
142. Bogdan, F.: "Perforated and Expanded Metals," *Machine Design*, Aug. 22, 1974, pp. 78–81.
143. Chase, H.: "Wire Forms Edge Up in High Production," *Wire and Wire Products*, Aug. 1969, pp. 43–46.
144. "Tips on the Fabrication of Steel Wire Products," *Wire and Wire Products*, Aug. 1971, pp. 50–56.
145. Taylor, F. W.: "On the Art of Cutting Metals," *Trans. ASME*, 28, 1907, pp. 31–350.
146. Strasser, F.: "Designing Parts that Are Easy to Machine," *Machine Design*, Aug. 7, 1975, pp. 65–69.
147. Branson, J. R.: "Materials Conservation and Improvement in Utilization by Computerised Techniques," *Sheet Metal Industries*, May 1976, pp. 280–285.
148. Kienzle, W.: "Scrap-Free Blanking," *Sheet Metal Industries*, Nov. 1965, pp. 839–845.
149. Branson, J. R.: "Pressworking Economics," *Sheet Metal Industries*, June 1974, pp. 309–313, July 1974, pp. 402–407.
150. Silverton, A.: "Effective Cost Reduction in the Press Shop," *Sheet Metal Industries*, Nov. 1976, pp. 365–367, 375.
151. MacGillavry, C. H.: "Fantasy and Symmetry: The Periodic Drawings of M. C. Escher," Published for the International Union of Crystallography by A. Oosthoek's Uitgeversmaatschappij, NV, Utrecht, 1965; reprinted in USA by H. N. Abrams, Inc., New York, 1976.
152. Ernst, B.: "The Magic Mirror of M. C. Escher," translated from the Dutch by J. E. Brigham, Ballantine Books, New York, 1976.
153. Coxeter, H. S. M. and W. O. J. Moser: "Generators and Relations for Discrete Groups," Springer-Verlag Company, Berlin, 1972.
154. Coxeter, H. S. M: "Introduction to Geometry," John Wiley & Sons, Inc., New York, 1965.
155. Heesch, H. and O. Kienzle: "Flachenschluss," Springer-Verlag Company, 1963, printed in Germany.
156. Gilmore, P. C. and R. E. Gomory: "Multistage Cutting Stock Problems of Two or More Dimensions," *Operations Research*, vol. 13, 1965, pp. 94–120.
157. Gilmore, P. C. and R. E. Gomory: "The Theory and Computation of Knapsack Functions," *Operations Research*, vol. 14, 1966, pp. 1045–1074.

158. Haims, M. J. and H. Freeman: "A Multistage Solution of the Template-Layout Problem," *IEEE Transactions on Systems Science and Cybernetics*, Apr. 1970, pp. 145–151.
159. Adamowicz M. and A. Albano: "A Solution of the Rectangular Cutting-Stock Problem," *IEEE Transactions on Systems, Man, and Cybernetics*, Apr. 1976, pp. 302–310.
160. Adamowicz, M. and A. Albano: "Nesting Two Dimensional Shapes in Rectangular Modules," *Computer Aided Design*, Jan. 1976, pp. 27–33.
161. Albano, A.: "A Method to Improve Two-Dimensional Layout," *Computer Aided Design*, Jan. 1977, pp. 48–52.

APPENDICES

APPENDIX A.1. VALUE ENGINEERING IDEA GENERATION CHECKLIST

Figure A-1 shows a three-dimensional matrix which involves all possible combinations of five materials, seven designs and five manufacturing methods. The designer should select the best combination from these 175 alternatives. Similar matrices can be constructed for various features of a product. Obviously, some combinations in the matrix are completely useless. We only need the best idea among the long list of alternatives.

In each product category, there are usually special areas where cost reduction is particularly fruitful. The design engineer should consult the records of cost reduction and product improvement. Value engineering case study can be used for the same purpose. The following list is representative of the questions the designer may have in the process of finalizing a design.

1. Can a material change improve the product performance?
2. Can lubricants and catalysts improve the performance of the product?
3. Should the appearance of the product be associated with major sports or cartoon figures?
4. Can a color change make the product more attractive? Day-glow color? Chrome plating? Glow-in-the-dark paints? Ultraviolet color? Reflective paints?

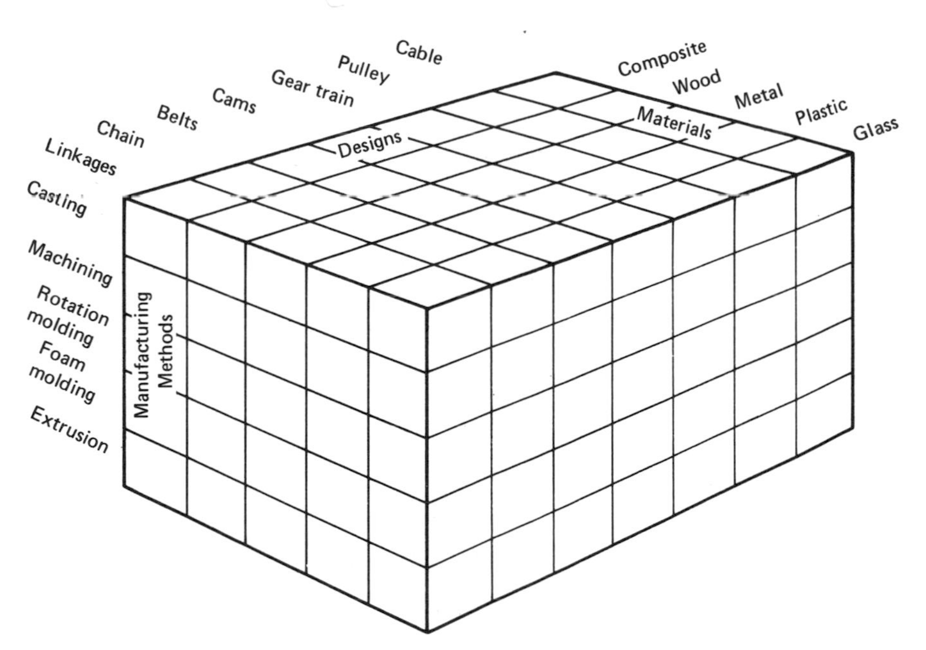

Fig. A-1. The alternative matrix.

5. Should fragrants be added to the product?
6. Would a change in surface texture improve the appeal of the product?
7. Can the product be made with a special click sound to demonstrate positive engagement?
8. Where two components joint or interact, can the male and female components be interchanged?
9. Can the handles and buttons be repositioned to improve man-machine interaction?
10. Can the order of operation, direction of cut, or direction of movement be improved?
11. Should the product be made foldable?
12. Can miniaturization enhance the product?
13. Can the number of components be reduced to improve strength and decrease assembly?
14. Is the packaging and labeling of the product appropriate in terms of cost and function?
15. Are all the components and manufacturing processes standard?

number of years n	interest rate i											
	1 %	2 %	3 %	5 %	7 %	10 %	12 %	15 %	20 %	25 %	30 %	35 %
1	0.9901	0.9804	0.9709	0.9524	0.9346	0.9091	0.8929	0.8696	0.8333	0.8000	0.7692	0.7407
2	0.9803	0.9612	0.9426	0.9070	0.8734	0.8264	0.7972	0.7561	0.6944	0.6400	0.5917	0.5487
3	0.9706	0.9423	0.9151	0.8638	0.8163	0.7513	0.7118	0.6575	0.5787	0.5120	0.4552	0.4064
4	0.9610	0.9238	0.8885	0.8227	0.7629	0.6830	0.6355	0.5718	0.4823	0.4096	0.3501	0.3011
5	0.9515	0.9057	0.8626	0.7835	0.7130	0.6209	0.5674	0.4972	0.4019	0.3277	0.2693	0.2230
6	0.9421	0.8880	0.8375	0.7462	0.6663	0.5645	0.5066	0.4323	0.3349	0.2621	0.2072	0.1652
7	0.9327	0.8706	0.8131	0.7107	0.6228	0.5132	0.4524	0.3759	0.2791	0.2097	0.1594	0.1224
8	0.9235	0.8535	0.7894	0.6768	0.5820	0.4665	0.4039	0.3269	0.2326	0.1678	0.1226	0.0906
9	0.9144	0.8368	0.7664	0.6446	0.5439	0.4241	0.3606	0.2843	0.1938	0.1342	0.0943	0.0671
10	0.9053	0.8204	0.7441	0.6139	0.5084	0.3855	0.3220	0.2472	0.1615	0.1074	0.0725	0.0497
11	0.8963	0.8043	0.7224	0.5847	0.4751	0.3505	0.2875	0.2149	0.1346	0.0859	0.0558	0.0368
12	0.8875	0.7885	0.7014	0.5568	0.4440	0.3186	0.2567	0.1869	0.1122	0.0687	0.0429	0.0273
13	0.8787	0.7730	0.6810	0.5303	0.4150	0.2897	0.2292	0.1625	0.0935	0.0550	0.0330	0.0202
14	0.8700	0.7579	0.6611	0.5051	0.3878	0.2633	0.2046	0.1413	0.0779	0.0440	0.0254	0.0150
15	0.8614	0.7430	0.6419	0.4810	0.3625	0.2394	0.1827	0.1229	0.0649	0.0352	0.0195	0.0111
16	0.8528	0.7285	0.6232	0.4581	0.3387	0.2176	0.1631	0.1069	0.0541	0.0281	0.0150	0.0082
17	0.8444	0.7142	0.6050	0.4363	0.3166	0.1978	0.1456	0.0929	0.0451	0.0225	0.0116	0.0061
18	0.8360	0.7002	0.5874	0.4155	0.2959	0.1799	0.1300	0.0808	0.0376	0.0180	0.0089	0.0045
19	0.8278	0.6864	0.5703	0.3957	0.2765	0.1635	0.1161	0.0703	0.0313	0.0144	0.0068	0.0033
20	0.8196	0.6730	0.5537	0.3769	0.2584	0.1486	0.1037	0.0611	0.0261	0.0115	0.0053	0.0025
25	0.7798	0.6095	0.4776	0.2953	0.1843	0.0923	0.0588	0.0304	0.0105	0.0038	0.0014	0.0006
30	0.7420	0.5521	0.4120	0.2314	0.1314	0.0573	0.0334	0.0151	0.0042	0.0012	0.0004	0.0001
35	0.7059	0.5000	0.3554	0.1813	0.0937	0.0356	0.0189	0.0075	0.0017	0.0004	0.0001	0.0000
40	0.6717	0.4529	0.3066	0.1421	0.0668	0.0221	0.0107	0.0037	0.0007	0.0001	0.0000	0.0000
45	0.6391	0.4102	0.2644	0.1113	0.0476	0.0137	0.0061	0.0019	0.0003	0.0000	0.0000	0.0000
50	0.6081	0.3715	0.2281	0.0872	0.0339	0.0085	0.0035	0.0009	0.0001	0.0000	0.0000	0.0000

P/F table

To obtain the present value, multiply the future value by the listed factor.

To obtain the future value, divide the present value by the listed factor.

INTEREST RATE TABLES

number of years n	interest rate i 1 %	2 %	3 %	5 %	7 %	10 %	12 %	15 %	20 %	25 %	30 %	35 %
1	0.990	0.980	0.971	0.952	0.935	0.909	0.893	0.870	0.833	0.800	0.769	0.741
2	1.970	1.941	1.913	1.859	1.808	1.736	1.690	1.626	1.528	1.440	1.361	1.289
3	2.941	2.884	2.829	2.723	2.624	2.487	2.402	2.283	2.106	1.952	1.816	1.696
4	3.901	3.808	3.717	3.546	3.387	3.170	3.037	2.855	2.589	2.362	2.166	1.997
5	4.853	4.713	4.580	4.329	4.100	3.791	3.605	3.352	2.991	2.689	2.436	2.220
6	5.795	5.601	5.417	5.076	4.766	4.355	4.111	3.784	3.326	2.951	2.643	2.385
7	6.727	6.472	6.230	5.786	5.389	4.868	4.564	4.160	3.605	3.161	2.802	2.508
8	7.651	7.325	7.020	6.463	5.971	5.335	4.968	4.487	3.837	3.329	2.925	2.598
9	8.565	8.162	7.786	7.108	6.515	5.759	5.328	4.772	4.031	3.463	3.019	2.665
10	9.470	8.982	8.530	7.722	7.024	6.145	5.650	5.019	4.192	3.571	3.092	2.715
11	10.366	9.786	9.252	8.306	7.499	6.495	5.938	5.234	4.327	3.656	3.147	2.752
12	11.254	10.575	9.954	8.863	7.943	6.814	6.194	5.421	4.439	3.725	3.190	2.779
13	12.132	11.348	10.635	9.393	8.358	7.103	6.424	5.583	4.533	3.780	3.223	2.799
14	13.002	12.106	11.296	9.899	8.745	7.367	6.628	5.724	4.611	3.824	3.249	2.814
15	13.863	12.849	11.938	10.380	9.108	7.606	6.811	5.847	4.675	3.859	3.268	2.825
16	14.716	13.577	12.561	10.838	9.447	7.824	6.974	5.954	4.730	3.887	3.283	2.834
17	15.560	14.291	13.166	11.274	9.763	8.022	7.120	6.047	4.775	3.910	3.295	2.840
18	16.396	14.991	13.753	11.689	10.059	8.201	7.250	6.128	4.812	3.928	3.304	2.844
19	17.224	15.678	14.323	12.085	10.336	8.365	7.366	6.198	4.843	3.942	3.311	2.848
20	18.043	16.351	14.877	12.462	10.594	8.514	7.469	6.259	4.870	3.954	3.316	2.850
25	22.021	19.523	17.413	14.094	11.654	9.077	7.843	6.464	4.948	3.985	3.329	2.856
30	25.805	22.396	19.600	15.372	12.409	9.427	8.055	6.566	4.979	3.995	3.332	2.857
35	29.405	24.998	21.487	16.374	12.948	9.644	8.176	6.617	4.992	3.998	3.333	2.857
40	32.831	27.355	23.114	17.159	13.332	9.779	8.244	6.642	4.997	3.999	3.333	2.857
45	36.091	29.489	24.518	17.774	13.606	9.863	8.283	6.654	4.999	4.000	3.333	2.857
50	39.192	31.423	25.729	18.256	13.801	9.915	8.305	6.661	4.999	4.000	3.333	2.857

P/A table

To obtain the present value, multiply the uniform annual payment by the listed factor.

To obtain the uniform annual payment, divide the present value by the listed factor.

APPENDIX A.3. STIFFNESS AND STRENGTH OF RIBBED SECTIONS

The moment of inertia I of a rectangular section and a trapezoidal section can be obtained from standard handbooks. These properties are combined by the parallel axis theorem to obtain the moment of inertia of the ribbed section.

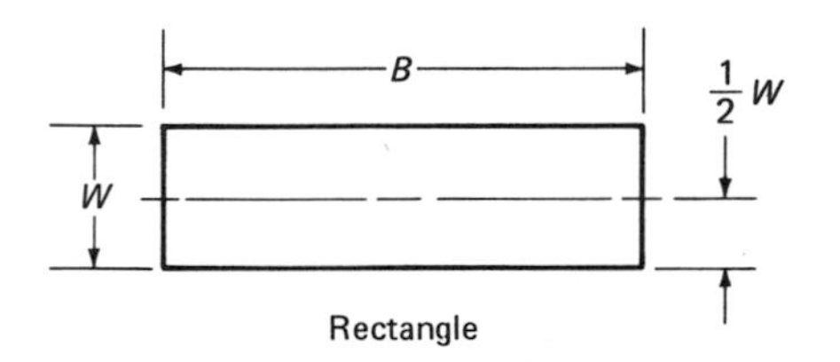

Rectangle

$$I_{\text{Rect.}} = \frac{BW^3}{12}$$

$$A_{\text{Rect.}} = BW$$

T

H

$$Y1 = \frac{H}{3}\left(\frac{2T+t}{T+t}\right)$$

t

Trapedzoid

$$I_{\text{Trap.}} = \frac{H^3}{36}\,\frac{T^2 + 4Tt + t^2}{T+t}$$

$$A_{\text{Trap.}} = \frac{H}{2}(t+T)$$

I_c

Y2

Y1

Y_c

Ribbed Section

$$Y2 = H + \frac{W}{2}$$

$$Y_c = \frac{A_{\text{Trap.}} \cdot Y1 + A_{\text{Rect.}} \cdot Y2}{A_{\text{Trap.}} + A_{\text{Rect.}}}$$

$$I_c = I_{\text{Trap.}} + A_{\text{Trap.}}(Y_c - Y1)^2 + I_{\text{Rect.}} + A_{\text{Rect.}}(Y_c - Y2)^2$$

$$\text{Stiffening factor} = \frac{I_c}{I_{\text{Rect.}}}$$

$$\text{Strengthening factor} = \frac{I_c}{I_{\text{Rect.}}} \cdot \frac{\left(\frac{W}{2}\right)}{Y_c}$$

$$\text{Rib material} = \frac{A_{\text{Trap.}}}{A_{\text{Rect.}}} = \frac{H}{2} \frac{(T+t)}{BW}$$

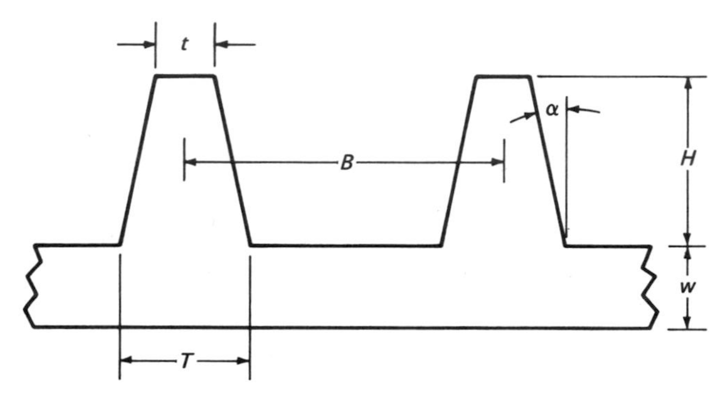

B = Rib spacing
H = Rib height
T = Rib thickness at root = $\frac{W}{2}$ or $\frac{3}{4}W$
t = Rib thickness at crest = $T - 2H \tan \alpha$
W = Wall thickness
α = Draft angle = $\frac{1}{2}^\circ$ or 1°

For calculation purpose,

$W = 1$

$\frac{B}{W} = B, \quad \frac{H}{W} = H$

$T = 0.5$ or 0.75

APPENDIX A.4. STIFFNESS AND STRENGTH OF CORRUGATIONS

The moment of inertia I is derived for a drawn corrugation and a square corrugation.

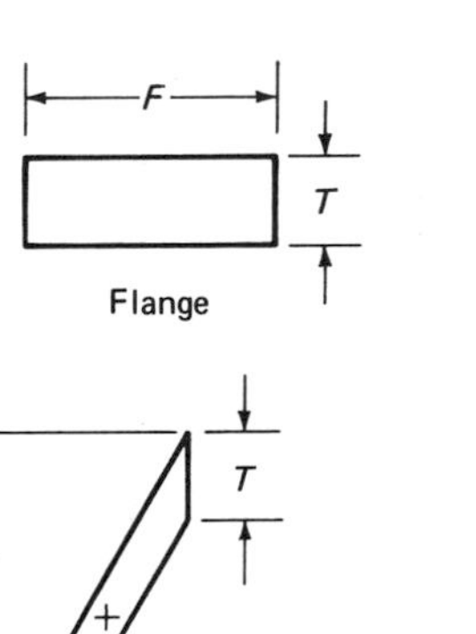

$$I_{\text{Flange}} = \frac{FT^3}{12}$$

$$A_{\text{Flange}} = FT$$

$$I_{\text{Web}} = \frac{TW}{12}(T^2 + H^2)$$

B = Wave length
F = Flange length

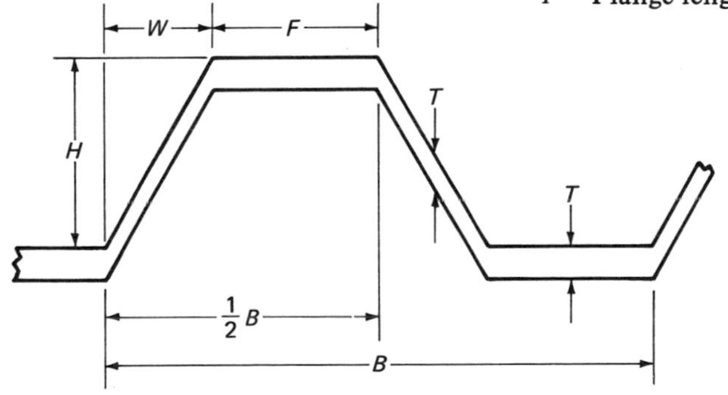

$$I_{\text{Corr.}} = 2 \times \left[I_{\text{Flange}} + A_{\text{Flange}} \left(\frac{H}{2} \right)^2 + I_{\text{Web}} \right]$$

$$I_{\text{Flat plate}} = \frac{BT^3}{12}$$

$$\text{Stiffening factor} = \frac{24}{BT^3} \left[FT \left(\frac{T^2}{12} + \frac{H^2}{4} \right) + \frac{TW}{12}(T^2 + H^2) \right]$$

$$= \left(\frac{H}{T} \right)^2 \left(1 + 4\frac{F}{B} \right) + 1$$

Strengthening factor

$$C_{\text{Corr.}} = \frac{H+T}{2}$$

$$C_{\text{Flat plate}} = \frac{T}{2}$$

$$\text{Strengthening factor} = \left[H^2\left(1 + 4\frac{F}{B}\right) + T^2\right] \frac{T}{T^2(H+T)}$$

$$= \frac{H^2(1 + 4F/B) + T^2}{T(H+T)}$$

Square corrugation

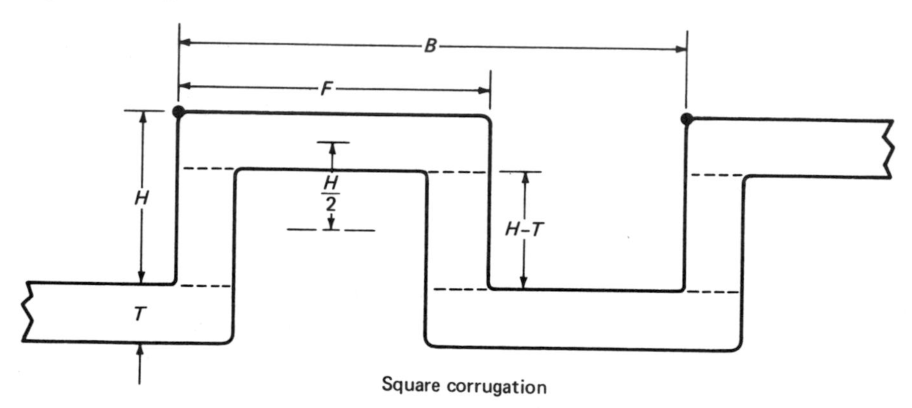

Square corrugation

$$I = \left[\frac{FT^3}{12} + FT\frac{H^2}{4} + \frac{(H-T)^3 T}{12}\right] \times 2$$

$$\text{Stiffening factor} = \frac{12}{BT^3}\left[\frac{FT^3}{6} + FT\frac{H^2}{2} + \frac{(H-T)^3 T}{6}\right]$$

$$= 2\frac{F}{B} + 6\frac{F}{B}\left(\frac{H}{T}\right)^2 + 2\frac{(H-T)^3}{BT^2}$$

$$\text{Strengthening factor} = \frac{2T}{H+T}\left[\frac{F}{B} + 3\frac{F}{B}\left(\frac{H}{T}\right)^2 + \frac{(H-T)^3}{BT^2}\right]$$

APPENDIX A.5. SNAP-IN AND SNAP-OUT FORCE CALCULATIONS

Refer to the free body diagram. The force acting at the interface consists of a normal component F_n and a tangential component F_f (friction).

Since all forces meet at one point, $\Sigma M = 0$

$$\Sigma\vec{F} = \vec{F_n} + \vec{F_f} + \vec{F_i} + \vec{F_s} = 0$$

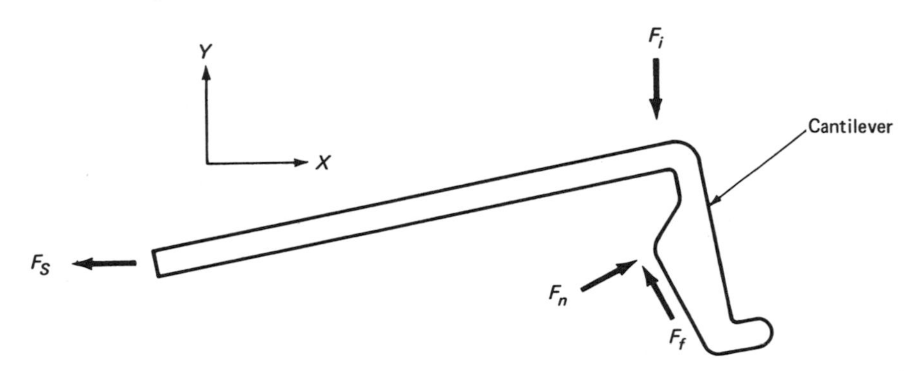

Free body diagram of the forces acting on a cantilever snap.

$$F_n(\cos\alpha\,\vec{i} + \sin\alpha\,\vec{j}) - F_f(\sin\alpha\,\vec{i} - \cos\alpha\,\vec{j}) - F_i\vec{j} - F_s\vec{i} = 0 \qquad \text{(A)}$$

where

F_i = snap-in force

F_s = snap force, the force needed to reduce interference to zero

F_n = normal force at the interference

F_f = friction force at the interference = $\mu \cdot F_n$

Consider the x-component of Eq. A,

$$F_n \cos\alpha - F_f \sin\alpha - F_s = 0$$

$$F_n = \frac{F_s}{\cos\alpha - \mu\sin\alpha} \qquad \text{(B)}$$

Consider the y-component of Eq. A,

$$F_n \sin\alpha + F_f \cos\alpha - F_i = 0$$

$$F_i = F_n(\sin\alpha + \mu\cos\alpha) \qquad \text{(C)}$$

Substitute B in C,

$$F_i = F_s \cdot \frac{(\sin\alpha + \mu\cos\alpha)}{(\cos\alpha - \mu\sin\alpha)} \qquad \text{(6.3)}$$

Equation 6.4 is similar to 6.3 with α replaced by β.
Equations 6.1 and 6.2 can be obtained from 6.3 and 6.4 by setting $\mu = 0$.

APPENDIX A.6. LINEAR PROGRAMMING FOR WOOD SELECTION

Linear programming optimization has been used to find the least cost mix of input lumber to satisfy a batch production of furniture. The following is a simplified example from Ref. [137]. The inputs are three different grades of wood A, B and C. Grade A costs more, but it has less defects and gives a better yield. The quality of lumber after cutting is uniform regardless of which grade was originally used because defects are eliminated during the cutting process. The outputs are three different lengths of plank: 8 MBF (thousand board feet) of 72-in. length wood, 6 MBF of 60-in. length and 1 MBF of 48-in. length. The objective is to minimize the total raw material cost:

$$\text{total cost} = 250\, X_A + 163\, X_B + 95\, X_C$$

where X_A, X_B, and X_C are the quantities of the three different grades of input lumber and \$250, \$163, and \$95 are the cost per MBF of the corresponding lumber.

The yield depends on the output length. If the output are short planks, the yield would be higher. A combination of short and long planks also results in a better yield than long planks only. Table 1 shows the yield as a function of the output length and input grade.

TABLE 1. YIELD FOR WOOD BOARD PRODUCTION*

	Input grades		
Output lengths	*A*	*B*	*C*
72-in. only	73.8%	49.2%	32.5%
72- and 60-in.	76.7%	56%	36.6%
72-, 60-, and 48-in.	82.6%	62.8%	46.1%

*Thomas, R. J. 1966. The yield of dimension stock from hardwood lumber. North Carolina Agri. Ext. Service, Misc. Extension Publication No. 16. Raleigh, N.C.

Additional variables are set-up to co-relate the input grades and the output lengths for purpose of writing the constraint equations. For example,

X_{A72} represents the amount of 72-in. planks produced from grade A wood.

X_{B60} represents the amount of 60-in. planks produced from grade B wood.

The constraint equations can now be expressed as follows:

$$X_{A72} \leqslant 0.738\, X_A$$
$$X_{A72} + X_{A60} \leqslant 0.767\, X_A$$
$$X_{A72} + X_{A60} + X_{A48} \leqslant 0.826\, X_A$$
$$X_{B72} \leqslant 0.492\, X_B$$
$$X_{B72} + X_{B60} \leqslant 0.56\, X_B$$

$$X_{B72} + X_{B60} + X_{B48} \leqslant 0.628\, X_B$$
$$X_{C72} \leqslant 0.325\, X_C$$
$$X_{C72} + X_{C60} \leqslant 0.366\, X_C$$
$$X_{C72} + X_{C60} + X_{C48} \leqslant 0.461\, X_C$$
$$X_{A72} + X_{B72} + X_{C72} \geqslant 8$$
$$X_{A60} + X_{B60} + X_{C60} \geqslant 6$$
$$X_{A48} + X_{B48} + X_{C48} \geqslant 1$$

The solution for this example is: $X_A = X_B = 0, X_C = 32.54$ MBF. The minimum cost is \$3091. Even though the better grades of lumber give high yields, for this set of data, grade C is the most economical. All 15 MBF of output would be cut from grade C at 46.1% yield (Table 1). Sensitivity analysis on this problem shows that the solution will not change if the prices of grades A and B, do not drop below \$199 and \$145, respectively, and the price of grade C does not raise above \$106.

TABLE 2. LUMBER PRICE RANGE WITHIN WHICH SOLUTION IS STABLE PRICES ARE LISTED IN \$ PER MBF.

	grade A	grade B	grade C
Upper value	∞	∞	\$106
Current value	\$250	\$163	\$ 95
Lower value	\$199	\$145	\$ 0

APPENDIX A.7. CONVERSION FACTORS

Length	1 mile = 1760 yards 1 yard = 3 feet 1 foot = 12 inches 1 inch = 10^6 micro-inch	1 meter = 100 cm = 1000 mm 1 cm = 10 mm 1 meter = 10^6 microns 1 meter = 10^{10} angstroms
	1 meter = 39.37 inches 1 inch = 25.4 mm	
Area	1 acre = 43, 560 sq ft 1 sq mile = 640 acres	
Volume	1 cu ft = 1728 cu in. 1 cu ft = 7.481 U.S. gallons 1 gallon = 4 quarts (fluid) 1 quart = 2 pints (fluid) 1 pint = 16 ounces (fluid)	1 liter = 1000 cu cm 1 cu meter = 1000 liters
	1 gallon = 3.785 liters 1 cu ft = 28.32 liters	
Mass	1 ton = 2000 pounds 1 pound = 16 ounces	1 Kg = 1000 gm
	1 Kg = 2.205 pounds 1 ounce = 28.35 gm	
Force	1 pound = 4.448 newtons 1 Kg = 2.205 pounds	1 Kg = 9.807 newtons 1 newton = 10^5 dynes
Velocity	15 mph = 22 ft/sec	1 meter/sec = 3.6 Km/hr
	1 mph = 1.609 Km/hr 1 meter/sec = 3.281 ft/sec	
Rotation velocity	1 rpm = 0.1047 rad/sec 1 rad/sec = 9.55 rpm	
Pressure	1 atmosphere = 14.7 psi = 760 mm Hg (0°C) 1 psi = 6895 newton/meter^2 1 psi = 51.71 mm Hg (0°C) psi (absolute) = psi (gage) + 14.7	
Density	1 gm/cu cm = 62.43 pound/cu ft = 8.345 pound/gallon = 0.03613 pound/cu in. = 0.5778 ounce/cu in. = 16.387 gm/cu in.	
Energy	1 Btu = 778.2 ft-pound 1 Btu = 252 calorie 1 ft-pound = 1.356 joule 1 calorie = 4.186 joule 1 Kilowatt-hr = 3.6×10^6 joule 1 joule (newton × meter) = 10^7 ergs (dyne × cm)	

Power	1 horsepower (HP) = 42.41 Btu/min
	1 HP = 745.7 watt
	1 HP = 550 ft-pound/sec
	$HP = \frac{\text{force} \times \text{velocity}}{33000}$ (pound) (ft/min)
	$HP = \frac{\text{angular velocity} \times \text{torque}}{63000}$ (rpm) (in-pound)
	watt = current × voltage (amphere) (volt)
	1 kilowatt = 239 calorie/sec
Temperature	degree kelvin ($^\circ$K) = degree celsius ($^\circ$C) + 273.16
	degree rankin ($^\circ$R) = degree fahrenheit ($^\circ$F) + 459.69
	degree celsius ($^\circ$C) = (degree fahrenheit ($^\circ$F) − 32) $\times \frac{5}{9}$
	degree fahrenheit ($^\circ$F) = $\left(\text{degree celsius } (^\circ C) \times \frac{9}{5}\right) + 32$
Angle	180 degrees = 3.1416 radians
	1 radian = 57.296 degrees
	1 degree = 0.01745 radians

Fractions and decimal conversions

1/64			=0.016	33/64			=0.516
	1/32		=0.031		17/32		=0.531
3/64			=0.047	35/64			=0.547
		1/16	=0.062			9/16	=0.562
5/64			=0.078	37/64			=0.578
	3/32		=0.094		19/32		=0.594
7/64			=0.109	39/64			=0.609
		1/8	=0.125			5/8	=0.625
9/64			=0.141	41/64			=0.641
	5/32		=0.156		21/32		=0.656
11/64			=0.172	43/64			=0.672
		3/16	=0.188			11/16	=0.688
13/64			=0.203	45/64			=0.703
	7/32		=0.219		23/32		=0.719
15/64			=0.234	47/64			=0.734
		1/4	=0.250			3/4	=0.750
17/64			=0.266	49/64			=0.766
	9/32		=0.281		25/32		=0.781
19/64			=0.297	51/64			=0.797
		5/16	=0.312			13/16	=0.812
21/64			=0.328	53/64			=0.828
	11/32		=0.344		27/32		=0.844
23/64			=0.359	55/64			=0.859
		3/8	=0.375			7/8	=0.875
25/64			=0.391	57/64			=0.891
	13/32		=0.406		29/32		=0.906
27/64			=0.422	59/64			=0.922
		7/16	=0.438			15/16	=0.938
29/64			=0.453	61/64			=0.953
	15/32		=0.469		31/32		=0.969
31/64			=0.484	63/64			=0.984
		1/2	=0.500				

INDEX